Charles Smith

Solutions of the Examples in an Elementary Treatise on Conic Sections

Charles Smith

Solutions of the Examples in an Elementary Treatise on Conic Sections

ISBN/EAN: 9783337256418

Printed in Europe, USA, Canada, Australia, Japan

Cover: Foto ©berggeist007 / pixelio.de

More available books at **www.hansebooks.com**

By the same Author.

AN ELEMENTARY TREATISE

ON

CONIC SECTIONS.

SIXTH EDITION.

Crown 8vo. 7*s.* 6*d.*

AN ELEMENTARY TREATISE

ON

SOLID GEOMETRY.

SECOND EDITION.

Crown 8vo. 9*s.* 6*d.*

ELEMENTARY ALGEBRA.

Globe 8vo. 4*s.* 6*d.*

A TREATISE ON ALGEBRA.

Crown 8vo. 7*s.* 6*d.*

A KEY

TO

TREATISE ON ALGEBRA.

[*Preparing.*]

SOLUTIONS OF THE EXAMPLES

IN

AN ELEMENTARY TREATISE

ON

CONIC SECTIONS.

SOLUTIONS OF THE EXAMPLES

IN

AN ELEMENTARY TREATISE

ON

CONIC SECTIONS

BY

CHARLES SMITH, M.A.,
FELLOW AND TUTOR OF SIDNEY SUSSEX COLLEGE, CAMBRIDGE.

London:
MACMILLAN AND CO.
AND NEW YORK.
1888

E.V.

Cambridge
PRINTED BY C. J. CLAY M.A. AND SONS
AT THE UNIVERSITY PRESS

PREFACE.

IT is hoped that the following Solutions of the Examples in my *Treatise on Conic Sections* will be of use to teachers, many of whom can ill afford time to write out detailed solutions of the questions which prove too difficult for their pupils; and that it will also be of service to those students who read the subject without the assistance of a teacher.

The solutions will, I trust, be found sufficiently full and clear. In some cases more than one solution is given.

My thanks are due to several of my friends, and particularly to Mr S. L. Loney, to Mr J. Bárnard and to Rev. H. T. Lewis, for their kindness in reading the proof sheets.

CHARLES SMITH.

SIDNEY SUSSEX COLLEGE,
July, 1888.

KEY

TO

CONIC SECTIONS.

CHAPTER I.

Page 5.

1. $\sqrt{\{(1+3)^2+(2+1)^2\}}=5$.

2. (i) $\sqrt{\{(1+1)^2+(-1-1)^2\}}=\sqrt{8}$; (ii) $\sqrt{\{(a+b)^2+(-a-b)^2\}}=(a+b)\sqrt{2}$; (iii) $\sqrt{\{(3+1)^2+(4-1)^2\}}=\sqrt{25}=5$.

3. The lengths of the lines joining the points are $\sqrt{\{(-1+\sqrt{3})^2+(-1-\sqrt{3})^2\}}$, $\sqrt{\{(-\sqrt{3}-1)^2+(\sqrt{3}-1)^2\}}$ and $\sqrt{\{(1+1)^2+(1+1)^2\}}$, each of which is equal to $\sqrt{8}$.

4. Call the points in the order given A, B, C, D. Then $AB=\sqrt{\{2^2+(-1-3)^2\}}=\sqrt{20}$, and $CD=\sqrt{\{(6-8)^2+(7-3)^2\}}=\sqrt{20}$. Again $BC=\sqrt{\{(-2-6)^2+(3-7)^2\}}=\sqrt{80}$, and $DA=\sqrt{\{8^2+(3+1)^2\}}=\sqrt{80}$. Since $AB=CD$ and $BC=DA$ the figure is a *parallelogram*. But $AC^2=6^2+(7+1)^2=100=AB^2+BC^2$, and $\therefore$ the angle ABC is a right angle.

5. $AB^2=(-2)^2+(-1-1)^2=8$, $BC^2=2^2+(1-3)^2=8$, $CD^2=2^2+(3-1)^2=8$, and $DA^2=(-2)^2+(1+1)^2=8$: the sides of $ABCD$ are therefore all equal. Also $AC^2=(-1-3)^2=16=AB^2+BC^2$, and therefore AB is perpendicular to BC. Hence $ABCD$ is a square.

In the second example, $AB^2=(2-4)^2+(1-3)^2=8$, $BC^2=(4-2)^2+(3-5)^2=8$, $CD^2=2^2+(5-3)^2=8$, and $DA^2=(-2)^2+(3-1)^2=8$: the sides are therefore all equal. Also $AC^2=(2-2)^2+(1-5)^2=16=AB^2+BC^2$, $\therefore$ AB is perpendicular to BC.

6. $AB^2=(2-5)^2+(1-4)^2=18$ and $CD^2=(4-1)^2+(7-4)^2=18$; $\therefore$ $AB=CD$. Again, $BC^2=(5-4)^2+(4-7)^2=10$ and $DA^2=(1-2)^2+(4-1)^2=10$; $\therefore BC=DA$. Hence opposite sides of $ABCD$ are equal, and therefore $ABCD$ is a parallelogram.

7. We have $(x-3)^2+(y-4)^2=(x-1)^2+(y+2)^2$, whence $4x+12y=20$.

Pages 7 and 8.

1. $\frac{1}{2}(2.3-4.1+4.5-2.3+2.1-2.5)=4.$

$\frac{1}{2}\{4(-6)-5(-5)+5.1-3(-6)+3(-5)-4.1\}=\frac{5}{2}.$

2. $\frac{1}{2}\{1.4-3.2+3.3-5.4+5.2-6.3+6.2-1.2\}=\frac{11}{2}.$

$\frac{1}{2}\{2.3-(-2)2+(-2)(-3)-(-3)3+(-3)(-2)-1(-3)+1.2-2(-2)\}$
$=20.$

Page 9.

1. Let (x, y) be any point on the locus; then we have

$$(x-3)^2+(y-4)^2=(x-5)^2+(y+2)^2,$$

whence $$4x-12y=4, \text{ or } x-3y=1.$$

2. Let (x, y) be any point on the locus; then we have

$$\{(x-a)^2+y^2\}+\{(x+a)^2+y^2\}=2c^2,$$

whence $$x^2+y^2=c^2-a^2.$$

3. Let (x, y) be any point on the locus; then

$$\{(x-a)^2+y^2\}\sim\{(x+a)^2+y^2\}=c^2,$$

whence $$4xy=\pm c^2.$$

4. Let the fixed points be (x_1, y_1) and (x_2, y_2), and let any point on the locus be (x, y); then the ratio of $(x-x_1)^2+(y-y_1)^2$ to $(x-x_2)^2+(y-y_2)^2$ is constant, $k:1$ suppose. Hence $(x-x_1)^2+(y-y_1)^2=k\{(x-x_2)^2+(y-y_2)^2\}$, which is the required equation.

5. Let (x, y) be any point on the locus; then the distance of this point from the origin is $\sqrt{(x^2+y^2)}$ and its distance from the axis of x is y. Hence $y=\frac{1}{2}\sqrt{(x^2+y^2)}$, whence $3y^2=x^2$.

6. Let (x, y) be any point on the locus; then we have

$$y=\sqrt{\{(x-1)^2+(y-1)^2\}},$$

whence $$x^2-2x-2y+2=0.$$

Page 11.

1. $x=1.\cos\frac{\pi}{2}=0,\ y=1.\sin\frac{\pi}{2}=1;\ \ x=2\cos\frac{\pi}{3}=1,\ y=2\sin\frac{\pi}{3}=\sqrt{3};$
$x=-4\cos\left(-\frac{\pi}{4}\right)=-2\sqrt{2},\ y=-4\sin\left(-\frac{\pi}{4}\right)=2\sqrt{2}.$

2. For the point $(-1, -1)$: $r=\sqrt{(x^2+y^2)}=\sqrt{\{(-1)^2+(-1)^2\}}=\pm\sqrt{2}$, and $\theta=\tan^{-1}\frac{y}{x}=\tan^{-1}\frac{-1}{-1}=\tan^{-1}1=\frac{\pi}{4}$ or $\frac{5\pi}{4}$. Now, in order that $r\cos\theta$

may be -1, if we consider that $r=+\sqrt{2}$ we must suppose $\theta=\frac{5\pi}{4}$; and if we take $r=-\sqrt{2}$ we must have $\theta=\frac{\pi}{4}$. We may therefore consider the polar coordinates to be either $r=\sqrt{2}$, $\theta=\frac{5\pi}{4}$; or $r=-\sqrt{2}$, $\theta=\frac{\pi}{4}$. [It will be obvious from a figure that these represent the same point.]

For the point $(-1, \sqrt{3})$: $r=\sqrt{(1+3)}=\pm 2$, and $\theta=\tan^{-1}\frac{\sqrt{3}}{-1}=\frac{2\pi}{3}$ or $\frac{5\pi}{3}$. If we take $r=+2$, we must take $\theta=\frac{2\pi}{3}$, since $x=r\cos\theta=-1$; so also, if we take $r=-2$ we must take $\theta=\frac{5\pi}{3}$. Hence the coordinates required are $r=2$, $\theta=\frac{2\pi}{3}$; *or* $r=-2$, $\theta=\frac{5\pi}{3}$.

For the point $(3, -4)$: $r=\sqrt{(3^2+4^2)}=\pm 5$ and $\theta=\tan^{-1}\frac{-4}{3}$; but, since $r\cos\theta$ is positive, r and θ must be so chosen that r and $\cos\theta$ may be both positive or both negative.

3. $\sqrt{\{2^2+4^2-2\,.\,2\,.\,4\cos 60^0\}}=\sqrt{12}$.

4. $\pm\frac{1}{2}\left\{1\,.\,1\sin\frac{\pi}{2}+1\,.\sqrt{2}\,.\sin\left(\frac{\pi}{4}-\frac{\pi}{2}\right)+\sqrt{2}\,.\,1\,.\sin\left(-\frac{\pi}{4}\right)\right\}=\frac{1}{2}$.

CHAPTER II.

Page 21.

1. Substituting the coordinates in the left-hand members, we have $2(3)-3(2)-1$ and $3(3)-2(2)-1$ which are respectively *negative* and *positive*.

2. Substituting $(2, -1)$ and $(1, 1)$ in $3x+4y-6$, we have $3(2)+4(-1)-6$ and $3(1)+4(1)-6$ which are of *contrary* signs.

3. Substitute the coordinates of the four points in the left-hand members of the two given equations: the signs of the results will then be found to be $+, +$; $-, -$; $-, +$; and $+, -$ respectively.

Pages 22—24.

1. The lines can be drawn by finding the intercepts on the axes, as in Art. 21. The intercepts are

(i) 2, 2. (ii) 4, -3. (iii) $-\frac{1}{4}, \frac{1}{3}$ and (iv) $-\frac{7}{2}, -\frac{7}{3}$.

2. (i) $\dfrac{y-3}{3-1}=\dfrac{x-2}{2-(-4)}$, or $x-3y+7=0$.

(ii) $\dfrac{x-a}{a-b}=\dfrac{y-b}{b-a}$, or $x+y-a-b=0$.

3. $\tan 150^0=-\dfrac{1}{\sqrt{3}}=-\tan 30^0$. Hence [Art. 23] the equations are respectively $y+1=-\dfrac{1}{\sqrt{3}}(x-1)$ and $y+1=\dfrac{1}{\sqrt{3}}(x-1)$.

4. See Art. 20.

5. The line is to pass through the point (4, 5) and to make an angle $\tan^{-1}\frac{2}{3}$ with the axis of x. Hence its equation is $y-5=\frac{2}{3}(x-4)$, that is $3y-2x-7=0$.

6. The equation of the line joining (2, 3) and (3, −1) is $\frac{y-3}{3+1}=\frac{x-2}{2-3}$; hence the line makes with the axis of x an angle $\tan^{-1}(-4)$. The equation of a parallel line through (2, 1) is therefore $y-1=-4(x-2)$, or $4x+y-9=0$.

7. Let a be the length of the intercept on either axis; then the equation of the line will be $\frac{x}{a}+\frac{y}{a}=1$. Now in order that the line may pass through (5, 6) we must have $\frac{5}{a}+\frac{6}{a}=1$, that is, $a=11$; and hence the required equation is $x+y=11$.

8. (i) $$\frac{x}{7.77-2(-99)}=\frac{y}{3(-99)-77.5}=\frac{1}{5.2-7.3},$$

that is $$\frac{x}{737}=-\frac{y}{682}=\frac{1}{-11}; \quad \therefore x=-67, \ y=62.$$

(ii) $$\frac{x}{-5.4-1.1}=\frac{y}{1.1-2.4}=\frac{1}{2.1-(-5)1},$$

that is $$\frac{x}{-21}=\frac{y}{-7}=\frac{1}{7}; \quad \therefore x=-3, \ y=-1.$$

(iii) $$\frac{x}{-\frac{1}{b}+\frac{1}{a}}=\frac{y}{-\frac{1}{b}+\frac{1}{a}}=\frac{1}{\frac{1}{a^2}-\frac{1}{b^2}}; \quad \therefore x=y=\frac{ab}{a+b}.$$

9. The first two lines meet in the point given by

$$\frac{x}{-30-28}=\frac{y}{-21+50}=\frac{1}{-20-9},$$

that is in the point $x=2$, $y=-1$; and (2, −1) is obviously on $x+2y=0$.

10. The equation of the line through the first two points is $\frac{x}{-2}=\frac{y-11}{11-3}$, that is $4x+y-11=0$; and the point (3, −1) is on the straight line $4x+y-11=0$ since $4(3)+(-1)-11=0$.

The equation of the line through $(3a, 0)$ and $(0, 3b)$ is $\frac{x}{3a}+\frac{y}{3b}=1$, and $(a, 2b)$ is on this line since $\frac{a}{3a}+\frac{2b}{3b}=1$.

11. The equations are $\frac{x-2}{2+3}=\frac{y-3}{3+5}$, $\frac{x+3}{-3-1}=\frac{y+5}{-5-2}$ and $\frac{x-1}{1-2}=\frac{y-2}{2-3}$ respectively. These reduce to $8x-5y-1=0$, $7x-4y+1=0$ and $x-y+1=0$.

12. The middle points of the sides opposite to (1, 2), (2, 3) and (−3, −5) are $\left(-\frac{1}{2}, -1\right)$, $\left(-1, -\frac{3}{2}\right)$ and $\left(\frac{3}{2}, \frac{5}{2}\right)$ respectively [Art. 5].

Hence the required equations are $\dfrac{x-1}{1+\frac{1}{2}}=\dfrac{y-2}{2+1}$, $\dfrac{x-2}{2+1}=\dfrac{y-3}{3+\frac{3}{2}}$ and $\dfrac{x+3}{-3-\frac{3}{2}}=\dfrac{y+5}{-5-\frac{5}{2}}$ respectively, that is $2x-y=0$, $3x-2y=0$ and $5x-3y=0$ respectively.

13. The extremities of one diagonal are (a, c), (b, d) and its equation is therefore $\dfrac{x-a}{a-b}=\dfrac{y-c}{c-d}$ which may be written $(d-c)x+(a-b)y+bc-ad=0$.

The extremities of the other diagonal are (a, d), (b, c) and its equation is therefore $\dfrac{x-a}{a-b}=\dfrac{y-d}{d-c}$ which may be written $(d-c)x+(b-a)y+ac-bd=0$.

14. The point of intersection of $3x+y-2=0$ and $2x-y-3=0$ is $(1, -1)$, and the point $(1, -1)$ will be on $ax+2y-3=0$ provided $a-2-3=0$; hence if the lines meet in a point a must be 5.

15. The equations of the lines are $x-3y+5=0$ and $x+y-5=0$, which meet in the point $x=y=\frac{5}{2}$; and the point $\left(\frac{5}{2}, \frac{5}{2}\right)$ is clearly midway between $(1, 2)$ and $(4, 3)$.

16. Substituting the coordinates of the two points in the left-hand member of $5y-6x+4=0$ the results are 7 and -4 which are of *different* signs, and hence the points are on *opposite* sides of the line.

17. The results of substitution are $+1$, and -1; hence the points are on *opposite* sides of the line.

18. In order that any particular point may be *within* a triangle it is necessary and sufficient that each angular point of the triangle and the point in question should be on the *same side* of the opposite side of the triangle. In the present case the angular points of the triangle will be found to be $(-1, -2)$, $(3, 4)$ and $(-4, 3)$. Now $(0, 0)$ and $(-1, -2)$ are both on the positive side of $x-7y+25=0$; also $(0, 0)$ and $(3, 4)$ are both on the positive side of $5x+3y+11=0$; and $(0, 0)$ and $(-4, 3)$ are both on the negative side of $3x-2y-1=0$. Hence the origin $(0, 0)$ is within the triangle.

Pages 30—31.

1. (i) $\tan^{-1}\dfrac{-2.1-1.3}{-2.3+1.1}=\tan^{-1}1=45^\circ$.

(ii) $\tan^{-1}\dfrac{2.2-(-1)1}{1.2+2(-1)}=\tan^{-1}\infty=90^\circ$.

(iii) $\tan^{-1}\dfrac{B(A+B)-(B-A)A}{A(A+B)+B(B-A)}=\tan^{-1}1=45^\circ$.

2. $7(x-3)-2(y-1)=0$, that is $7x-2y=19$.

3. The line through (0, 0) perpendicular to $3x+2y-5=0$ is $2x-3y=0$ and the foot of the perpendicular is $\left(\frac{15}{13}, \frac{10}{13}\right)$.

The line through (0, 0) perpendicular to $4x+3y-7=0$ is $3x-4y=0$ and the foot of the perpendicular is $\left(\frac{28}{25}, \frac{21}{25}\right)$.

Hence the line joining these points is $\dfrac{x-\frac{15}{13}}{\frac{15}{13}-\frac{28}{25}} = \dfrac{y-\frac{10}{13}}{\frac{10}{13}-\frac{21}{25}}$, that is $\frac{13x-15}{11}=\frac{13y-10}{-23}$, which reduces to $23x+11y=35$.

4. $\dfrac{4.2+3.3-7}{\sqrt{(4^2+3^2)}}=\frac{10}{5}=2$, $\dfrac{5.2+12.3-20}{\sqrt{(5^2+12^2)}}=\frac{26}{13}=2$, $\dfrac{3.2+4.3-8}{\sqrt{(3^2+4^2)}}$
$=\frac{10}{5}=2$.

5. The lines are $3(x-1)+4(y-1)=0$ and $3(x+2)+4(y+1)=0$, that is $3x+4y-7=0$ and $3x+4y+10=0$. The distance between two parallel lines is the length of the perpendicular on either line drawn from any point on the other; and the perpendicular from $(-2, -1)$ on $3x+4y-7=0$ is $\frac{3(-2)+4(-1)-7}{5}=-\frac{17}{5}$, so that the distance required is $\frac{17}{5}$.

6. The equations are of the form $y-3=m(x-2)$, where m is found from $\pm 1=\dfrac{-\frac{1}{2}-m}{1+\left(-\frac{1}{2}\right)m}=\frac{2m+1}{m-2}$. The two values of m are -3 and $\frac{1}{3}$, and the corresponding equations are $y-3+3(x-2)=0$ and $3(y-3)-(x-2)=0$, that is $3x+y-9=0$ and $3y-x-7=0$.

7. Any straight line parallel to $x+7y+2=0$ is $x+7y+c=0$; and the lines whose perpendicular distances from $(1, -1)$ are ± 1 are those for which c is given by $\frac{1-7+c}{\sqrt{50}}=\pm 1$. Hence the equations required are $x+7y+6\pm\sqrt{50}=0$.

8. Any line through the intersection of the given lines is

$$x-4y-7+\lambda(y+2x-1)=0 \text{ [Art. 33]};$$

this will go through the origin if $\lambda=-7$, and hence the required equation is $13x+11y=0$.

Or thus:

The point of intersection of the lines $x-4y-7=0$ and $y+2x-1=0$ is $\left(\frac{11}{9}, -\frac{13}{9}\right)$; $\therefore$ the line required is $\frac{x}{-\frac{11}{9}}=\frac{y}{\frac{13}{9}}$, or $13x+11y=0$.

9. Any line through the intersection is given by

$$3x+4y-2+\lambda(x-2y+5)=0.$$

This will pass through (1, 1), if $5+\lambda(4)=0$, that is if $\lambda=-\frac{5}{4}$.

$$\therefore \qquad 4(3x+4y-2)-5(x-2y+5)=0,$$

or $7x+26y-33=0$, is the equation required.

10. Any line through the intersection is given by

$$y-4x-1+\lambda(2x+5y-6)=0.$$

This will be perpendicular to $3y+4x=0$, if $4(-4+2\lambda)+3(1+5\lambda)=0$, that is if $23\lambda-13=0$. Therefore the required equation is

$$23(y-4x-1)+13(2x+5y-6)=0, \text{ or } 88y-66x-101=0.$$

11. The equations of the sides are

$$\frac{x-3}{3+1}=\frac{y-2}{2+1}, \quad \frac{x+1}{-1-2}=\frac{y+1}{-1-1} \text{ and } \frac{x-2}{2-3}=\frac{y-1}{1-2},$$

that is $3x-4y-1=0$, $2x-3y-1=0$ and $x-y-1=0$.

Hence the lengths of the perpendiculars on the lines from (0, 0) are respectively $\frac{1}{5}$, $\frac{1}{\sqrt{13}}$ and $\frac{1}{\sqrt{2}}$.

12. The equations of the bisectors are

$$\frac{4y+3x-12}{\sqrt{(4^2+3^2)}}=\pm\frac{3y+4x-24}{\sqrt{(3^2+4^2)}}.$$

which reduce to $y-x+12=0$ and $7y+7x-36=0$.

13. The lines $x+3y-10=0$ and $3x-y+5=0$ will be found to intersect in the point $\left(-\frac{1}{2}, \frac{7}{2}\right)$, and the lines $x+3y-20=0$ and $3x-y-5=0$ in the point $\left(\frac{7}{2}, \frac{11}{2}\right)$. Hence the equation of one diagonal is $2x-4y+15=0$.

Again, the lines $x+3y-10=0$ and $3x-y-5=0$ will be found to intersect in the point $\left(\frac{5}{2}, \frac{5}{2}\right)$, and the lines $x+3y-20=0$ and $3x-y+5=0$ in the point $\left(\frac{1}{2}, \frac{13}{2}\right)$. Hence the equation of the other diagonal is $4x+2y-15=0$.

Also the two diagonals $2x-4y+15=0$ and $4x+2y-15=0$ will be found to meet in the point $\left(\frac{3}{2}, \frac{9}{2}\right)$.

14. The angular points are $(c, -c)$, (c, c) and $(0, 0)$. Hence, from Art. 6, the area $= \frac{1}{2}\{c \cdot c - c(-c)\} = c^2$.

The result will be obvious if the three lines are drawn in a figure.

15. The angular points of the triangle will be found to be $(-2, -6)$, $\left(-\frac{4}{3}, -\frac{8}{3}\right)$ and $(0, 0)$. Hence the area

$$= \pm\frac{1}{2}\left\{(-2)\left(-\frac{8}{3}\right) - \left(-\frac{4}{3}\right)(-6)\right\} = \frac{4}{3}.$$

16. The angular points are $(7, -8)$, $\left(-\frac{5}{3}, \frac{2}{3}\right)$ and $\left(-\frac{3}{7}, \frac{22}{7}\right)$; and therefore the area of the triangle is

$$\pm\frac{1}{2}\left\{7 \cdot \frac{2}{3} - \left(-\frac{5}{3}\right)(-8) + \left(-\frac{5}{3}\right)\frac{22}{7} - \left(-\frac{3}{7}\right)\frac{2}{3} + \left(-\frac{3}{7}\right)(-8) - 7 \cdot \frac{22}{7}\right\} = \frac{338}{21}.$$

17. The lines intercept a length $c_2 - c_1$ on the axis of y, and the abscissa of their point of intersection is $\frac{c_2 - c_1}{m_1 - m_2}$; hence the area of the triangle is

$$\frac{1}{2}(c_2 - c_1)\frac{c_2 - c_1}{m_1 - m_2}.$$

18. If the lines be called respectively BCP, CAQ and ABR, where P, Q, R are the points of intersection with the axis of y; then, it will be seen from a figure that

$$\triangle ACB = \triangle AQR + \triangle BRP + \triangle CPQ,$$

provided that the area of a triangle is considered to be positive or negative according as the triangle lies on the left-hand or on the right-hand in going round in the given order of the angular points.

Hence, from the preceding example,

$$2\triangle ACB = \frac{(c_2 - c_3)^2}{m_2 - m_3} + \frac{(c_3 - c_1)^2}{m_3 - m_1} + \frac{(c_1 - c_2)^2}{m_1 - m_2}.$$

19. Let the equations of the two given straight lines be

$$x\cos\alpha + y\sin\alpha - p = 0 \text{ and } x\cos\beta + y\sin\beta - q = 0.$$

Then, if (x', y') be any point on the locus, we have

$$x'\cos\alpha + y'\sin\alpha - p + x'\cos\beta + y'\sin\beta - q = \text{constant} = k \text{ suppose}.$$

Hence (x', y') must always lie on the straight line whose equation is

$$x(\cos\alpha + \cos\beta) + y(\sin\alpha + \sin\beta) - p - q = k.$$

Pages 34—35.

1. The lines are [Art. 36] inclined at an angle

$$\tan^{-1}\frac{2\sqrt{(\sec^2\theta-1)}}{2}=\tan^{-1}(\tan\theta)=\theta.$$

2. The equation is equivalent to $(x+3y-2)(x-2y+9)=0$, and therefore represents the two lines $x+3y-2=0$, $x-2y+9=0$, which are inclined [Art. 29, iii] at any angle $\tan^{-1}\frac{1(-2)-1.3}{1.1+3(-2)}=\tan^{-1}1=45^\circ$.

3. (i) The lines are $x=a$ and $y=a$, which are obviously at right angles to one another.

(ii) The equation is equivalent to $(x-2y)(x+2y)=0$ and therefore represents the lines $x-2y=0$, $x+2y=0$: these are inclined at an angle

$$\tan^{-1}\frac{1.2-1(-2)}{1.1+2(-2)}=\tan^{-1}-\frac{4}{3}.$$

(iii) The lines are the axes $x=0$, $y=0$.

(iv) The equation may be written $(x-3)(y-2)=0$ and therefore represents the lines $x-3=0$, $y-2=0$ which are obviously at right angles to one another.

(v) The equation may be written $(x-4y)(x-y)=0$, and therefore represents the lines $x-4y=0$ and $x-y=0$ which are inclined at an angle

$$\tan^{-1}\frac{1(-4)-1(-1)}{1.1+(-4)(-1)}=\tan^{-1}\frac{-3}{5}.$$

(vi) The equation may be written $(x-4y+4)(x-y-1)=0$ and therefore represents the lines $x-4y+4=0$, $x-y-1=0$; the lines are parallel to the lines in (v) and are inclined at an angle $\tan^{-1}\frac{-3}{5}$.

(vii) Writing the equation in the form $(x+y\cot 2a)^2-(y^2+y^2\cot^2 2a)=0$, that is $(x+y\cot 2a)^2-y^2\operatorname{cosec}^2 2a=0$, we see that the equation represents the two lines $x+y\cot 2a\pm y\operatorname{cosec} 2a=0$, that is $x\sin a+y\cos a=0$ and $x\cos a-y\sin a=0$.

4. From Art. 37 (iii), the condition is

$$12.2.\lambda-12.\left(-\frac{5}{2}\right)^2-2\left(\frac{11}{2}\right)^2-\lambda(-5)^2+2\left(-\frac{5}{2}\right)\left(\frac{11}{2}\right)(-5)=0,$$

whence $\lambda=2$.

The lines are parallel to $12x^2-10xy+2y^2=0$ and are therefore inclined at an angle $\tan^{-1}\frac{2\sqrt{(25-24)}}{12+2}=\tan^{-1}\frac{1}{7}$.

5. From Art. 37 (iii) the condition is

$$12.2.2-12\left(-\frac{5}{2}\right)^2-2\left(\frac{11}{2}\right)^2-2\left(\frac{\lambda}{2}\right)^2+2\left(-\frac{5}{2}\right)\left(\frac{11}{2}\right)\frac{\lambda}{2}=0,$$

whence $2\lambda^2+55\lambda+350=0$, so that $\lambda=-10$ or $\lambda=-17\frac{1}{2}$.

6. The condition is

$$12.\lambda.3-12.3^2-\lambda.3^2-3.18^2+2.3.3.18=0,$$

whence $\lambda=28$.

The lines are imaginary since $18^2-12.28$ is negative.

7. The condition is $-2\left(\frac{\lambda}{2}\right)^2+2.\frac{3}{2}.\frac{5}{2}.\frac{\lambda}{2}=0$, whence $\lambda=\frac{15}{2}$. [For the value $\lambda=0$ the equation would represent *one* straight line.]

8. The equation of the lines is $3x^2+5xy-3y^2+(2x+3y)(3x-2y)=0$, that is $9x^2+10xy-9y^2=0$; and the pair of lines $9x^2+10xy-9y^2=0$ are at right angles since the sum of the coefficients of x^2 and y^2 is zero.

Page 39.

1. Put $a=-1$, $b=1$, $h=0$ in the formula of Art. 44.

2. The equation is of the form $a(x-1)+b(y-2)=0$ and if the line is perpendicular to $x+2y=0$ we have [Art. 42, (iii)] $a+2b-(2a+b)\cos 60^0=0$, whence $b=0$, so that the required equation is $x-1=0$.

3. $$\tan^{-1}\frac{5\sin\omega}{1+5\cos\omega}=\tan^{-1}\frac{5.\frac{4}{5}}{1+5.\frac{3}{5}}=\tan^{-1}1=45^0.$$

4. The lines $y=m_1x$, $y=m_2x$ make *equal* angles with $y=0$ if

$$\frac{m_1\sin\omega}{1+m_1\cos\omega}=-\frac{m_2\sin\omega}{1+m_2\cos\omega},$$

whence $m_1+m_2+2m_1m_2\cos\omega=0$.

5. $m_1+m_2=-\frac{2B}{C}$ and $m_1m_2=\frac{A}{C}$; hence, from 4, $-B+A\cos\omega=0$.

6. The lines are inclined at an angle

$$\tan^{-1}2\sin\omega\sqrt{(\cos^2\omega-\cos 2\omega)}/(1+\cos 2\omega-2\cos^2\omega)=\tan^{-1}\infty=\frac{\pi}{2}.$$

7. The polar equation of the line joining (r_1, θ_1), (r_2, θ_2) is

$$r_1r_2\sin(\theta_2-\theta_1)+r_2r\sin(\theta-\theta_2)+rr_1\sin(\theta_1-\theta)=0,$$

or $r\cos\theta(r_1\sin\theta_1-r_2\sin\theta_2)+r\sin\theta(r_2\cos\theta_2-r_1\cos\theta_1)-r_1r_2\sin(\theta_1-\theta_2)=0.$

Comparing with $p=r\cos(\theta-\alpha)$, that is with

$$r\cos\theta\cos\alpha+r\sin\theta\sin\alpha-p=0, \text{ we have}$$

$$\frac{\cos\alpha}{r_1\sin\theta_1-r_2\sin\theta_2}=\frac{\sin\alpha}{r_2\cos\theta_2-r_1\cos\theta_1}=\frac{p}{r_1r_2\sin(\theta_1-\theta_2)};$$

$$\therefore\quad p=r_1r_2\sin(\theta_1-\theta_2)/\sqrt{\{(r_1\sin\theta_1-r_2\sin\theta_2)^2+(r_2\cos\theta_2-r_1\cos\theta_1)^2\}}$$
$$=r_1r_2\sin(\theta_1-\theta_2)/\sqrt{\{r_1^2+r_2^2-2r_1r_2\cos(\theta_1-\theta_2)\}},$$

and $$\alpha=\tan^{-1}\frac{r_2\cos\theta_2-r_1\cos\theta_1}{r_1\sin\theta_1-r_2\sin\theta_2}.$$

The required polar coordinates are p and α.

Pages 43—46.

1. Let the equation of the straight line in any one of its possible positions be $\frac{x}{h}+\frac{y}{k}=1$; then we have $\frac{1}{h}+\frac{1}{k}=\text{const.}=\frac{1}{a}$ suppose, that is $\frac{a}{h}+\frac{a}{k}=1$, from which it is clear that the line always passes through the fixed point (a, a).

2. From Art. 29 it follows that two lines $y=m_1x$ and $y=m_2x$ are at right angles when the value of $\frac{y}{x}$ for one line is equal to the value of $-\frac{x}{y}$ for the other. Hence the two lines through the origin perpendicular to

$$ax^2+2hxy+by^2=0,$$

that is to $a+2h\frac{y}{x}+b\left(\frac{y}{x}\right)^2$ are given by

$$a+2h\left(-\frac{x}{y}\right)+b\left(-\frac{x}{y}\right)^2=0,$$

that is $ay^2-2hxy+bx^2=0$.

3. The line through (a, b) perpendicular to

$$y-mx=0 \text{ is } (x-a)+m(y-b)=0,$$

so that the value of $\frac{y}{x}$ for one line is equal to the value of $-\frac{x-a}{y-b}$ for the perpendicular line through (a, b). Hence the lines through (a, b) perpendicular to the lines

$$p_0y^n+p_1y^{n-1}x+p_2y^{n-2}x^2+\ldots+p_nx^n=0,$$

that is to $$p_0\left(\frac{y}{x}\right)^n+p_1\left(\frac{y}{x}\right)^{n-1}+p_2\left(\frac{y}{x}\right)^{n-2}+\ldots+p_n=0$$

are given by $$p_0\left(-\frac{x-a}{y-b}\right)^n+\ldots\ldots+p_n=0,$$

that is $$p_n(y-b)^n-\ldots\ldots+(-1)^np_0(x-a)^n=0.$$

4. The angles the lines make with the axis of x are given by

$$1+3\tan\theta-3\tan^2\theta-\tan^3\theta=0,$$

whence $\tan 3\theta=-1$. Hence the lines make angles

$$\frac{\pi}{4},\ \frac{\pi}{4}+\frac{2\pi}{3}\text{ and }\frac{\pi}{4}+\frac{4\pi}{3}$$

with the axis of x, and therefore are equally inclined to one another.

5. Take OA for the axis of x and OB for the axis of y. Let $OA=a$, $OB=b$, and let $AP:BQ=m:n$. Then if $AP=mk$, BQ will be nk. The coordinates of P will be $a+mk$, 0 and those of Q will be 0, $b+nk$. Hence, if (x, y) be the middle point of PQ, we have

$$x=\frac{1}{2}(a+mk),\ y=\frac{1}{2}(b+nk);\quad \therefore \frac{2x-a}{m}=\frac{2y-b}{n},$$

which is the equation of the required locus.

6. Let the equation of the straight line in any one of its possible positions be $x\cos\alpha+y\sin\alpha-p=0$, where α and p are unknown; and let (a_1, b_1), (a_2, b_2), &c. be the given fixed points. Then

$$(a_1\cos\alpha+b_1\sin\alpha-p)+(a_2\cos\alpha+b_2\sin\alpha-p)+\ldots\ldots=0\ldots\ldots\ldots\text{(i)}.$$

Let $\Sigma a=a_1+a_2+\ldots$, and $\Sigma b=b_1+b_2+\ldots$, and let there be n of the given fixed points; then (i) may be written $\Sigma a\,.\cos\alpha+\Sigma b\,.\sin\alpha-np=0$, that is $\frac{\Sigma a}{n}\,.\cos\alpha+\frac{\Sigma b}{n}\,.\sin\alpha-p=0$; whence it follows that the fixed point $\left(\frac{\Sigma a}{n}, \frac{\Sigma b}{n}\right)$ is always on the straight line. [The fixed point through which the line always passes is the centre of mean position of the given points.]

7. Let the fixed lines OM, ON be taken as axes, and let ω be the angle between them.

Let x, y be the coordinates of P and ξ, η the coordinates of Q. Then it will be seen from a figure that $OM=x+y\cos\omega$, and $ON=y+x\cos\omega$. Hence, as $OM=\xi$ and $ON=\eta$, we have

$$\xi=x+y\cos\omega\text{ and }\eta=y+x\cos\omega;$$

$$\therefore\ x\sin^2\omega=\xi-\eta\cos\omega\text{ and }y\sin^2\omega=\eta-\xi\cos\omega\ \ldots\ldots\ldots\ldots\text{(i)}.$$

Hence, if $Ax+By+C=0$ be the equation of the straight line on which P moves, we have, by substituting from (i),

$$A(\xi-\eta\cos\omega)+B(\eta-\xi\cos\omega)+C\sin^2\omega=0,$$

which is a relation of the first degree in ξ and η, and therefore shews that $Q(\xi, \eta)$ is always on a fixed straight line.

8. Take the point O for pole and any straight line through O for initial line, and let the equations of the fixed straight lines be $p_1=r\cos(\theta-\alpha_1)$ and $p_2=r\cos(\theta-\alpha_2)$. Then, if the line OPQ make any angle θ with the initial line, we have

$$p_1=OP\cos(\theta-\alpha_1)\text{ and }p_2=OQ\cos(\theta-\alpha_2);$$

$$\therefore \quad \frac{1}{OP} + \frac{1}{OQ} = \frac{\cos(\theta - \alpha_1)}{p_1} + \frac{\cos(\theta - \alpha_2)}{p_2},$$

so that $$\frac{2}{OR} = \frac{\cos(\theta - \alpha_1)}{p_1} + \frac{\cos(\theta - \alpha_2)}{p_2} = A\cos\theta + B\sin\theta,$$

where $$A = \frac{\cos\alpha_1}{p_1} + \frac{\cos\alpha_2}{p_2} \quad \text{and} \quad B = \frac{\sin\alpha_1}{p_1} + \frac{\sin\alpha_2}{p_2}.$$

Hence the equation of the locus of R is

$$2 = A \,.\, OR\cos\theta + B \,.\, OR\sin\theta$$
$$= Ax + By,$$

where x, y are the rectangular coordinates of R.

9. One diagonal is the line joining the point of intersection of $\alpha = 0$ and $\alpha' = 0$ to the point of intersection of $\alpha = c$ and $\alpha' = c$.

Now any line through the point of intersection of $\alpha = 0$ and $\alpha' = 0$ is given by

$$\alpha - \lambda\alpha' = 0 \text{..............................} \quad \text{(i) [Art. 33].}$$

The equation (i) will therefore be the equation of one of the diagonals provided λ be so chosen that (i) goes through the point of intersection of $\alpha = c$ and $\alpha' = c$, the condition for which is $c - \lambda c = 0$, so that $\lambda = 1$.

Hence the equation of one diagonal is $\alpha - \alpha' = 0$. Similarly the equation of the other diagonal is $\alpha + \alpha' - c = 0$.

10. Let $AB = a$, $AD = b$, and angle $BAD = \alpha$.

Then the equation of AB is $\theta = 0$, and of AD is $\theta = \alpha$.

The perpendicular from A on BC is $a\sin\alpha$, and it makes an angle $\alpha - \frac{\pi}{2}$ with AB; hence the equation of BC is $a\sin\alpha = r\cos\left(\theta - \alpha + \frac{\pi}{2}\right)$.

The perpendicular from A on CD is $b\sin\alpha$ and it makes an angle $\frac{\pi}{2}$ with AB; hence the equation of CD is

$$b\sin\alpha = r\cos\left(\theta - \frac{\pi}{2}\right).$$

The line AC makes with AB an angle $\tan^{-1} b\sin\alpha / (a + b\cos\alpha)$, and therefore its equation is $\theta = \tan^{-1} b\sin\alpha / (a + b\cos\alpha)$.

The equation of BD referred to AB and AD as oblique axes is obviously $x/a + y/b = 1$.

But $x/\sin(\alpha - \theta) = y/\sin\theta = r/\sin\alpha$; hence the polar equation of BD is

$$r\sin(\alpha - \theta)/a + r\sin\theta/b = \sin\alpha.$$

11. Let L be the point (h, k), and let M, N be the feet of the perpendiculars on the axes, and LK the perpendicular from L to MN.

Then $LK \,.\, MN = 2\triangle MLN = ML \,.\, NL\sin\omega$. But $ML = k\sin\omega$ and $NL = h\sin\omega$; also, since O, M, L, N are on a circle whose diameter is OL,

$$MN = OL\sin\omega = \sin\omega\sqrt{(h^2 + k^2 + 2hk\cos\omega)}.$$

Hence $$LK = hk\sin^2\omega/\sqrt{(h^2+k^2+2hk\cos\omega)}.$$

Again, since $OM = h + k\cos\omega$ and $ON = k + h\cos\omega$, the equation of MN is

$$\frac{x}{h+k\cos\omega} + \frac{y}{k+h\cos\omega} = 1.$$

Also the equation of LK is of the form

$$A(x-h) + B(y-k) = 0 \quad \text{.........................} \quad \text{(i)}.$$

But LK is perpendicular to MN; hence [Art. 42]

$$A(k+h\cos\omega) + B(h+k\cos\omega) - A\cos\omega(h+k\cos\omega) - B\cos\omega(k+h\cos\omega) = 0,$$

that is $$Ak\sin^2\omega + Bh\sin^2\omega = 0 \quad \text{.........................} \quad \text{(ii)}.$$

From (i) and (ii) it follows that the equation of LK is

$$h(x-h) - k(y-k) = 0.$$

12. Let $y = mx$ be the equation of a straight line through (0, 0) which is at a distance δ from (x_1, y_1); then

$$\frac{y_1 - mx_1}{\sqrt{(1+m^2)}} = \delta; \quad \therefore (y_1 - mx_1)^2 = \delta^2(1+m^2).$$

But $m = \frac{y}{x}$; and therefore

$$\left(y_1 - \frac{y}{x}x_1\right)^2 = \delta^2\left(1 + \frac{y^2}{x^2}\right),$$

that is $$(xy_1 - x_1y)^2 = \delta^2(x^2+y^2).$$

13. Take AB for the axis of x and AC for the axis of y, and let $AB = a$, $AC = b$.

Then F is $(a, -a)$ and C is $(0, b)$; $\therefore$ equation of FC is $\frac{x-a}{a} = \frac{y+a}{-a-b}$, that is $(a+b)x + ay - ab = 0$. Similarly the equation of KB is

$$(a+b)y + bx - ab = 0.$$

Hence $$(a+b)x + ay - ab - \{(a+b)y + bx - ab\} = 0,$$

that is $ax - by = 0$, is the equation of a line through the intersection of FC and KB. But $ax - by = 0$ is obviously perpendicular to $\frac{x}{a} + \frac{y}{b} = 1$, that is to BC; hence $ax - by = 0$ is the equation of AL, so that AL passes through the intersection of FC and AB.

14. The equation of the line joining (3, 4) and (1, −1) is $5x - 2y - 7 = 0$. A side of the square makes an angle $\frac{\pi}{4}$ with the diagonal, and hence if a side be parallel to $y = mx$, we have $\pm 1 = \dfrac{m - \frac{5}{2}}{1 + \frac{5}{2}m}$, whence $m = -\frac{7}{3}$ or $m = \frac{3}{7}$.

Hence the required equations are $3(x-3) - 7(y-4) = 0$, $3(x-1) - 7(y+1) = 0$, $7(x-3) + 3(y-4) = 0$ and $7(x-1) + 3(y+1) = 0$.

15. Take A for origin and ABX for the axis of x, and let $AB=c$.

Then if $y=mx$ be the equation of AC and $y=m'(x-c)$ the equation of BC; then $m=\tan XAC$ and $m'=\tan XBC$. But since $ABC\sim BAC$ is constant, it follows that the sum of the angles XAC and XBC is constant, and therefore $\frac{m+m'}{1-mm'}=$const. Hence, putting $\frac{y}{x}$ for m and $\frac{y}{x-c}$ for m', we have for the required equation $\dfrac{\frac{y}{x}+\frac{y}{x-c}}{1-\frac{y}{x}\cdot\frac{y}{x-c}}=$const. $=k$ suppose; therefore $y(2x-c)=k(x^2-y^2-cx)$.

16. Take the given line as axis of x and any perpendicular line as axis of y. Let AB and CD be the fixed lengths, and let $OA=a$, $OB=b$, $OC=c$ and $OD=d$.

Then, if (x', y') be any point P on the locus, the equations of PA, PB, PC and PD will be respectively

$$\frac{x-x'}{x'-a}=\frac{y-y'}{y'},\quad \frac{x-x'}{x'-b}=\frac{y-y'}{y'},\quad \frac{x-x'}{x'-c}=\frac{y-y'}{y'} \text{ and } \frac{x-x'}{x'-d}=\frac{y-y'}{y'}.$$

Hence the angle APB is $\tan^{-1}\left(\frac{y'}{x'-a}-\frac{y'}{x'-b}\right)\Big/\left\{1+\frac{y'^2}{(x'-a)(x'-b)}\right\}$ $=\tan^{-1}(a-b)\,y'/\{(x'-a)(x'-b)+y'^2\}$; and so also the angle CPD is $\tan^{-1}(c-d)\,y'/\{(x'-c)(x'-d)+y'^2\}$. But the angles APB and CPD are equal; hence

$$(a-b)\,y'/\{(x'-a)(x'-b)+y'^2\}=(c-d)\,y'/\{(x'-c)(x'-d)+y'^2\}.$$

Hence, rejecting the solution $y=0$ for which each of the angles APB and CPD is zero, the equation of the required locus is

$$(a-b)\{(x-c)(x-d)+y^2\}=(c-d)\{(x-a)(x-b)+y^2\}.$$

17. If (x', y') be any point which satisfies the given condition, we have

$$(x'\cos\theta+y'\sin\theta-a)(x'\cos\phi+y'\sin\phi-a)$$

$$=\left(x'\cos\frac{\theta+\phi}{2}+y'\sin\frac{\theta+\phi}{2}-a\cos\frac{\theta-\phi}{2}\right)^2;$$

$$\therefore\ x'^2\left(\cos\theta\cos\phi-\cos^2\frac{\theta+\phi}{2}\right)+y'^2\left(\sin\theta\sin\phi-\sin^2\frac{\theta+\phi}{2}\right)$$

$$+x'y'\left(\sin\theta\cos\phi+\cos\theta\sin\phi-2\sin\frac{\theta+\phi}{2}\cos\frac{\theta+\phi}{2}\right)$$

$$-ax'\left(\cos\theta+\cos\phi-2\cos\frac{\theta+\phi}{2}\cos\frac{\theta-\phi}{2}\right)$$

$$-ay'\left(\sin\theta+\sin\phi-2\sin\frac{\theta+\phi}{2}\cos\frac{\theta-\phi}{2}\right)+a^2\sin^2\frac{\theta-\phi}{2}=0.$$

$$\therefore \sin^2\frac{\theta-\phi}{2}(x'^2+y'^2-a^2)=0,$$

the coefficients of $x'y'$, x' and y' being zero.

Hence (x', y') is always on the locus whose equation is $x^2+y^2-a^2=0$.

18. Take the given straight line for the axis of y and the line AB for the axis of x, and let A be $(a, 0)$ and B be $(b, 0)$.

Then, if (x', y') be any point P on the locus, the equations of PA, PB will be $\frac{x-a}{a-x'}=\frac{y}{-y'}$ and $\frac{x-b}{b-x'}=\frac{y}{-y'}$.

The intercepts made on the axis of y by PA, PB respectively are therefore $\frac{ay'}{a-x'}$ and $\frac{by'}{b-x'}$. Hence $\frac{ay'}{a-x'}-\frac{by'}{b-x'}=\text{constant}=l$ suppose.

Hence (x', y') is on the locus whose equation is

$$\frac{ay}{a-x}-\frac{by}{b-x}=l,$$

or

$$(b-a)\,xy=l\,(a-x)\,(b-x).$$

19. The area of a parallelogram is easily seen to be equal to $p_1p_2 \operatorname{cosec}\theta$, where p_1, p_2 are the perpendicular distances between the pairs of parallel sides, and θ is the angle between two intersecting sides. In the present case we have

$$p_1=\frac{7a_1-7a_2}{\sqrt{(3^2+4^2)}}=\frac{7}{5}(a_1-a_2), \text{ and } p_2=\frac{7}{5}(b_1-b_2);$$

also

$$\theta=\tan^{-1}\frac{4.4-3.3}{3.4+4.3}=\tan^{-1}\frac{7}{24}=\operatorname{cosec}^{-1}\frac{25}{7}$$

Hence the area

$$=\frac{7}{5}(a_1-a_2)\cdot\frac{7}{5}(b_1-b_2)\cdot\frac{25}{7}=7\,(a_1-a_2)\,(b_1-b_2).$$

20. Let the two straight lines given by $ax^2+2hxy+by^2=0$ be $y=m_1x$ and $y=m_2x$; then $m_1+m_2=-\frac{2h}{b}$, $m_1m_2=\frac{a}{b}$.

Now $y=m_1x$ and $y=m_2x$ meet $lx+my+n=0$ in the points

$$\left(-\frac{n}{l+mm_1}, -\frac{nm_1}{l+mm_1}\right) \text{ and } \left(-\frac{n}{l+mm_2}, -\frac{nm_2}{l+mm_2}\right)$$

respectively. Hence the area of the triangle is

$$\frac{1}{2}\frac{n^2(m_2-m_1)}{(l+mm_1)\,(l+mm_2)}=\frac{1}{2}\frac{n^2\sqrt{\{(m_1+m_2)^2-4m_1m_2\}}}{l^2+lm\,(m_1+m_2)+m^2m_1m_2}$$

$$=\frac{1}{2}\frac{n^2\sqrt{\left\{\frac{4h^2}{b^2}-4\frac{a}{b}\right\}}}{l^2-2lm\frac{h}{b}+m^2\frac{a}{b}}=\frac{n^2\sqrt{(h^2-ab)}}{bl^2-2hlm+am^2}.$$

21. The equation of the bisectors of the angles between the lines $ax^2+2hxy+by^2=0$ is

$$\frac{x^2-y^2}{a-b}=\frac{xy}{h}\ \text{[Art. 39.]}$$

Also the equation of the bisectors of the angles between the lines

$$ax^2+2hxy+by^2+\lambda(x^2+y^2)=0 \text{ is}$$

$$\frac{x^2-y^2}{(a+\lambda)-(b+\lambda)}=\frac{xy}{h}.$$

Hence the two pairs of lines have the same bisectors, whence the result obviously follows.

22. If one of the lines given by $ax^2+2hxy+by^2=0$ is the same as one of the lines given by $a'x^2+2h'xy+b'y^2=0$, then the two equations

$$a+2hm+bm^2=0 \text{ and } a'+2h'm+b'm^2=0$$

will both be satisfied by some value of m, and this value of m is given by

$$\frac{m^2}{ah'-a'h}=\frac{2m}{ba'-b'a}=\frac{1}{hb'-h'b},$$

whence it follows that

$$4(ah'-a'h)(hb'-h'b)=(ba'-b'a)^2,$$

which is the condition required.

23. The two lines through the origin perpendicular to $ax^2+2hxy+by^2=0$ are given by $bx^2-2hxy+ay^2=0$. [See question 2.]

The condition required is therefore the condition that one of the lines $bx^2-2hxy+ay^2=0$ may coincide with one of the lines $a'x^2+2h'x'y'+b'y^2=0$. Hence, from the result of the preceding example, the condition required is

$$4(bh'+a'h)(-hb'-h'a)=(aa'-bb')^2.$$

24. The perpendiculars on the sides of a triangle from the centre of any one of the four circles which touch the sides are equal in *magnitude*.

Hence the four centres are given by

$$\frac{4y+3x}{5}=\pm\frac{12y-5x}{13}=\pm\frac{y-15}{1}.$$

Now the angular points of the triangle are (36, 15), (−20, 15) and (0, 0); and when the coordinates of these angular points are respectively substituted in the left-hand members of the equations $4y+3x=0$, $12y-5x=0$ and $y-15=0$, the results are +, +, −1. These must also be the signs of the results when the coordinates of the *inscribed* circle are substituted, since the centre of the inscribed circle is *within* the triangle. [See solution of **18**, page 24.] Hence the coordinates of the centre of the inscribed circle satisfy the relations

$$\frac{4y+3x}{5}=\frac{12y-5x}{13}=-\frac{y-15}{1};$$

$\therefore\ 9y+3x=75$ and $25y-5x=195$, whence $x=1$ and $y=8$.

25. The equations of the sides are

$$2x+y-7=0,\ x+2y-5=0,\ x-y+1=0;$$

and when the coordinates of the angular points are substituted in the left-hand members of the equations of the opposite sides of the triangle the results are −, +, + respectively. These must also be the signs of the results when the coordinates of the inscribed circle are substituted, since the centre of the inscribed circle is *within* the triangle. Hence the coordinates of the inscribed circle satisfy the relations

$$-\frac{2x+y-7}{\sqrt{5}}=\frac{x+2y-5}{\sqrt{5}}=\frac{x-y+1}{\sqrt{2}},$$

$$\text{whence } x=\frac{1}{6}(8+\sqrt{10}),\ y=\frac{1}{6}(16-\sqrt{10}).$$

The centre of the escribed circles corresponding to the points (1, 2), (2, 3) and (3, 1) are given respectively by the relations

$$\frac{2x+y-7}{\sqrt{5}}=\frac{x+2y-5}{\sqrt{5}}=\frac{x-y+1}{\sqrt{2}},$$

$$-\frac{2x+y-7}{\sqrt{5}}=-\frac{x+2y-5}{\sqrt{5}}=\frac{x-y+1}{\sqrt{2}},$$

$$\text{and} -\frac{2x+y-7}{\sqrt{5}}=\frac{x+2y-5}{\sqrt{5}}=-\frac{x-y+1}{\sqrt{2}}.$$

26. Putting $y=x\tan\theta$, we have

$$1-3\tan^2\theta=m(\tan^3\theta-3\tan\theta);$$

$$\therefore\ \tan 3\theta=-\frac{1}{m}.$$

Hence, if a be any one value of $\frac{1}{3}\tan^{-1}\left(-\frac{1}{m}\right)$, $a+\frac{2\pi}{3}$ and $a+\frac{4\pi}{3}$ will also be values, and the three lines represented by the given equation will be those for which $\theta=a$, $\theta=a+\frac{2\pi}{3}$ and $\theta=a+\frac{4\pi}{3}$.

27. Let the lines be $y-m_1x=0$ and $y-m_2x=0$; then $m_1+m_2=-2\frac{h}{b}$, and $m_1m_2=\frac{a}{b}$.

The product of the perpendiculars from (x', y') on the two lines will be

$$\frac{y'-m_1x'}{\sqrt{(1+m_1^2)}}\times\frac{y'-m_2x'}{\sqrt{(1+m_2^2)}}=\frac{y'^2-(m_1+m_2)x'y'+m_1m_2x'^2}{\sqrt{\{1+m_1^2+m_2^2+m_1^2m_2^2\}}}$$

$$=\frac{y'^2+\frac{2h}{b}x'y'+\frac{a}{b}x'^2}{\sqrt{\left\{1+\frac{4h^2}{b^2}-\frac{2a}{b}+\frac{a^2}{b^2}\right\}}}$$

$$=\frac{by'^2+2hx'y'+ax'^2}{\sqrt{\{(a-b)^2+4h^2\}}}.$$

28. Let $ax^2+2hxy+by^2=b(y-m_1x)(y-m_2x)$, so that the lines are $y-m_1x=0$ and $y-m_2x=0$.

Then $$p_1^2+p_2^2=\frac{(y-m_1x)^2}{1+m_1^2}+\frac{(y-m_2x)^2}{1+m_2^2}$$

$$=\frac{2y^2+y^2(m_1^2+m_2^2)-2xy(1+m_1m_2)(m_1+m_2)+x^2(m_1^2+m_2^2+2m_1^2m_2^2)}{1+m_1^2+m_2^2+m_1^2m_2^2}$$

$$=\frac{2y^2+y^2\left(\frac{4h^2}{b^2}-2\frac{a}{b}\right)+2xy\left(1+\frac{a}{b}\right)\frac{2h}{b}+x^2\left(\frac{4h^2}{b^2}-2\frac{a}{b}+2\frac{a^2}{b^2}\right)}{1+\frac{4h^2}{b^2}-2\frac{a}{b}+\frac{a^2}{b^2}},$$

$$\therefore\ (p_1^2+p_2^2)\{(a-b)^2+4h^2\}=y^2(2b^2-2ab+4h^2)+4h(a+b)xy$$
$$+x^2(2a^2-2ab+4h^2).$$

29. If the straight lines be $y-m_1x=0$, $y-m_2x=0$, $y-m_3x=0$; then

$$k^6=\frac{(y-m_1x)^2(y-m_2x)^2(y-m_3x)^2}{(1+m_1^2)(1+m_2^2)(1+m_3^2)}.$$

But $ay^3+bxy^2+cx^2y+dx^3=a(y-m_1x)(y-m_2x)(y-m_3x)$.

Hence $$m_1+m_2+m_3=-\frac{b}{a},$$

$$m_2m_3+m_3m_1+m_1m_2=\frac{c}{a},$$

and $$m_1m_2m_3=-\frac{d}{a}.$$

Hence $(1+m_1^2)(1+m_2^2)(1+m_3^2)$

$$=1+(m_1+m_2+m_3)^2-2(m_2m_3+m_3m_1+m_1m_2)+(m_2m_3+m_3m_1+m_1m_2)^2$$
$$-2m_1m_2m_3(m_1+m_2+m_3)+m_1^2m_2^2m_3^2$$

$$=1+\frac{b^2}{a^2}-2\frac{c}{a}+\frac{c^2}{a^2}-2\frac{bd}{a^2}+\frac{d^2}{a^2}$$

$$=\frac{1}{a^2}\{(a-c)^2+(b-d)^2\}.$$

Hence $$k^3=\frac{ay^3+bxy^2+cx^2y+dx^3}{\sqrt{\{(a-c)^2+(b-d)^2\}}}.$$

30. If the three lines be $y-m_1x=0$, $y-m_2x=0$, and $y-m_3x=0$; then

$$m_1m_2m_3=-\frac{A}{D}.$$

And, if the first two lines are at right angles, $m_1m_2=-1$; $\therefore\ m_3=\frac{A}{D}$. But m_3 is a root of the equation $A+3Bm+3Cm^2+Dm^3=0$;

$$\therefore\ A+3B.\frac{A}{D}+3C.\frac{A^2}{D^2}+D.\frac{A^3}{D^3}=0,$$

that is $$AD^2+3ABD+3A^2C+A^3=0.$$

31. The equations of any two pairs of perpendicular lines through the origin are of the form

$$x^2+\lambda xy-y^2=0 \text{ and } x^2+\mu xy-y^2=0.$$

Hence we have to shew that for some values of λ and μ the equation

$$a(x^2+\lambda xy-y^2)(x^2+\mu xy-y^2)=0$$

must be identical with the given equation.

Comparing the coefficients of x^3y, x^2y^2 and xy^3 we must have

$$-4b=a(\lambda+\mu),$$

$$6c=a(\lambda\mu-2),$$

and

$$-4b=a(\lambda+\mu).$$

Since there are only *two* different conditions to be satisfied by λ and μ, these conditions can always be satisfied.

The two pairs of lines will coincide if $\lambda=\mu$; and in this case $a\lambda=-2b$ and $a\lambda^2=2a+6c$; $\therefore 2b^2=a^2+3ac$.

32. The equation of any pair of perpendicular lines through the origin can be put into the form $y^2+\lambda xy-x^2=0$. Hence the necessary and sufficient condition that two of the lines represented by the given equation may be at right angles is that the left-hand member of the given equation may have a factor of the form $y^2+\lambda xy-x^2$.

Hence

$$ay^4+bxy^3+cx^2y^2+dx^3y+ex^4=(y^2+\lambda xy-x^2)(ay^2+\mu xy-ex^2);$$

$$\therefore\ b=a\lambda+\mu,\ c=\lambda\mu-a-e \text{ and } d=-\mu-e\lambda.$$

Eliminating λ and μ we obtain the required result.

33. The locus of the equation

$$g'(ax^2+2hxy+by^2+2gx)-g(a'x^2+2h'xy+b'y^2+2g'x)=0\ldots\ldots\text{(i)}$$

clearly passes through all points which satisfy both $ax^2+2hxy+by^2+2gx=0$, and $a'x^2+2h'xy+b'y^2+2g'x=0$.

But (i) is a homogeneous equation of the second degree, and therefore represents two straight lines through the origin. Hence (i) is the equation of the straight lines joining the origin to the points of intersection of the given curves. The condition that the lines represented by (i) are at right angles is [Art. 36]

$$ag'-a'g+bg'-b'g=0,$$

or

$$g'(a+b)=g(a'+b').$$

34. Let ABC, $A'B'C'$ be the two triangles, and let A, B, C, be (a_1, b_1), (a_2, b_2) and (a_3, b_3) respectively, and A', B', C' be (α_1, β_1) (α_2, β_2) and (α_3, β_3) respectively.

The equations of the three perpendiculars from A', B', C' on BC, CA, AB respectively are

$$x(a_2-a_3)+y(b_2-b_3)-\alpha_1(a_2-a_3)-\beta_1(b_2-b_3)=0 \quad\ldots\ldots\ldots\ldots\text{(i)},$$

$$x(a_3-a_1)+y(b_3-b_1)-\alpha_2(a_3-a_1)-\beta_2(b_3-b_1)=0 \quad\ldots\ldots\ldots\ldots\text{(ii)},$$

and $$x(a_1-a_2)+y(b_1-b_2)-\alpha_3(a_1-a_2)-\beta_3(b_1-b_2)=0 \quad\ldots\ldots\ldots\ldots\text{(iii)}.$$

From (i) and (ii) by addition, we have

$$x(a_2-a_1)+y(b_2-b_1)-\alpha_1(a_2-a_3)-\beta_1(b_2-b_3)$$
$$-\alpha_2(a_3-a_1)-\beta_2(b_3-b_1)=0 \quad\ldots\ldots\ldots\ldots\text{(iv)}.$$

Now (iv) is the equation of a straight line through the point of intersection of (i) and (ii) [Art. 33], and this straight line is clearly parallel to (iii). Hence in order that the lines (i), (ii), (iii) should meet in a point it is necessary and sufficient that (iv) and (iii) should represent the *same* straight line; and the condition for this is that

$$\alpha_1(a_2-a_3)+\beta_1(b_2-b_3)+\alpha_2(a_3-a_1)+\beta_2(b_3-b_1)+\alpha_3(a_1-a_2)+\beta_3(b_1-b_2)=0,$$

which may be written in the form

$$a_1(\alpha_2-\alpha_3)+b_1(\beta_2-\beta_3)+a_2(\alpha_3-\alpha_1)+b_2(\beta_3-\beta_1)+a_3(\alpha_1-\alpha_2)+b_3(\beta_1-\beta_2)=0,$$

and this is the necessary and sufficient condition that the perpendiculars from A, B, C to the sides $B'C'$, $C'A'$, $A'B'$ respectively should meet in a point.

35. Let $A'B'C'$ be any triangle having its angular points A', B', C' on the fixed straight lines OA, OB, OC respectively; and let the side $A'B'$ pass through the fixed point P and the side $B'C'$ through the fixed point Q.

Take OA for axis of x, and OC for axis of y, and let the equation of OB be $y=mx$, and let P be (α, β) and Q be (γ, δ).

Then, if the co-ordinate of B' be (x', mx'), the equations of $B'PA'$ and $B'QC'$ will be respectively

$$\frac{x-\alpha}{\alpha-x'}=\frac{y-\beta}{\beta-mx'} \text{ and } \frac{x-\gamma}{\gamma-x}=\frac{y-\delta}{\delta-mx'};$$

$$\therefore\ OA'=\alpha-\frac{\beta(\alpha-x')}{\beta-mx'}=\frac{x'(\beta-m\alpha)}{\beta-mx'}.$$

and $$OC'=\delta-\frac{\gamma(\delta-mx')}{\gamma-x'}=x'\frac{m\gamma-\delta}{\gamma-x'}.$$

Hence the equation of $A'C'$ is

$$\frac{x(\beta-mx')}{x'(\beta-m\alpha)}+\frac{y(\gamma-x')}{x'(m\gamma-\delta)}=1,$$

or $$\frac{x\beta}{\beta-m\alpha}+\frac{y\gamma}{m\gamma-\delta}-x'\left(\frac{mx}{\beta-m\alpha}+\frac{y}{m\gamma-\delta}+1\right)=0,$$

which shews that $A'C'$ always passes through a fixed point for all values of x', namely through the point of intersection of the straight lines

$$\frac{x\beta}{\beta-m\alpha}+\frac{y\gamma}{m\gamma-\delta}=0 \text{ and } \frac{mx}{\beta-m\alpha}+\frac{y}{m\gamma-\delta}+1=0.$$

CHAPTER III.

Page 49.

2. The equation becomes $(x-1)^2-(y+2)^2+2(x-1)+4(y+2)=0$, that is $x^2-y^2+3=0$.

3. Referred to parallel axes through the point (α, β), the equation will be $6(x+\alpha)^2+5(x+\alpha)(y+\beta)-6(y+\beta)^2-17(x+\alpha)+7(y+\beta)+5=0$, that is

$$6x^2+5xy-6y^2+x(12\alpha+5\beta-17)+y(5\alpha-12\beta+7)+6\alpha^2+5\alpha\beta-6\beta^2 -17\alpha+7\beta+5=0.$$

If α and β have the values given by $12\alpha+5\beta-17=0$ and $5\alpha-12\beta+7=0$, namely the values $\alpha=\beta=1$, the coefficients of x and y in the transformed equation will be zero; and we have to shew that the terms which do not contain x or y will also vanish for the values $\alpha=1$, $\beta=1$, and this is at once seen to be true.

4. The equation becomes

$$4\left(\frac{x\sqrt{3}-y}{2}\right)^2+2\sqrt{3}\cdot\frac{x\sqrt{3}-y}{2}\cdot\frac{x+y\sqrt{3}}{2}+2\left(\frac{x+y\sqrt{3}}{2}\right)^2=1,$$

that is $5x^2+y^2=1$.

5. Transform to parallel axes through $(-1, 0)$; then the equation will become

$$(x-1)^2-2(x-1)y+y^2+(x-1)-3y=0,$$

that is

$$x^2-2xy+y^2-x-y=0.$$

Now turn the axes through 45°; then the equation will become

$$\left(\frac{x-y}{\sqrt{2}}\right)^2-2\frac{x-y}{\sqrt{2}}\cdot\frac{x+y}{\sqrt{2}}+\left(\frac{x+y}{\sqrt{2}}\right)^2-\frac{x-y}{\sqrt{2}}-\frac{x+y}{\sqrt{2}}=0,$$

that is $\sqrt{2}y^2-x=0$.

6. The transformed equation will be

$$\left(\frac{x-y}{\sqrt{2}}\right)^2+c\frac{x-y}{\sqrt{2}}\cdot\frac{x+y}{\sqrt{2}}+\left(\frac{x+y}{\sqrt{2}}\right)^2=a^2,$$

that is $(2+c)x^2+(2-c)y^2=2a^2$.

Page 52.

1. By turning the axes through any angle θ, $ax^2+2hxy+by^2$ will become

$$a(x'\cos\theta-y'\sin\theta)+2h(x'\cos\theta-y'\sin\theta)(x'\sin\theta+y'\cos\theta)$$
$$+b(x'\sin\theta+y'\cos\theta)^2, \text{ or}$$
$$x'^2(a\cos^2\theta+2h\sin\theta\cos\theta+b\sin^2\theta)+2x'y'\{(b-a)\sin\theta\cos\theta+h(\cos^2\theta-\sin^2\theta)\}$$
$$+y'^2(a\sin^2\theta-2h\sin\theta\cos\theta+b\cos^2\theta).$$

Hence $a'+b'=(a\cos^2\theta+2h\sin\theta\cos\theta+b\sin^2\theta)+(a\sin^2\theta-2h\sin\theta\cos\theta$
$+b\cos^2\theta)=a+b.$

Also $h'^2-a'b'=\{(b-a)\sin\theta\cos\theta+h(\cos^2\theta-\sin^2\theta)\}^2$

$$-(a\cos^2\theta+2h\sin\theta\cos\theta+b\sin^2\theta)(a\sin^2\theta-2h\sin\theta\cos\theta+b\cos^2\theta)$$
$=(h^2-ab)(\sin^2\theta+\cos^2\theta)^2$, the other terms all vanishing,
$=h^2-ab.$

2. Since $x^2+y^2+2xy\cos\omega$ is the square of the distance of the point (x, y) from the origin,

$x^2+y^2+2xy\cos\omega$ must be changed into $x'^2+y'^2+2x'y'\cos\omega'$.

But $x^2+y^2+2xy\cos\omega$ will become

$$(mx'+ny')^2+(m'x'+n'y')^2+2(mx'+ny')(m'x'+n'y')\cos\omega$$
$$=x'^2(m^2+m'^2+2mm'\cos\omega)$$
$$+y'^2(n^2+n'^2+2nn'\cos\omega)$$
$$+2x'y'\{mn+m'n'+(mn'+m'n)\cos\omega\}.$$

Hence $$m^2+m'^2+2mm'\cos\omega=1,$$
and $$n^2+n'^2+2nn'\cos\omega=1;$$
$$\therefore \frac{m^2+m'^2-1}{mm'}=\frac{n^2+n'^2-1}{nn'}.$$

Page 59.

1. $PQ\,.\,RS+PR\,.\,SQ+PS\,.\,QR$
$=(PS+SQ)RS+(PS+SR)SQ+PS\,.\,QR$
$=PS(RS+SQ+QR)+SQ(RS+SR)$
$=0$, since $RS+SQ+QR=0$ and $RS+SR=0$.

2. $\{PQRS\}=PQ\,.\,RS/PS\,.\,RQ.$
$\{QPSR\}=QP\,.\,SR/QR\,.\,SP$
$=(-PQ)(-RS)/(-RQ)(-PS)$
$=PQ\,.\,RS/PS\,.\,RQ.$
$\{RSPQ\}=RS\,.\,PQ/RQ\,.\,PS.$
$\{SRQP\}=SR\,.\,QP/SP\,.\,QR=RS\,.\,PQ/RQ\,.\,PS.$

3. $\{PQRS\} = PQ \, . \, RS / PS \, . \, RQ$,

and $\{PSRQ\} = PS \, . \, RQ / PQ \, . \, RS$;

$$\therefore \{PQRS\} = \frac{1}{\{PSRQ\}}.$$

Again $$\{PQRS\} = \frac{PQ \, . \, RS}{PS \, . \, RQ} \text{ and } \{PRQS\} = \frac{PR \, . \, QS}{PS \, . \, QR};$$

$$\therefore \{PQRS\} + \{PRQS\} = \frac{PQ \, . \, RS + PR \, . \, SQ}{PS \, . \, RQ}$$

$$= 1, \text{ (from question 1)}.$$

4. The four points can be taken in 24 different orders. From question 2 if the points be taken in any particular order there are three other orders for which the cross ratio is the same, and therefore there can only be $24 \div 4$, that is 6, different cross ratios; and it follows from question 3 that three of the six different cross ratios are the reciprocals of the other three.

5. From 2 it follows that $\{SRQP\}$ is always equal to $\{PQRS\}$; and from 3 it follows that $\{PQRS\} + \{PRQS\} = 1$.

Hence, when $\{PQRS\} = -1$, $\{SRQP\} = -1$, and $\{PRQS\} = 2$.

Also $$\{PSQR\} = \frac{PS \, . \, QR}{PR \, . \, QS} = \frac{1}{\{PRQS\}};$$

hence when $$\{PRQS\} = 2, \{PSQR\} = \frac{1}{2}.$$

6. We have $$PQ \, . \, RS = PS \, . \, QR;$$

$$\therefore (PO + OQ)(OS - OR) = (PO + OS)(OR - OQ);$$

or, since $PO = OR$,

$$(OR + OQ)(OS - OR) = (OR + OS)(OR - OQ);$$

whence $$OQ \, . \, OS = OR^2 = OP^2.$$

7. We have $$PQ \, . \, RS = -PS \, . \, RQ;$$

$$\therefore PQ(PS - PR) = -PS(PQ - PR);$$

$$\therefore 2PQ \, . \, PS = PQ \, . \, PR + PR \, . \, PS;$$

$$\therefore \frac{2}{PR} = \frac{1}{PQ} + \frac{1}{PS}.$$

So also from $$PQ \, . \, RS = -PS \, . \, RQ,$$

we have $$RS(-RP + RQ) = -RQ(-RP + RS);$$

$$\therefore 2RQ \, . \, RS = RP \, . \, RS + RP \, . \, RQ;$$

$$\therefore \frac{2}{RP} = \frac{1}{RQ} + \frac{1}{RS}.$$

CHAPTER IV.

Pages 63—64.

1. (i) We have $\left(x-\frac{1}{2}\right)^2+\left(y-\frac{1}{2}\right)^2=\frac{1}{2}$. Hence the centre is $\left(\frac{1}{2}, \frac{1}{2}\right)$, and the radius $=\frac{1}{2}\sqrt{2}$.

(ii) We have $x^2+y^2+x-2y+\frac{3}{4}=0$, or $\left(x+\frac{1}{2}\right)^2+(y-1)^2=\frac{1}{4}+1-\frac{3}{4}$. Hence the centre is $\left(-\frac{1}{2}, 1\right)$, and the radius $=\frac{1}{2}\sqrt{2}$.

2. The general equation of a circle is $x^2+y^2+2gx+2fy+c=0$. This circle will pass through the given points provided $c=0$, $a^2+2ga+c=0$ and $b^2+2fb+c=0$. Hence $2g=-a$, $2f=-b$, and $c=0$; and therefore the required equation is $x^2+y^2-ax-by=0$.

3. Substitute successively the co-ordinates of the points $(a, 0)$, $(-a, 0)$ and $(0, b)$ in the general equation of a circle, namely in the equation $x^2+y^2+2gx+2fy+c=0$; then we have $a^2+2ga+c=0$, $a^2-2ga+c=0$, $b^2+2fb+c=0$.

Hence $g=0$, $c=-a^2$, $2f=\frac{a^2-b^2}{b}$: the required equation is therefore

$$x^2+y^2+\frac{a^2-b^2}{b}y-a^2=0.$$

5. Let (ξ, η) be the co-ordinates of any point P on the circle; then the equations of the lines joining P to the given points (x', y'), (x'', y'') will be respectively

$$\frac{x-x'}{x'-\xi}=\frac{y-y'}{y'-\eta} \text{ and } \frac{x-x''}{x''-\xi}=\frac{y-y''}{y''-\eta}.$$

Since these lines are at right angles, we have [Art. 42, (iii)]

$$(x'-\xi)(x''-\xi)+(y'-\eta)(y''-\eta)+\{(x'-\xi)(y''-\eta)+(x''-\xi)(y'-\eta)\}\cos\omega=0.$$

Thus the equation of the locus of (ξ, η) is

$$(x-x')(x-x'')+(y-y')(y-y'')+\{(x-x')(y-y'')+(x-x'')(y-y')\}\cos\omega=0.$$

6. Let the centre be (d, e) and the radius a; then [Art. 67, (ii)] the given equation is the same as

$$x^2+y^2+2xy\cos\omega-2x(d+e\cos\omega)-2y(d\cos\omega+e)+d^2+e^2+2de\cos\omega-a^2=0.$$

Hence $2\cos\omega=1$, $d+e\cos\omega=-1=d\cos\omega+e$, and $d^2+e^2+2de\cos\omega-a^2=0$;

$$\therefore\ \omega=60^0,\ d=e=-\frac{2}{3},\ a=\frac{2}{3}\sqrt{3}.$$

7. Let the equation of the circle be

$$x^2+y^2+2gx+2fy+c=0 \quad\text{.........................(i).}$$

Then, since (x', y'), (x'', y'') and (x''', y''') are all on the circle, we have

$$x'^2+y'^2+2gx'+2fy'+c=0 \quad\text{.........................(ii),}$$

$$x''^2+y''^2+2gx''+2fy''+c=0 \quad\text{.....................(iii),}$$

$$x'''^2+y'''^2+2gx'''+2fy'''+c=0 \quad\text{.....................(iv).}$$

Eliminating g, f, c from the equations (i), (ii), (iii), (iv), [Smith's Algebra, Art. 428] we have the required equation, namely

$$\begin{vmatrix} x^2+y^2, & x, & y, & 1 \\ x'^2+y'^2, & x', & y', & 1 \\ x''^2+y''^2, & x'', & y'', & 1 \\ x'''^2+y'''^2, & x''', & y''', & 1 \end{vmatrix}=0.$$

Pages 69—70.

1. The abscissae of the common points are given by

$$x^2+(2x+1)^2=2;\ \therefore x=-1 \text{ or } x=\frac{1}{5}.$$

If $x=-1$, $y=2x+1=-1$; and if $x=\frac{1}{5}$, $y=2x+1=\frac{7}{5}$. Hence the points are $(-1, 1)$ and $\left(\frac{1}{5}, \frac{7}{5}\right)$.

2. The abscissae of the points of intersection are given by

$$x^2+\left(\frac{3x}{2}\right)^2-3x+3x=0,$$

whence $x^2=0$; hence the two values of x, and therefore also the two values of y, are equal, so that the line touches the circle.

3. It is obvious that the point $(1, 1)$ is on both circles; and the tangent to the first circle at $(1, 1)$ is $x+y=2$; and the tangent to the second circle is $x+y-3(x+1)-3(y+1)+10=0$, which reduces to $x+y=2$. Hence the two circles have the same tangent at $(1, 1)$.

4. To find where the circle cuts the axis of x, put $y=0$; then we have $x^2-2ax+a^2=0$. Hence the circle cuts the axis of x in coincident points, and so also for the axis of y.

5. Let the equation of the circle be $x^2+y^2+2gx+2fy+k=0$. Since $x=0$ is a tangent, the roots of $y^2+2fy+k=0$ must be equal; $\therefore k=f^2$. Again, since $y=0$ is a tangent, the roots of $x^2+2gx+k=0$ must be equal; $\therefore k=g^2$. Also, since $x-c=0$ is a tangent, the roots of $c^2+y^2+2gc+2fy+k=0$ must be equal; $\therefore c^2+2gc+k=f^2$. From the equations $c^2+2gc+k=f^2$, $k=g^2$ and $k=f^2$, we have $g=-\frac{c}{2}$, $k=\frac{c^2}{4}$, $f=\pm\frac{c}{2}$.

Hence the *two* circles which satisfy the conditions are given by the equation $x^2+y^2-cx\pm cy+\frac{c^2}{4}=0$.

6. Let the equation of the circle be $x^2+y^2+2gx+2fy+c=0$. Since the circle touches $x=0$, the roots of $y^2+2fy+c=0$ must be equal, and therefore $c=f^2$...(i).

Since the circle touches $x=a$, the roots of $a^2+y^2+2ga+2fy+c=0$ must be equal, and therefore $f^2=a^2+2ga+c$...(ii).

Since the circle touches $3x+4y+5a=0$, the roots of

$$x^2+\left(\frac{3x+5a}{4}\right)^2+2gx-2f\left(\frac{3x+5a}{4}\right)+c=0,$$

that is of

$$25x^2+(30a+32g-24f)\,x+25a^2-40af+16c=0$$

must be equal, and therefore

$$(15a+16g-12f)^2=25\,(25a^2-40af+16c)\ldots\ldots\ldots\ldots\ldots\text{(iii)}.$$

From (i) and (ii) we have $c=f^2$, $2g=-a$; and therefore (iii) becomes

$$(7a-12f)^2=25\,(25a^2-40af+16f^2);$$

or

$$576a^2-832af+256f^2=0,$$

that is

$$(a-f)\,(576a-256f)=0;$$

$$\therefore\ f=a,\ \text{or}\ f=\frac{9}{4}a.$$

Hence the equations of the two possible circles are

$$x^2+y^2-ax+2ay+a^2=0,\ \text{and}$$

$$x^2+y^2-ax+\frac{9}{2}ay+\frac{81}{16}a^2=0.$$

Or thus:—

Having found, as above, $c=f^2$...(i) and $f^2=a^2+2ga+c$...(ii), the equation of the circle may be written in the form $x^2+y^2-ax+2fy+f^2=0$, that is $\left(x-\frac{a}{2}\right)^2+(y+f)^2=\frac{a^2}{4}$, so that the centre of the circle is $\left(\frac{a}{2},\,-f\right)$ and the radius $\frac{a}{2}$.

Hence the perpendicular distance of the line $3x+4y+5a=0$ from the point $\left(\frac{a}{2}, -f\right)$ is $\pm\frac{a}{2}$, and therefore $\dfrac{3\left(\frac{a}{2}\right)+4(-f)+5a}{5}=\pm\frac{a}{2}$, whence we we find $f=a$ or $f=\frac{9}{4}a$.

7. Writing the equation of the circle in the form $(x-a)^2+y^2=a^2$, we see that the centre is $(a, 0)$ and the radius a.

Now any straight line whose perpendicular distance from the centre of a circle is equal to the radius, will touch the circle; and the perpendicular distance of $(a, 0)$ from $y-m(x-a)-a\sqrt{(1+m^2)}=0$ is a, so that the requisite condition is satisfied.

8. Let (ξ, η) be the co-ordinates of the point of intersection of the straight lines; then the equations of the lines will be

$$\frac{x-a}{a-\xi}=\frac{y}{-\eta} \text{ and } \frac{x+a}{-a-\xi}=\frac{y}{-\eta}.$$

$$\text{Hence } \tan\theta=\frac{\dfrac{\eta}{\xi-a}\sim\dfrac{\eta}{\xi+a}}{1+\dfrac{\eta^2}{(\xi-a)(\xi+a)}}=\frac{\pm 2a\eta}{\xi^2+\eta^2-a^2}.$$

Hence the locus of (ξ, η) is the circle

$$x^2+y^2\mp 2ay\cot\theta-a^2=0.$$

9. Take the given perpendicular straight lines for the axes of x and y respectively.

Let $x^2+y^2+2gx+2fy+c=0$ be the equation of the circle in any one of its possible positions; then, since the circle touches the axis of x, the roots of $x^2+2gx+c=0$ must be equal, and therefore $c=g^2$...(i)

The circle cuts the axis of y where $y^2+2fy+c=0$, and the difference of the roots of this equation is $\sqrt{4f^2-4c}$; hence $4l^2=4f^2-4c=4f^2-4g^2$, or $f^2-g^2=l^2$.

But, if (x, y) be the centre of the circle $x=-g$ and $y=-f$, and therefore the locus of the centres is given by $y^2-x^2=l^2$.

10. Let the equation of the straight line be $x\cos\alpha+y\sin\alpha-p=0$; then we have

$$a\cos\alpha-p+(-a\cos\alpha-p)=\text{constant}.$$

Hence p is constant, which shews that the line always touches a fixed circle whose centre is at the origin.

11. The equation of any line which makes an angle of 60° with the axis of x is $y=\sqrt{3}x+c$. This line meets the circle $x^2+y^2=3$ where

$$x^2+(\sqrt{3}x+c)^2=3,$$

or

$$4x^2+2c\sqrt{3}x+c^2-3=0.$$

The line will touch the circle if the roots of the above quadratic equation be equal, the condition for which is $4(c^2-3)=3c^2$, or $c=\pm 2\sqrt{3}$. Hence the required equations are $y=\sqrt{3}(x\pm 2)$.

12. The angular points of the triangle opposite to the sides $x-1=0$, $2y-5=0$, $3x-4y-5=0$ are $\left(5, \frac{5}{2}\right)$, $\left(1, -\frac{1}{2}\right)$ and $\left(1, \frac{5}{2}\right)$ respectively; and the results of substituting the co-ordinates of the angular points in the left-hand members of the equations of the opposite sides are $+, -, -$ respectively.

Now, if (α, β) be the centre of any one of the four circles which touch the sides, then $\alpha-1$, $\frac{2\beta-5}{2}$ and $\frac{3\alpha-4\beta-5}{5}$ are equal in absolute magnitude; and for the centre of the *inscribed* circle these quantities are respectively $+, -, -$.

Hence the centre of the inscribed circle is given by

$$\alpha-1=-\frac{2\beta-5}{2}=-\frac{3\alpha-4\beta-5}{5},$$

whence

$$\alpha=2,\ \beta=\frac{3}{2}.$$

Thus the centre of the inscribed circle is $\left(2, \frac{3}{2}\right)$, and its radius is the perpendicular from $\left(2, \frac{3}{2}\right)$ on $x-1=0$, so that the radius is 1. Hence the required equation is $(x-2)^2+\left(y-\frac{3}{2}\right)^2=1$.

13. Writing the equations of the circles in the forms

$$(x-a)^2+(y-b)^2=(a+b)^2 \text{ and } (x+b)^2+(y+a)^2=(a+b)^2,$$

we see that the square of the distance between their centres is

$$(a+b)^2+(a+b)^2,$$

which is equal to the sum of the squares of the radii of the circles. Hence the lines drawn from a common point to the centres of the circles are at right angles, and therefore the tangents to the circles at a common point are at right angles.

14. The equations of the circles may be written

$$(x+d)^2+y^2=d^2-k^2, \text{ and } x^2+(y+d')^2=d'^2+k^2.$$

The square of the distance between the centres of the two circles is $d^2+d'^2$, and this is equal to $(d^2-k^2)+(d'^2+k^2)$ which is the sum of the squares of the radii of the circles. Hence the circles cut orthogonally.

Page 73.

1. (i) $2x+3y=4$. (ii) $3x-y=4$. (iii) $x-y=4$.

2. (i) If $4x+6y-7=0$ is the same as $xx'+yy'-35=0$, we have

$$\frac{x'}{4}=\frac{y'}{6}=\frac{35}{7};$$

$\therefore x'=20$ and $y'=30$. Thus the pole is (20, 30).

(ii) If $3x-2y-5=0$ is the same as $xx'+yy'-35=0$, we have

$$\frac{x'}{3}=\frac{y'}{-2}=\frac{35}{5};\ \therefore \text{the pole is } (21,\ -14).$$

(iii) If $ax+by-1=0$ is the same as $xx'+yy'-35=0$, we have $\frac{x'}{a}=\frac{y'}{b}=35$; $\therefore$ the pole is $(35a,\ 35b)$.

3. At the intersection we have $16+y^2=4$; $\therefore y=\pm\sqrt{-12}$. Thus the points of intersection are $(4,\ \pm\sqrt{-12})$, and therefore the tangents are

$$4x+\sqrt{-12}\,y=4 \text{ and } 4x-\sqrt{-12}\,y=4,$$

which clearly intersect in the point (1, 0).

4. The polar of $(x',\ y')$ with respect to the circle

$$x^2+y^2=a^2 \text{ is } xx'+yy'-a^2=0.$$

This line will touch the circle $(x-a)^2+y^2=a^2$ if its distance from the centre $(a,\ 0)$ is equal to the radius; the condition that this should be the case is

$$\frac{(ax'-a^2)^2}{x'^2+y'^2}=a^2, \text{ or } y'^2+2ax'=a^2.$$

Pages 79—80.

1. From Art. 82, the lengths of the tangents are

$$\sqrt{(2^2+5^2-2\,.\,2-3\,.\,5-1)}=3,$$

and
$$\sqrt{\left(4^2+1^2-\frac{3}{4}\,.\,4-\frac{1}{4}\,.\,1-\frac{7}{4}\right)}=\sqrt{12}.$$

2. The circle will be $x^2+y^2+2gx+2fy+c=0$, where $g,\ f,\ c$ satisfy the relations $9+6g+c=0$, $4+4f+c=0$ and $1+1-2g+2f+c=0$.

Hence $c=-\frac{18}{5}$, $f=-\frac{1}{10}$ and $g=-\frac{9}{10}$, so that the equation of the circle is

$$x^2+y^2-\frac{9}{5}x-\frac{1}{5}y-\frac{18}{5}=0.$$

The rectangle of the segments of any chord through the origin is found by substituting the coordinates 0, 0 in the equation of the circle; this rectangle is therefore $-\frac{18}{5}$.

3. The equation of the radical axis is

$$x^2+y^2+2x+3y-7-(x^2+y^2-2x-y+1)=0,$$

that is
$$4x+4y-8=0, \text{ or } x+y-2=0.$$

4. The equation of the radical axis is

$$x^2+y^2+bx+by-c-\left(x^2+y^2+ax+\frac{b^2}{a}y\right)=0,$$

that is
$$(b-a)x+\left(b-\frac{b^2}{a}\right)y-c=0,$$

or
$$ax-by+\frac{ac}{a-b}=0.$$

5. The equation of the radical axis is

$$x^2+y^2+ax+by+c-(x^2+y^2+bx+ay+c)=0,$$

that is
$$x-y=0.$$

The line $x-y=0$ meets either circle where $x^2+x^2+ax+bx+c=0$, whence $(x_1-x_2)^2=(x_1+x_2)^2-4x_1x_2=\left(\frac{a+b}{2}\right)^2-4\frac{c}{2}$.

Now the length of the common chord

$$=\sqrt{\{(x_1-x_2)^2+(y_1-y_2)^2\}}=\sqrt{\{2(x_1-x_2)^2\}}=\sqrt{\left\{\frac{1}{2}(a+b)^2-4c\right\}}.$$

6. The radical axis of the first and second circles is $x-2y-4=0$. The radical axis of the second and third circles is $2x-4y-8=0$, which is the same as the radical axis of the first and second circles, from which it follows that the three circles have a common radical axis.

7. The radical axis of the first and second circles is

$$x^2+y^2+4x+7-\left(x^2+y^2+\frac{3}{2}x+\frac{5}{2}y+\frac{9}{2}\right)=0,$$

that is $x-y+1=0$.

The radical axis of the first and third circles is $4x-y+7=0$.

Hence the radical centre is the point of intersection of the lines $x-y+1=0$ and $4x-y+7=0$, and this point is $(-2, -1)$.

8. Any tangent to $x^2+y^2-4=0$ is $x\cos\alpha+y\sin\alpha-2=0$; and the condition that this line should touch $(x-4)^2+y^2-1=0$ is $4\cos\alpha-2=\pm1$, whence $\cos\alpha=\frac{3}{4}$ or $\cos\alpha=\frac{1}{4}$.

If $\cos\alpha=\frac{3}{4}$, $\sin\alpha=\pm\frac{1}{4}\sqrt{7}$; and, if $\cos\alpha=\frac{1}{4}$, $\sin\alpha=\pm\frac{1}{4}\sqrt{15}$.

Hence the tangents are $3x\pm y\sqrt{7}-8=0$, and $x\pm y\sqrt{15}-8=0$.

The centres of similitude are the points where the common tangents cut the line joining the centres of the circles, and the line joining the centres is the axis of x; hence the centres of similitude are the points (8, 0) and $\left(\frac{8}{3}, 0\right)$.

9. The square of the tangent from (f, g) to the first circle is f^2+g^2-6, and the square of the tangent from (f, g) to the second circle is

$$f^2+g^2+3f+3g.$$

Hence
$$f^2+g^2-6=4\,(f^2+g^2+3f+3g);$$
$$\therefore\; f^2+g^2+4f+4g+2=0.$$

10. If (x, y) be any point which satisfies the condition we have

$$\sqrt{(x^2+y^2+2x)}=3\sqrt{(x^2+y^2-4)};$$
$$\therefore\; x^2+y^2+2x=9\,(x^2+y^2-4),$$

or
$$4x^2+4y^2-x-18=0.$$

11. The equation of any circle through the intersection of the two given circles is found by giving a suitable value to λ in the equation

$$x^2+y^2+2x+3y-7-\lambda\,(x^2+y^2+3x-2y-1)=0. \quad \text{[Art. 86.]}$$

The above circle will pass through the point (1, 2) provided

$$1+4+2+6-7-\lambda\,(1+4+3-4-1)=0,$$

or
$$\lambda=2.$$

Hence the required equation is

$$x^2+y^2+2x+3y-7-2\,(x^2+y^2+3x-2y-1)=0,$$

or
$$x^2+y^2+4x-7y+5=0.$$

12. The equation of any circle through the points of intersection of the given circles is given by

$$\lambda\,(x^2+y^2-4)+x^2+y^2-2x-4y+4=0.$$

The above circle will touch $x+2y=0$ provided that

$$\lambda\,(4y^2+y^2-4)+4y^2+y^2+4y-4y+4,$$

that is $5\,(\lambda+1)\,y^2-4\,(\lambda-1)$, is a perfect square, the condition for which is $\lambda=1$. Hence the equation required is

$$x^2+y^2-4+x^2+y^2-2x-4y+4=0,$$

or
$$x^2+y^2-x-2y=0.$$

Pages 83—87.

1. Let the fixed point be (h, k), and the fixed line

$$x\cos\alpha+y\sin\alpha-p=0.$$

Then
$$(x-h)^2+(y-k)^2 \propto (x\cos\alpha+y\sin\alpha-p).$$

Hence the equation of the locus is

$$(x-h)^2+(y-k)^2=c\,(x\cos\alpha+y\sin\alpha-p),$$

where c is some constant: the locus is therefore a circle.

2. Take axes parallel to the sides and midway between them; then, if $2a$ be the length of a side of the square, the equations of the sides will be

$$x-a=0,\ x+a=0,\ y-a=0,\ y+a=0.$$

Hence $(x-a)^2+(x+a)^2+(y-a)^2+(y+a)^2=\text{constant}=4c^2$ suppose;

$$\therefore\ x^2+y^2=c^2-a^2.$$

3. Let the n given points be (a_1, b_1) (a_2, b_2), &c.; and let (x, y) be any point on the locus; then we have

$$\Sigma\{(x-a_1)^2+(y-b_1)^2\}=\text{constant};$$

$$\therefore\ n(x^2+y^2)-2x\Sigma a_1-2y\Sigma b_1+\Sigma a_1^2+\Sigma b_1^2=\text{const.};$$

$$\therefore\ \left(x-\frac{\Sigma a_1}{n}\right)^2+\left(y-\frac{\Sigma b_1}{n}\right)^2=\text{constant}.$$

Hence the required locus is a circle whose centre is the point

$$\left(\frac{\Sigma a_1}{n}, \frac{\Sigma b_1}{n}\right).$$

4. Take the line joining the two given points for axis of x, and its middle point for the origin, and let the given points be $(a, 0)$, $(-a, 0)$. Then, if (x, y) be any point on the locus, we have

$$(x-a)^2+y^2=n^2\{(x+a)^2+y^2\};$$

$$\therefore\ x^2+y^2-2ax\frac{1+n^2}{1-n^2}+a^2=0.$$

Thus the locus, for any particular value of n, is a circle; and it is obvious that all the circles, obtained by giving different values to n, will cut $x=0$ in the same (imaginary) points.

5. Take the base and the bisector of the vertical angle for axes; then the equations of the three sides are of the form (i) $y=0$, (ii) $\frac{x}{a}+\frac{y}{d}=1$, (iii) $-\frac{x}{a}+\frac{y}{d}=1$. Hence, if (x, y) be any point on the locus we have

$$y^2=\frac{\frac{x}{a}+\frac{y}{d}-1}{\sqrt{\left(\frac{1}{a^2}+\frac{1}{d^2}\right)}}\cdot\frac{-\frac{x}{a}+\frac{y}{d}-1}{\sqrt{\left(\frac{1}{a^2}+\frac{1}{d^2}\right)}};$$

$$x^2+y^2+2y\frac{a^2}{d}-a^2=0\ldots\ldots\ldots\ldots\ldots\text{(i)}.$$

The locus is therefore a circle.

The circle (i) cuts $y=0$ where $x=\pm a$; and the tangents at the points $(\pm a, 0)$ are $\pm ax+y\frac{a^2}{d}-a^2=0$, that is $\pm\frac{x}{a}+\frac{y}{d}=1$; hence the circle (i) touches the sides of the triangle at the extremities of the base.

6. The angular points of the triangle will be found to be $(1, 2)$, $(7, -1)$ and $(-2, 8)$.

Hence $x^2+y^2+2gx+2fy+c=0$ will be the equation of the circle, provided

$$1+4+2g+4f+c=0,\ 49+1+14g-2f+c=0$$

and $4+64-4g+16f+c=0$, whence $2g=-17$, $2f=-19$, $c=50$, so that the required equation is

$$x^2+y^2-17x-19y+50=0.$$

7. The common chord is $2x+1=0$, which meets the circles where

$$y^2+3y+\frac{1}{4}=0,\ 2x+1=0.$$

Now the circle on (x_1, y_1), (x_2, y_2) as diameter is

$$(x-x_1)(x-x_2)+(y-y_1)(y-y_2)=0,$$

or

$$x^2-x(x_1+x_2)+x_1x_2+y^2-y(y_1+y_2)+y_1y_2=0.$$

Hence the equation required is

$$x^2+x+\frac{1}{4}+y^2+3y+\frac{1}{4}=0.$$

8. The equation of the lines is

$$x^2+y^2-(2x+2y)\frac{x+2y}{3}=0 \qquad \text{[Art. 38];}$$

and the lines are at right angles since the sum of the coefficients of x^2 and y^2 is zero.

9. Taking the point O for pole, and the initial line perpendicular to the fixed straight line on which P moves, the equation of the locus of P is $r\cos\theta=a$, where a is the length of the perpendicular ON.

Let the coordinates of Q be ρ, θ; then $\rho r=k^2$; and therefore

$$\rho\frac{a}{\cos\theta}=k^2, \text{ or } \rho=\frac{k^2}{a}\cos\theta.$$

Hence, from Art. 81 (iii), the locus of Q is a circle whose diameter is $\frac{k^2}{a}$.

[The result is easily proved geometrically: for if A be the point in the perpendicular ON such that $OA \cdot ON=k^2=OP \cdot OQ$, it follows that A, N, P, Q are on a circle, and therefore the angle OQA is a right angle, whence Q is always on the fixed circle whose diameter is OA.]

10. Take O for pole and the initial line through the centre of the circle on which P moves, and let a be the radius of the circle and c the distance of its centre from O. Then the equation of the locus of P is [Art. 81]

$$r^2 - 2cr\cos\theta + c^2 - a^2 = 0.$$

Hence, if ρ, θ be the coordinates of Q corresponding to the coordinates r, θ of P, we have since $r\rho = k^2$

$$\frac{k^4}{\rho^2} - 2c \cdot \frac{k^2}{\rho}\cos\theta + c^2 - a^2 = 0,$$

$$\therefore\ \rho^2 - 2\frac{ck^2}{c^2 - a^2}\rho\cos\theta + \frac{k^4}{c^2 - a^2} = 0,$$

which shews that the locus of Q is a circle whose centre is on the initial line at a distance $\frac{ck^2}{c^2 - a^2}$ from the pole.

11 and 12. The tangents to all the circles of a co-axial system drawn from any point T on their radical axis are all equal. Hence, if P, Q be the points of contact of a line through T which touches two of the circles, and if O be either of the two point-circles of the system, we have $TP = TQ = TO$, so that PQ is bisected at T, and O is on the circle whose diameter is PQ.

13. The equations of any two circles take the required form when the line of centres is the axis of x and the radical axis is the axis of y. [See Art. 85.]

If one of the circles is within the other (1) their radical axis must cut them in imaginary points, and therefore b must be positive; and (2) the centres of the circles must be on the same side of the radical axis, and therefore a and a' must be both positive or both negative. Hence the *necessary* conditions are that b and aa' should both be positive, and these conditions are also easily seen to be *sufficient*.

14. Let P, Q be the two points, and let their coordinates be x_1, y_1 and x_2, y_2.

Then the ratio of the distances P, Q from the centre is

$$\sqrt{x_1^2 + y_1^2} : \sqrt{x_2^2 + y_2^2}.$$

The polars of P, Q are $xx_1 + yy_1 - a^2 = 0$ and $xx_2 + yy_2 - a^2 = 0$ respectively; and therefore the distance of P from the polar of Q is $\frac{x_1x_2 + y_1y_2 - a^2}{\sqrt{(x_2^2 + y_2^2)}}$, and the distance of Q from the polar of P is $\frac{x_2x_1 + y_2y_1 - a^2}{\sqrt{(x_1^2 + y_1^2)}}$. Hence the ratio of these two distances is also equal to $\sqrt{(x_1^2 + y_1^2)} : \sqrt{(x_2^2 + y_2^2)}$, which proves the proposition.

15. Let the equations of any two circles be

$$x^2 + y^2 - a^2 = 0 \text{ and } (x - a)^2 + y^2 - b^2 = 0.$$

Then if (x, y) be any point such that the tangents from it to the two circles are in the ratio of the radii, we have

$$x^2+y^2-a^2 : (x-a)^2+y^2-b^2=a^2 : b^2.$$

Hence the locus of points which satisfy this relation is the circle whose equation is

$$b^2(x^2+y^2-a^2)=a^2\{(x-a)^2+y^2-b^2\}\text{.....................}(i).$$

Now the centre of the circle (i) is clearly on the axis of x; also the circle meets the axis of x in points given by $x=\frac{a}{a\mp b}a$, and these are the points which divide the line joining the centres of the given circles internally and externally in the ratio of the radii. Hence (i) is the equation of a circle having the centres of similitude as extremities of a diameter.

16. Let the circles be $x^2+y^2-a^2=0$ and $x^2+y^2-b^2=0$; then we have

$$x^2+y^2-a^2 : x^2+y^2-b^2 :: b^2 : a^2;$$
$$\therefore\ a^2(x^2+y^2-a^2)=b^2(x^2+y^2-b^2);$$
$$\therefore\ x^2+y^2=a^2+b^2.$$

17. Let T be the middle point of AB, and let P be one of the points of intersection of the circles whose diameters are AB and CD. Then [see Ex. 6 page 59] $TC \,.\, TD=TB^2=TP^2$. Hence TP touches the circle CPD at P; but the tangent at P to the circle APB is perpendicular to TP; the two circles must therefore cut orthogonally.

18. Let T be the centre of one circle, and let any diameter ATB of that circle cut the other circle in C and D. Then, if P be one of the points of intersection of the circles, TP is a tangent to the circle CPD, and therefore $TC \,.\, TD=TP^2=AT^2$.

Now A, B, C, D is an harmonic range provided $AC \,.\, BD=AD \,.\, CB$, or $(AT+TC)(TD-AT)=(AT+TD)(AT-TC)$, or $AT^2=TC \,.\, TD$.

19. We may take

$$x\cos a+y\sin a-a=0,$$

$$x\cos\left(a+\frac{2\pi}{n}\right)+y\sin\left(a+\frac{2\pi}{n}\right)-a=0,$$

$$\cdots\cdots\cdots\cdots\cdots\cdots\cdots\cdots\cdots\cdots$$

$$x\cos\left\{a+\frac{2\pi(n-1)}{n}\right\}+y\sin\left\{a+\frac{2\pi(n-1)}{n}\right\}-a=0,$$

for the equations of the sides.

The sum of the squares of the perpendiculars from (x, y) is the sum of the squares of the left-hand members of the above equations; and in this sum the coefficients of x^2 and y^2 are equal, since

$$\cos 2a+\cos 2\left(\ +\frac{2\pi}{n}\right)+\ldots\ldots+\cos 2\left\{a+\frac{2\pi(n-1)}{n}\right\}=0;$$

also the coefficient of xy is zero, since

$$\sin 2a + \sin 2\left(a + \frac{2\pi}{n}\right) + \ldots\ldots + \sin 2\left\{a + \frac{2\pi(n-1)}{n}\right\} = 0.$$

Hence the required locus is a circle.

20. Take the point O for origin and the fixed straight lines for axes. Then the equation of every circle through the origin is

$$x^2 + y^2 + 2gx + 2fy = 0 \ldots\ldots\ldots\ldots\ldots\ldots\ldots\ldots\ldots\ldots \text{(i)}.$$

This circle cuts the axis in the points $(-2g,\ 0)$, $(0, -2f)$. Hence the equation of PQ is

$$\frac{x}{-2g} + \frac{y}{-2f} = 1.$$

But PQ passes through a fixed point, (α, β) suppose.

Hence $$\frac{\alpha}{g} + \frac{\beta}{f} + 2 = 0 \ldots\ldots\ldots\ldots\ldots\ldots\ldots\ldots\ldots\ldots \text{(ii)}.$$

Now the centre of the circle (i) is the point $x = -g$, $y = -f$. Hence, from (ii), the equation of the required locus is $\frac{\alpha}{x} + \frac{\beta}{y} - 2 = 0$.

21. Let A be the point (a, α) and B the point (b, β) and let (r, θ) be any point P on the locus. Then, since $AP^2 + BP^2 = AB^2$, we have from Art. 10

$$\{a^2 + r^2 - 2ar\cos(\theta - \alpha)\} + \{b^2 + r^2 - 2br\cos(\theta - \beta)\} = a^2 + b^2 - 2ab\cos(\alpha - \beta):$$

$$\therefore \quad r^2 - r\{a\cos(\theta - \alpha) + b\cos(\theta - \beta)\} + ab\cos(\alpha - \beta) = 0.$$

22. We have to eliminate θ between the given equations. We have

$$r\cos\theta\cos\beta + r\sin\theta\sin\beta = p;$$

$$\therefore \quad (r\cos\theta\cos\beta - p)^2 = r^2\sin^2\beta\,(1 - \cos^2\theta);$$

$$(r^2\cos\beta - 2ap)^2 = r^2\sin^2\beta\,(4a^2 - r^2),$$

or $$r^4 - 4a\,(p\cos\beta + a\sin^2\beta)\,r^2 + 4a^2p^2 = 0,$$

the equation required.

The line will touch the circle when the two values of r^2 are equal, and the condition for this is that

$$4a^2p^2 = 4a^2\,(p\cos\beta + a\sin^2\beta)^2,$$

or $$\pm p = p\cos\beta + a\sin^2\beta;$$

$$p = 2a\cos^2\frac{\beta}{2} \text{ or } p = -2a\sin^2\frac{\beta}{2}.$$

23. The angular points are $(24, 7)$, $\left(-9, -\frac{27}{4}\right)$ and $(0, 0)$; and when the coordinates of the angular points are substituted in the left-hand members of the equations of the opposite sides of the triangle the results are $+$, $+$, $-$ respectively. These must also be the signs of the results when the coordinates of the centre of the *inscribed* circle are substituted.

Hence the coordinates of the centre of the inscribed circle satisfy the relations

$$\frac{3x-4y}{\sqrt{(3^2+4^2)}}=\frac{7x-24y}{\sqrt{(7^2+24^2)}}=-\frac{5x-12y-36}{\sqrt{(5^2+12^2)}},$$

whence $x=\frac{5}{8}$, $y=-\frac{5}{4}$.

24. Let the equations of the circles be

$$x^2+y^2+2gx+c=0 \text{ and } x^2+y^2+2g'x+c=0.$$

Then the polars of (x', y') with respect to the circles are

$$xx'+yy'+g(x+x')+c=0, \text{ and } xx'+yy'+g'(x+x')+c=0.$$

If these lines make a constant angle α with one another, we have

$$\{(x'+g)(x'+g')+y'^2\}\tan\alpha=\pm y'(g'-g).$$

Hence the point (x', y') must be on one or other of the two circles

$$x^2+y^2+(g+g')x+gg'\pm(g'-g)y\cot\alpha=0.$$

25. Let A, B be the centres of the two circles and a, b their radii. Let T be any point on the radical axis, and let TP touch the circle whose centre is A and TQ the circle whose centre is B, and let AP, BQ meet in R; then, since $TP=TQ$ and the angles TPR, TQR are right angles, $RP=RQ$.

Hence $RA\pm a=RB\pm b$.

If therefore A be $(\alpha, 0)$ and B be $(\beta, 0)$, the equation of the locus of R is given by $\sqrt{\{(x-\alpha)^2+y^2\}}\pm a=\sqrt{\{(x-\beta)^2+y^2\}}\pm b.$

26. Let $Ax+By+C=0$ cut the first circle in the points P, P', and $A'x+B'y+C'=0$ cut the second circle in Q, Q', and let T be the point of intersection of the lines. Then in order that P, P', Q, Q' may lie on a circle it is necessary and sufficient that $TP\,.\,TP'=TQ\,.\,TQ'$; hence T must lie on the radical axis of the given circles, and the equation of this radical axis is

$$(a-a')x+(b-b')y+c-c'=0.$$

Thus the points will lie on a circle provided the three lines

$$Ax+By+C=0 \quad\text{(i)},$$

$$A'x+B'y+C'=0 \quad\text{(ii)},$$

and $$(a-a')x+(b-b')y+c-c'=0 \quad\text{(iii)},$$

meet in a point; that is provided the equations (i), (ii), (iii) are simultaneously true, the condition for which is [Algebra, Art. 428]

$$\begin{vmatrix} A, & B, & C \\ A', & B', & C' \\ a-a', & b-b', & c-c' \end{vmatrix}=0.$$

27. Take the line joining the two given points and the perpendicular line which is midway between the points for axes; then the equation of any circle of the system will be

$$x^2+y^2+2gx-c^2=0,$$

where c is the same for all the circles.

Now the tangent at the point (x', y') is

$$xx' + yy' + g(x + x') - c^2 = 0 \text{..............(i)},$$

where $$x'^2 + y'^2 + 2gx' - c^2 = 0 \text{..............(ii)}.$$

Now, if the tangent at (x', y') be parallel to the fixed line $y + mx = 0$, we have from (i) $m = \dfrac{x' + g}{y'}$, or $g = my' - x'$.

Substitute for g in (ii), and we have

$$x'^2 + y'^2 + 2(my' - x')x' - c^2 = 0.$$

Thus the equation of the required locus is

$$y^2 - x^2 + 2mxy - c^2 = 0.$$

28. The equations of the system of co-axial circles can be expressed in the forms

$$x^2 + y^2 + 2g_1x + c = 0 \text{..............(i)},$$

$$x^2 + y^2 + 2g_2x + c = 0 \text{..............(ii)},$$

$$x^2 + y^2 + 2g_3x + c = 0 \text{..............(iii)}.$$

The centres are the points $(-g_1, 0)$, $(-g_2, 0)$ and $(-g_3, 0)$ respectively.

Hence $BC = -g_3 + g_2$, $CA = -g_1 + g_3$ and $AB = -g_2 + g_1$.

Also, if (x, y) be the coordinates of any point

$$t_1^2 = x^2 + y^2 + 2g_1x + c,$$

$$t_2^2 = x^2 + y^2 + 2g_2x + c,$$

and $$t_3^2 = x^2 + y^2 + 2g_3x + c.$$

Hence $(g_2 - g_3)t_1^2 + (g_3 - g_1)t_2^2 + (g_1 - g_2)t_3 = 0$.

29. Let the equations of the three circles be of the form

$$x^2 + y^2 + 2g_1x + 2f_1y + c_1 = 0, \text{ \&c.}$$

Then, if t_1, t_2, t_3 be the lengths of the tangents from (x, y) to the three circles the equation of the locus of (x, y) which satisfies the relation

$$At_1^2 + Bt_2^2 + Ct_3^2 - D = 0,$$

is $$A(x^2 + y^2 + 2g_1x + 2f_1y + c_1) + B(x^2 + y^2 + 2g_2x + 2f_2y + c_2)$$
$$+ C(x^2 + y^2 + 2g_3x + 2f_3y + c_3) - D = 0 \text{.........(i)}.$$

The equation (i) represents a circle for all values of A, B, C, D except those for which $A + B + C = 0$, in which case (i) clearly represents a straight line. If $A + B + C \neq 0$, there are three quantities at our disposal, namely the ratios $A : B : C : D$, and these may be so chosen as to make (i) pass through any three points and therefore represent *any* given circle. There is however one case of exception, for if the original circles have their centres on a straight line, every circle included in (i) will also have its centre on that straight line. [This is obvious if the line on which the centres lie is the axis of x, so that $f_1=f_2=f_3=0$.]

30. If two circles cut one another orthogonally the radius of either circle must be equal to the length of the tangent drawn from its centre to the other; and, conversely, a circle whose radius is equal to the tangent drawn from its centre to another circle will cut the other circle orthogonally.

Hence, if a circle cut two given circles orthogonally, the tangents to the two given circles from its centre must both be equal to its radius; and therefore the centre of the orthogonal circle must be on the radical axis of the two given circles. Hence the tangents from the centre of the orthogonal circle to all the circles of the coaxial system defined by the two given circles will all be equal to the radius of the circle which is orthogonal to the two given circles, and hence all the circles of the co-axial system will be cut orthogonally.

Or thus:

Let the equations of the two given circles, referred to their line of centres and their radical axis as axes, be

$$x^2+y^2+2g_1x+c=0 \qquad\text{(i)},$$

$$x^2+y^2+2g_2x+c=0 \qquad\text{(ii)}.$$

Then any other circle of the coaxial system will be

$$x^2+y^2+2gx+c=0 \qquad\text{(iii)}.$$

Now the conditions that the circle

$$x^2+y^2+2Gx+2Fy+C=0,$$

should cut both (i) and (ii) orthogonally are [Art. 88, Ex. 1]

$$2Gg_1-C-c=0 \text{ and } 2Gg_2-C-c=0.$$

Hence $G=0$ and $C=-c$; from which it follows that

$$x^2+y^2+2Fy-c=0$$

will cut (iii) at right angles for all values of g.

31. As in 30, all the circles given by the equation

$$x^2+y^2+2gx+c=0 \qquad\text{(i)},$$

g being supposed to have any value whatever but c being the same for all the circles, are cut orthogonally by any one of the circles

$$x^2+y^2+2Fy-c=0 \qquad\text{(ii)}.$$

Thus all the circles of the co-axial system (i) are cut orthogonally by any circle of the co-axial system (ii).

32. Let A, B be the centres of the two fixed circles, and let O be the centre of any circle which touches the fixed circles in P and Q respectively. Let PQ cut the circle whose centre is B in the point Q'; then since $OP=OQ$ and $BQ=BQ'$, it follows that OPA is parallel to BQ'.

Hence PQ, produced if necessary, will cut the line AB in a point K which divides AB internally or externally in the ratio of the radii, that is to say PQ always passes through one or other of the centres of similitude. Now, for the same centre of similitude, $KQ'.KQ$ is constant, and $KQ':KP$ is constant, and therefore $KQ.KP$ is constant; hence the tangent from K to the touching circle is of constant length, and therefore all the circles which touch the two given circles so that the line joining the points of contact passes through either of the centres of similitude are cut orthogonally by a fixed circle whose centre is that centre of similitude.

33. Take the tangents to the two circles at one of their common points for axes; then the equations of the circles will be

$$x^2+y^2-2ax=0 \text{ and } x^2+y^2-2by=0.$$

The centres of the circles are $(a, 0)$ and $(0, b)$ respectively, and therefore the common diameter of the circles is $\frac{x}{a}+\frac{y}{b}=1$.

Any line perpendicular to the common diameter is $ax-by=c$; and the poles, (x', y') and (x'', y''), of this line with respect to the two circles are found by comparing $ax-by-c=0$ with $xx'+yy'-a(x+x')=0$ and with $xx''+yy''-b(y+y'')=0$.

Hence $\frac{x'-a}{a}=\frac{y'}{-b}=\frac{ax'}{c}$, and $\frac{x''}{a}=\frac{y''-b}{-b}=\frac{by''}{c}$.

Thus pairs of points (x', y') and (x'', y'') can be found satisfying the required conditions for all values of c; moreover we have $\frac{x'}{y'}=-\frac{c}{ab}=-\frac{y''}{x''}$, and therefore $x'x''+y'y''=0$, which shews that the lines from the origin to (x', y') and (x'', y'') are at right angles.

34. Let the equations of the circles $S=0$ and $S'=0$ be respectively

$$(x-a)^2+y^2-a^2=0 \quad\text{and}\quad (x+a)^2+y^2-a'^2=0.$$

Then the circles $\frac{S}{a}\pm\frac{S'}{a'}=0$ are

$$a'\{(x-a)^2+y^2-a^2\}\pm a\{(x+a)^2+y^2-a'^2\}=0,$$

or
$$x^2+y^2-2ax\frac{a'-a}{a'+a}+a^2-aa'=0,$$

and
$$x^2+y^2-2ax\frac{a'+a}{a'-a}+a^2+aa'=0.$$

The condition that these two circles should cut orthogonally is [Art. 88 (I)]

$$2a\frac{a'-a}{a'+a}\cdot a\frac{a'+a}{a'-a}-(a^2-aa')-(a^2+aa')=0,$$

which is clearly satisfied.

35. The equation of any tangent to the first circle is

$$(x-a)\cos\alpha + y\sin\alpha = b \quad\text{.........(i).}$$

Hence any perpendicular tangent to the second circle is

$$(x+a)\cos\left(\alpha \pm \frac{\pi}{2}\right) + y\sin\left(\alpha \pm \frac{\pi}{2}\right) = c,$$

that is

$$y\cos\alpha - (x+a)\sin\alpha = \pm c \quad\text{.........(ii).}$$

From (i) and (ii) we have

$$\frac{\cos\alpha}{\mp cy - b(x+a)} = \frac{\sin\alpha}{-by \pm c(x-a)} = \frac{1}{a^2 - x^2 - y^2};$$

$$\therefore \{\mp cy - b(x+a)\}^2 + \{-by \pm c(x-a)\}^2 = (a^2 - x^2 - y^2)^2.$$

Now the bisectors of the angles between the lines (i) and (ii) are

$$(x-a)\cos\alpha + y\sin\alpha - b \pm \{y\cos\alpha - (x+a)\sin\alpha \mp c\} = 0,$$

that is

$$x(\cos\alpha + \sin\alpha) + (y+a)(\sin\alpha - \cos\alpha) - b \pm c = 0,$$

and

$$x(\cos\alpha - \sin\alpha) + (y-a)(\sin\alpha + \cos\alpha) - b \mp c = 0,$$

and writing these in the form

$$x\cos\left(\alpha - \frac{\pi}{4}\right) + (y+a)\sin\left(\alpha - \frac{\pi}{4}\right) = \frac{1}{\sqrt{2}}(b \mp c),$$

and

$$x\cos\left(\alpha + \frac{\pi}{4}\right) + (y-a)\sin\left(\alpha + \frac{\pi}{4}\right) = \frac{1}{\sqrt{2}}(b \pm c),$$

we see that every bisector touches one or other of the *four* circles

$$x^2 + (y+a)^2 = \frac{1}{2}(b \mp c)^2,$$

$$x^2 + (y-a)^2 = \frac{1}{2}(b \pm c)^2.$$

36. Take the lines AB, AC for axes of x and y. Then the equations of the three escribed circles will be easily seen to be respectively

$$x^2 + y^2 + 2xy\cos A - 2sx - 2sy + s^2 = 0 \quad\text{.........(i),}$$

$$x^2 + y^2 + 2xy\cos A + 2(s-c)x - 2(s-c)y + (s-c)^2 = 0 \quad\text{.........(ii),}$$

$$x^2 + y^2 + 2xy\cos A - 2(s-b)x + 2(s-b)y + (s-b)^2 = 0 \quad\text{.........(iii).}$$

The radical axis of (i) and (ii) is

$$2x(a+b) + 2cy - (a+b)c = 0.$$

The radical axis of (ii) and (iii) is

$$2x - 2y + b - c = 0.$$

Hence the radical centre of the three circles is the point $\left\{\frac{c(a+c)}{4s}, \frac{b(a+b)}{4s}\right\}$ in which the two radical axes meet.

Now the radius of the circle orthogonal to three given circles is the length of the tangent from their radical centre to either of the three given circles; and hence the square of the required radius is found by substituting the coordinates of the radical centre in the left-hand member of (i).

Hence the square of the diameter

$$= \frac{1}{4s^2}\{c^2(a+c)^2 + b^2(a+b)^2 + (a+b)(a+c)(b^2+c^2-a^2)\}$$
$$-2\{c(a+c)+b(a+b)\}+4s^2$$
$$=\frac{1}{2s}\{\Sigma a^2b + abc\},$$

after reduction.

Now $\dfrac{a^2}{\sin^2 A}\{1+\Sigma\cos B\cos C\}$

$$=\frac{a^2b^2c^2}{4s(s-a)(s-b)(s-c)}\left\{1+\Sigma\frac{(a^2-b^2+c^2)(a^2+b^2-c^2)}{4a^2bc}\right\}$$
$$=\frac{1}{16s(s-a)(s-b)(s-c)}[4a^2b^2c^2+\Sigma bc(a^2-b^2+c^2)(a^2+b^2-c^2)].$$

Now it will be found that

$$8(s-a)(s-b)(s-c) = -\Sigma a^3 + \Sigma a^2b - 2abc,$$

and $4a^2b^2c^2 + \Sigma bc(a^2-b^2+c^2)(a^2+b^2-c^2) = -\Sigma a^5b + \Sigma a^4bc$
$$+2\Sigma a^3b^3 + 4a^2b^2c^2 = (-\Sigma a^3 + \Sigma a^2b - 2abc).(\Sigma a^2b + abc).$$

Hence $\dfrac{a^2}{\sin^2 A}\{1+\Sigma\cos B\cos C\} = \dfrac{1}{2s}(\Sigma a^2b + abc)$, which proves the proposition.

[Another solution will be found in the 'Solutions of the Cambridge Problems and Riders for 1875,' page 61.]

37. Let the fixed points be $(\mp a, 0)$, and let the centres of the circles be (x', y') and (x'', y''); then, if (x, y) be the point of contact, we have

$$2x = x' + x'',\ 2y = y' + y'' \quad\text{.........(i).}$$

Also
$$(x'+a)^2 + y'^2 = c^2 \quad\text{.........(ii),}$$
$$(x''-a)^2 + y''^2 = c^2 \quad\text{.........(iii),}$$
and
$$(x'-x'')^2 + (y'-y'')^2 = 4c^2 \quad\text{.........(iv).}$$

From (ii) and (iii)
$$x(x'-x'') + y(y'-y'') + 2ax = 0 \quad\text{.........(v).}$$

Subtract (iv) from twice the sum of (ii) and (iii); then

$$(x'+x'')^2 + (y'+y'')^2 + 4a(x'-x'') + 4a^2 = 0,$$

i.e.
$$x^2 + y^2 + a^2 + a(x'-x'') = 0 \quad\text{.........(vi).}$$

Then, from (v),

$$y(y'-y'')-\frac{x}{a}(x^2+y^2-a^2)=0 \quad\text{..................... (vii).}$$

Hence substituting the values of $x'-x''$ and $y'-y''$ given in (vi) and (vii) in (iv) we have the equation of the required locus, namely

$$y^2(x^2+y^2+a^2)^2+x^2(x^2+y^2-a^2)^2-4a^2c^2y^2=0.$$

If $a=c$, the above equation is equivalent to

$$\{(x^2+y^2)^2-a^2(x^2-3y^2)\}(x^2+y^2-a^2)=0.$$

CHAPTER V.

Pages 96—97.

1. The extremities of the latus-rectum are $(a, 2a)$ and $(a, -2a)$. Hence [Art. 95] the equations of the tangents at the points are $\pm 2ay = 2a(x+a)$, i.e. $\pm y = x + a$. Also, by Art. 97, the equations of the normals at their points are

$$(y \mp 2a)\,2a \pm 2a\,(x-a) = 0, \text{ i.e. } y \pm x \mp 3a = 0.$$

2. Substituting for y, we have $(3x-a)^2 - 4ax = 0$, that is

$$9x^2 - 10ax + a^2 = 0; \;\therefore\; x = a \text{ or } x = \frac{a}{9}.$$

Where $x = a$, $y = 3a - a = 2a$; and where $x = \frac{a}{9}$, $y = \frac{a}{3} - a = -\frac{2}{3}a$.

Hence the two points of intersection are $(a, 2a)$ and $\left(\frac{a}{9}, -\frac{2a}{3}\right)$.

3. The two tangents are $yy' = 2a(x+x')$ and $-\frac{4a^2}{y'}y = 2a\left(x - \frac{a^2}{x'}\right)$. These are at right angles from Art. 29.

4. At points common to $y = 2x + \frac{a}{2}$ and $y^2 - 4ax = 0$ we have $\left(2x + \frac{a}{2}\right)^2 - 4ax = 0$, that is $\left(2x - \frac{a}{2}\right)^2 = 0$: thus the two values of x, and therefore also the two values of y, are equal, and hence the points of intersection are coincident.

At points common to $y = 2x + \frac{a}{2}$ and $20x^2 + 20y^2 = a^2$ we have $20x^2 + 20\left(2x + \frac{a}{2}\right)^2 = a^2$; $\therefore$ $(10x + 2a)^2 = 0$: thus the two values of x, and therefore also the two values of y, are equal, and hence the points of intersection are coincident.

5. Let $y = mx + c$ be the equation of the common tangent line; then $x^2 + (mx+c)^2 - 2a^2 = 0$ must have equal roots; therefore $(1+m^2)(c^2 - 2a^2) = m^2c^2$, that is

$$c^2 - 2a^2 - 2a^2m^2 = 0 \quad\text{..............................}\quad \text{(i)}.$$

Again, the equation $(mx+c)^2=8ax$ must have equal roots; and therefore $mc=2a$...(ii).

From (i) and (ii) $4a^2-2a^2m^2-2a^2m^4=0$; $\therefore m^2=1$ or $m^2=-2$. Rejecting the imaginary values $\pm\sqrt{-2}$, we have $m=\pm 1$ and the corresponding values of c are $\pm 2a$. Hence $y=\pm(x+2a)$ are the real common tangents.

6. The ordinates of the points common to $7x+6y=13$ and $y^2-7x-8y+14=0$ are given by $y^2-2y+1=0$, i.e. $(y-1)^2$. Hence the two values of y are equal, and therefore also the two values of x, so that the straight line must cut the curve in two coincident points.

7. The equation may be written in the form $(x+2a)^2=-2a(y-2a)$. Hence, if the origin be changed to the point $(-2a, 2a)$, the equation will become $x'^2=-2ay'$, which shews that the curve is a parabola whose vertex is at the new origin and whose axis lies along the axis of y but in the *negative* direction.

8. The equation of any parabola whose vertex is at the origin and whose axis is the axis of y must be of the form $x^2=4ay$; hence the equation of any parabola whose vertex is at any point (h, k) and whose axis is parallel to the axis of y must be $(x-h)^2=4a(y-k)$. The equation

$$x^2+2Ax+2By+C=0$$

will therefore represent any parabola of the system if we take $A=-h$, $B=-2a$ and $C=h^2+4ak$.

9 and 10. (i) We have $y^2=5(x+2)$, which represents a parabola whose vertex is the point $(-2, 0)$, whose axis is the line $y=0$, and whose latus-rectum is 5.

The focus is on the axis and at a distance from the vertex equal to one-quarter of the latus-rectum; hence the focus is the point

$$x=-2+\frac{5}{4}=-\frac{3}{4},\quad y=0.$$

The directrix is perpendicular to the axis and cuts it at a distance from the vertex, measured in the negative direction along the axis equal to one-quarter of the latus-rectum; hence the equation of the directrix is

$$x=-2-\frac{5}{4},\ \text{or } 4x+13=0.$$

(ii) We have $(x-2)^2=-2(y-2)$. Hence if the origin be changed to the point (2, 2), the equation will become $x^2=-2y$, which represents a parabola whose vertex is the new origin, whose axis is along the negative part of the axis of y, and whose latus-rectum is 2.

The focus is at a distance $\frac{1}{2}$ from the vertex measured along the axis of the parabola; hence, referred to the new origin, the focus is $\left(0, -\frac{1}{2}\right)$, and therefore referred to the original axes the focus is $\left(2, 2-\frac{1}{2}\right)$.

The directrix is perpendicular to the axis of the parabola and cuts the axis at a distance from the vertex equal to $\frac{1}{2}$ measured in the negative direction along the axis; hence the equation of the directrix referred to the new origin is $y=\frac{1}{2}$, and therefore the equation when referred to the original origin is $y=2\frac{1}{2}$.

(iii) The equation $(y-2)^2=5(x+4)$, represents a parabola the axis being $y-2=0$ and the tangent at the vertex $x+4=0$, the vertex is $(-4, 2)$ and the latus-rectum 5.

The focus is along $y-2=0$ at a distance $\frac{5}{4}$ from the vertex, and is therefore at the point $x=-4+\frac{5}{4}=-\frac{11}{4}$, $y=2$.

The directrix is perpendicular to the axis and at a distance $\frac{5}{4}$ from the vertex along the axis backwards. Hence the directrix is $x=-4-\frac{5}{4}$, or $4x+21=0$.

(iv) We have $(x+2)^2=\frac{8}{3}\left(y+\frac{3}{2}\right)$, which represents a parabola whose axis is $x+2=0$, the tangent at the vertex being $y+\frac{3}{2}=0$. The vertex is therefore $\left(-2, -\frac{3}{2}\right)$ and the latus-rectum is $\frac{8}{3}$.

The focus is along the axis $x+2=0$ at a distance $\frac{2}{3}$ from the vertex measured along the axis, and is therefore at the point $\left(-2, -\frac{3}{2}+\frac{2}{3}\right)$.

The directrix is parallel to $y+\frac{3}{2}=0$ and at a distance $\frac{2}{3}$ from $\left(-2, -\frac{3}{2}\right)$; its equation is therefore $y=-\frac{3}{2}-\frac{2}{3}$ or $6y+13=0$.

11. Let (x, y) be any point on the curve; then its distance from the origin is $\sqrt{x^2+y^2}$ and its distance from the straight line $2x-y-1=0$ is $\frac{2x-y-1}{\sqrt{5}}$. Hence the required equation is $x^2+y^2=\frac{1}{5}(2x-y-1)^2$.

The line whose equation is $2y=4x-1$, i.e. $2x-y-\frac{1}{2}=0$ is parallel to the directrix $2x-y-1=0$ and is midway between it and the focus $(0, 0)$; hence the line is the tangent at the vertex of the parabola.

12. If the ordinates of P, P' be y_1, y_2 respectively, the equation of PP' will be $y(y_1+y_2)-4ax-y_1y_2=0$. If this pass through the fixed point $(c, 0)$

on the axis of the parabola, we have $y_1y_2+4ac=0$. Hence the product of the ordinates of P and P' is constant.

Since $y_1y_2=-4ac$, $16a^2c^2=y_1^2y_2^2=16a^2x_1x_2$; $\therefore x_1x_2=c^2$.

13. The point of intersection is given by $x=\frac{a}{mm'}$, $y=\frac{a}{m}+\frac{a}{m'}$. Hence, if mm' is constant, the abscissa of the point of intersection is constant; and when $mm'=-1$, $x=-a$ which is the equation of the directrix.

14. At the points of intersection of the straight line and the curve we have $\left\{m(x+a)+\frac{a}{m}\right\}^2=4a(x+a)$, i.e. $\left\{m(x+a)-\frac{a}{m}\right\}^2=0$. Hence the two values of x, and therefore also the two values of y, are equal, and hence the line cuts the curve in coincident points.

15. From 14, the equation of any tangent to $y^2=4a(x+a)$ is of the form

$$y=mx+ma+\frac{a}{m}\quad\text{.......................(i)},$$

and of any tangent to $y^2=4a'(x+a')$ of the form

$$y=m'x+m'a'+\frac{a'}{m'}.$$

Hence the equation of a tangent to the second parabola perpendicular to (i) is

$$y=-\frac{1}{m}x-\frac{a'}{m}-a'm\quad\text{.......................(ii)}.$$

Where (i) and (ii) intersect we have

$$0=\left(m+\frac{1}{m}\right)(x+a+a').$$

Hence the equation of the required locus is $x+a+a'=0$.

16. Let the two points be $(a+d, 0)$ $(a-d, 0)$.

The equation $y=mx+\frac{a}{m}$ is the general equation of a tangent to the parabola.

Hence the difference of the squares of the perpendiculars is

$$\frac{\left\{m(a+d)+\frac{a}{m}\right\}^2}{1+m^2}\sim\frac{\left\{m(a-d)+\frac{a}{m}\right\}^2}{1+m^2}=\frac{4md\left(ma+\frac{a}{m}\right)}{1+m^2}=4ad=\text{constant}$$

for all values of m.

17. Let P be (x_1, y_1) and Q (x_2, y_2); then the equation of PQ will be $y(y_1+y_2)-4ax-y_1y_2=0$. Hence PQ cuts the axis in the point $\left(-\frac{y_1y_2}{4a}, 0\right)$.

Now the equation of AP is $\frac{x}{y}=\frac{x_1}{y_1}=\frac{4ax_1}{4ay_1}=\frac{y_1}{4a}$; and similarly the equation of AQ is $\frac{x}{y}=\frac{y_2}{4a}$.

Hence, if AP and BQ are at right angles, we have $16a^2+y_1y_2=0$. Hence PQ cuts the axis in the fixed point $(4a, 0)$.

18. The ordinates of the points common to $y^2-4ax=0$ and

$$x^2+y^2+Ax+By+C=0$$

are given by $\left(\frac{y^2}{4a}\right)^2+y^2+A\left(\frac{y^2}{4a}\right)+By+C=0$.

This is an equation of the fourth degree in which the coefficient of y^3 is zero, from which it follows that the sum of the roots is zero.

19. Let $y=mx+\frac{a}{m}$ be any tangent to the parabola. Then T is the point $\left(-\frac{a}{m^2}, 0\right)$, and Y is the point $\left(0, \frac{a}{m}\right)$. Hence the coordinates of Q are $x=-\frac{a}{m^2}$, $y=\frac{a}{m}$; therefore the locus of Q for different values of m is the parabola $y^2+ax=0$.

20. The tangents at (x_1, y_1) and (x_3, y_3) are $yy_1=2a(x+x_1)$, and $yy_3=2a(x+x_3)$; and these tangents meet where

$x=\frac{y_1x_3-x_1y_3}{y_3-y_1}=\frac{y_1y_3}{4a}$. Hence, if $y_1y_3=y_2^2$, $x=\frac{y_2^2}{4a}=x_2$.

21. $$2\Delta=\begin{vmatrix} x_1 & y_1 & 1 \\ x_2 & y_2 & 1 \\ x_3 & y_3 & 1 \end{vmatrix}=\frac{1}{4a}\begin{vmatrix} y_1^2 & y_1 & 1 \\ y_2^2 & y_2 & 1 \\ y_3^2 & y_3 & 1 \end{vmatrix}$$

$$=-\frac{1}{4a}(y_2-y_3)(y_3-y_1)(y_1-y_2).$$

Pages 105—111.

[The equation of a parabola is always supposed to be $y^2-4ax=0$.]

1. Let O be (x', y'), then its polar is $yy'-2a(x+x')=0$, and $\therefore$ T is $(-x', 0)$. The equation of OM is $2a(y-y')+y'(x-x')$, which meets the axis in G, the point $(x'+2a, 0)$. Now $TS=a+x'=SG=SP$; and, since $TS=SG$ and TMG is a right angle, $TS=TM$.

2. The curves $y^2=ax$ and $x^2=by$ meet in the point $(a^{\frac{1}{3}}b^{\frac{2}{3}},\ a^{\frac{2}{3}}b^{\frac{1}{3}})$, and the tangents to the two parabolas at this point are

$$2y\,a^{\frac{2}{3}}b^{\frac{1}{3}}=a\,(x+a^{\frac{1}{3}}b^{\frac{2}{3}}) \text{ and } 2x\,a^{\frac{1}{3}}b^{\frac{2}{3}}=b\,(y+a^{\frac{2}{3}}b^{\frac{1}{3}});$$

hence the angle between the tangents is

$$\tan^{-1}(\tfrac{1}{2}a^{\frac{1}{3}}b^{-\frac{1}{3}}\sim 2a^{\frac{1}{3}}b^{-\frac{1}{3}})/(1+a^{\frac{2}{3}}b^{-\frac{2}{3}})=\tan^{-1}3a^{\frac{1}{3}}b^{\frac{1}{3}}/(2a^{\frac{2}{3}}+2b^{\frac{2}{3}}).$$

3. If the ordinates of P, Q be y_1, y_2 respectively, the equation of PQ will be $y\,(y_1+y_2)-4ax-y_1y_2=0$; and if PQ pass through the focus, $(a, 0)$, we have $y_1y_2+4a^2=0$ (i). Equation of AP is $y=\dfrac{y_1}{x_1}x=\dfrac{4a}{y_1}x$, which meets the directrix $x=-a$ where $y=-\dfrac{4a^2}{y_1}=y_2$ from (i).

4. The tangents at y_1, y_2 meet where $4ax=y_1y_2$, $y=\frac{1}{2}(y_1+y_2)$. Hence, if $y_2=ky_1$, we have $\dfrac{4ax}{k}=\left(\dfrac{2y}{1+k}\right)^2$.

5. The directions of the tangents from (x, y) are given by $y=mx+\dfrac{a}{m}$, that is $m^2x-my+a=0$. Hence, if m_1, m_2 be the roots, we have $m_1+m_2=\dfrac{y}{x}$ and $m_1\,m_2=\dfrac{a}{x}$. (i) $m_1+m_2=k$; $\therefore\ \dfrac{y}{x}=k$, so that the locus is the straight line $y=kx$. (ii) $m_1^2+m_2^2=k$, that is $(m_1+m_2)^2-2m_1m_2=k$; $\therefore$ locus is

$$\frac{y^2}{x^2}-2\,\frac{a}{x}=k.$$

6. The angle between the tangents from (x, y) is $\tan^{-1}(m_1-m_2)/(1+m_1m_2)$, where m_1, m_2 are the roots of $m^2x-my+a=0$. Hence equation of required locus is $1=\sqrt{\left(\dfrac{y^2}{x^2}-\dfrac{4a}{x}\right)}\Big/\left(1+\dfrac{a}{x}\right)$, which reduces to $y^2=x^2+6ax+a^2$.

7. Let the tangents be $y=m_1x+\dfrac{a}{m_1}$, $y=m_2x+\dfrac{a}{m_2}$.

Then $\dfrac{a}{m_1}-\dfrac{a}{m_2}=\text{constant}=c$ suppose. Now the tangents meet where $y=a\left(\dfrac{1}{m_1}+\dfrac{1}{m_2}\right)$ and $x=\dfrac{a}{m_1m_2}$, whence $y^2-4ax=a^2\left(\dfrac{1}{m_1}-\dfrac{1}{m_2}\right)^2=c^2$.

8. The two tangents being $y=m_1x+\dfrac{a}{m_1}$ and $y=m_2x+\dfrac{a}{m_2}$, we have $m_1m_2=1$ since the tangents make equal angles with the axis and directrix respectively but are not at right angles. But $x=\dfrac{a}{m_1m_2}$ at the point of intersection of the tangents; $\therefore$ the tangents intersect in the line $x=a$.

9. The tangents at the ends of the latus-rectum are $y=x+a$ and $y=-x-a$. The perpendiculars on these lines from the point (a, k) are

$$y-k=-x+a \text{ and } y-k=x-a$$

respectively.

Hence the feet of the perpendiculars are $\left(\frac{k}{2}, \frac{k}{2}+a\right)$ and $\left(-\frac{k}{2}, \frac{k}{2}-a\right)$,

and the equation of the line joining the feet is therefore $2ax-ky+\frac{k^2}{2}=0$,

or $yk=2a\left(x+\frac{k^2}{4a}\right)$, which is the equation of the tangent at $\left(\frac{k^2}{4a}, k\right)$.

10. The equation of the chord of contact of the tangents drawn from $(-4a, k)$ is

$$yk-2a(x-4a)=0 \quad \text{..................................(i).}$$

The equation of the lines joining the vertex to the points where (i) cuts $y^2-4ax=0$ is [Art. 38]

$$y^2+4ax\,\frac{yk-2ax}{8a^2}=0, \text{ or}$$

$y^2-x^2+\frac{k}{2a}xy=0$, which obviously is the equation of two perpendicular lines.

11. Let T be the point (ξ, η); then the equation of the chord of contact will be

$$y\eta-2a(x+\xi)=0.$$

Hence $TN=(\eta^2-4a\xi)/\sqrt{(\eta^2+4a^2)}$.

The equation of the line TNM is $2a(y-\eta)+\eta(x-\xi)=0$; $\therefore$ M is the point $(\xi+2a, 0)$, and $TM=\sqrt{\{\eta^2+(\xi+2a-\xi)^2\}}=\sqrt{\{\eta^2+4a^2\}}$.

Hence if $TN.TM=\text{constant}=c^2$ suppose, we have $\eta^2-4a\xi=c^2$, so that the locus of T is the parabola $y^2-4a\left(x+\frac{c^2}{4a}\right)=0$.

Again, if $TN:TM=\text{constant}=\lambda$, we have $\eta^2-4a\xi=\lambda(\eta^2+4a^2)$, so that the locus of T is the parabola whose equation is

$$(1-\lambda)y^2-4ax-4a^2\lambda=0.$$

12. Take the common tangent at the vertices for axis of y, and the axis of x midway between the axes of the two parabolas; then the equations of the parabolas will be $(y-b)^2=4ax$ and $(y+b)^2=4ax$. Now the line $y=k$ will meet the curves in points whose abscissa are $\frac{(k-b)^2}{4a}$ and $\frac{(k+b)^2}{4a}$, and therefore at the middle point of the intercept $x=\frac{k^2+b^2}{4a}$ and $y=k$.

Hence, for all values of k, we have $y^2+b^2=4ax$, which is the equation of an equal parabola.

13. Take for axes the common tangent and the diameter through its point of contact; and let the equations of the parabolas be $y^2-4ax=0$ and $y^2-4bx=0$. The equations of the tangents to the two parabolas at (x', y') and (x'', y'') respectively will be

$$yy'-2a(x+x')=0 \text{ and } yy''-2b(x+x'')=0,$$

and these tangents meet $x=0$ where

$$y=\frac{2ax'}{y'}=\frac{1}{2}y' \text{ and } y=\frac{1}{2}y''$$

respectively. Hence, if the tangents meet $x=0$ in the same point, $y'=y''$, and therefore the line joining (x', y'), (x'', y'') is parallel to the axis.

14. Let the equations of the two parabolas be

$$x^2=4b(x+c)\text{..............(i)},$$

and $$y^2=4ax\text{.....................(ii)}.$$

Then, if $(x'\ y')$ be any point on (i), we have $y'^2-4b(x'+c)=0$, and the equation of the chord of contact of the tangents from (x', y') to (ii) is

$$yy'=2a(x+x')\text{ ....(iii)}.$$

Now, if (ξ, η) be the coordinates of the middle point of any chord of the parabola (ii), the chord will from Art. 102 make an angle $\tan^{-1}\frac{2a}{\eta}$ with the axis of x and therefore the equation of the chord will be

$$\eta(y-\eta)=2a(x-\xi)\text{.............................(iv)}.$$

Hence if (ξ, η) be the middle point of the chord (iii) we have by comparing (iii) and (iv)

$$\eta=y' \text{ and } \eta^2-2a\xi=2ax'.$$

But $y'^2-4b(x'+c)=0$; $\therefore\ \eta^2-\frac{2b}{a}(\eta^2-2a\xi)-4bc=0$,

shewing that (ξ, η) is always on the parabola whose equation is

$$y^2=\frac{4ab}{2b-a}(x-c).$$

15. The chord whose middle point is (ξ, η) makes an angle $\tan^{-1}\frac{2a}{\eta}$ with the axis [Art. 102, (iii.)]; and therefore the equation of the chord is

$$\eta(y-\eta)-2a(x-\xi)=0.$$

If therefore the chord pass through the fixed point (f, g), we have

$$\eta(g-\eta)-2a(f-\xi)=0,$$

shewing that the locus of (ξ, η) is the parabola

$$y(y-g)-2a(x-f)=0,$$

or thus:—

Let (ξ, η) be the middle point of the chord whose equation is

$$y-g=m(x-f).$$

Then at the extremities of the chord we have

$$y-g=m\left(\frac{y^2}{4a}-f\right), \text{ and } \{g+m(x-f)\}^2=4ax.$$

Hence $$2\eta=y_1+y_2=\frac{4a}{m},$$

and $$2\xi=x_1+x_2=\frac{4a}{m^2}+\frac{2m(mf-g)}{m^2}.$$

Hence, by eliminating m, we have $2a(\xi-f)+g\eta-\eta^2=0$.

16. Let the pole of the chord be (ξ, η); then the equation of the chord will be $y\eta-2a(x+\xi)=0$.

This meets $y^2=4ax$, where $4a^2(x+\xi)^2=4a\eta^2x$.

But, if x_1 and x_2 be roots of the above quadratic equation in x, we have by supposition $\frac{x_1+x_2}{2}=\text{constant}=c$ suppose.

Hence $\eta^2-2a\xi=2ca$, so that the pole of the chord is on the parabola

$$y^2=2a(x+c).$$

17. The equation of the chord of contact of the tangents drawn from $T(x', y')$ is $$yy'-2a(x+x')=0.$$

The equation of the lines joining A to the points where the chord cuts the parabola is therefore

$$y^2-4ax\frac{yy'-2ax}{2ax'}=0. \qquad \text{[Art. 38.]}$$

Hence the lines PA, QA cut the directrix in points p, q whose ordinates are given by

$$y^2+4a^2\frac{yy'+2a^2}{2ax'}=0,$$

the ordinate of the middle point of pq is therefore

$$\frac{1}{2}(y_1+y_2)=-\frac{ay'}{x'}.$$

Also the equation of AT is $\frac{x}{x'}=\frac{y}{y'}$ hence the ordinate of t is $-\frac{ay'}{x'}$. Hence t coincides with the middle point of pq.

18. Take for axes the tangent at O and the diameter through O, and let $y^2-4ax=0$ be the equation of the parabola.

Then if y_1, y_2 be the ordinates of the extremities of any chord, the equation of the chord will be $y(y_1+y_2)-4ax-y_1y_2=0$.

Hence $$OP=\frac{y_1y_2}{4a}.$$

The equations of the tangents at the extremities of the chord will be

$$yy_1 - 2a\left(x + \frac{y_1^2}{4a}\right) = 0$$

and
$$yy_2 - 2a\left(x + \frac{y_2^2}{4a}\right) = 0.$$

Hence $$OQ \,.\, OQ_1 = \left(-\frac{y_1^2}{4a}\right)\left(-\frac{y_2^2}{4a}\right) = \left(\frac{y_1 y_2}{4a}\right)^2 = OP^2.$$

19. Take the vertex for origin and axes parallel and perpendicular to the fixed straight line on which the base moves, and let the equation of the fixed straight line be $x = a$.

Let the equation of the circumscribing circle be

$$x^2 + y^2 + 2gx + 2fy = 0 \quad \text{.........................(i)}.$$

Then the intercept on the line $x = a$ must be of given length, $2l$ suppose.

Hence $(y_1 - y_2)^2 = 4l^2$, where y_1 y_2 are the roots of

$$a^2 + y^2 + 2ga + 2fy = 0.$$

Hence $$4f^2 - 4(a^2 + 2ga) = 4l^2.$$

But the centre of the circle (i) is $(-g, -f)$.

Hence the locus of the centre is the parabola whose equation is

$$y^2 = -2ax + a^2 + l^2.$$

20. The polar of (h, k) with respect to

$$x^2 + y^2 + 2ax - 3a^2 = 0$$

is
$$hx + ky + a(x + h) - 3a^2 = 0.$$

Now this touches $y^2 + 4ax = 0$ if the equation

$$(h + a)y^2 - 4aky - 4a^2(h - 3a) = 0$$

has equal roots, *i.e.* if

$$k^2 = -(h + a)(h - 3a)$$

or
$$h^2 + k^2 - 2ah - 3a^2 = 0,$$

i.e. if (h, k) lies on the circle

$$x^2 + y^2 - 2ax - 3a^2 = 0.$$

21. Let P be (x_1, y_1) and P' be (x_2, y_2); then the equation of PP' will be

$$y(y_1 + y_2) - 4ax - y_1 y_2 = 0.$$

Hence, as PP' is a focal chord, $y_1 y_2 = -4a^2$, whence also $x_1 x_2 = a^2$.

The coordinates of V are $\frac{1}{2}(x_1 + x_2)$, $\frac{1}{2}(y_1 + y_2)$.

Therefore the equation of VO is

$$\left\{x-\frac{1}{2}(x_1+x_2)\right\}(y_1+y_2)+4a\left\{y-\frac{1}{2}(y_1+y_2)\right\}=0.$$

Hence, at O, $$x=\frac{1}{2}(x_1+x_2)+2a;$$

$$\therefore\ SO=\frac{1}{2}(x_1+x_2)+a=\frac{1}{2}\{a+x_1+a+x_2\}=\frac{1}{2}(SP+SP').$$

Again, $$VO^2=\frac{1}{4}(y_1+y_2)^2+\left\{\frac{1}{2}(x_1+x_2)+2a-\frac{1}{2}(x_1+x_2)\right\}^2$$

$$=\frac{1}{4}\{4ax_1+4ax_2+2y_1y_2+16a^2\}$$

$$=(ax_1+ax_2+a^2+x_1x_2),$$

since $$y_1y_2=-4a^2 \text{ and } x_1x_2=a^2$$

$$=(a+x_1)(a+x_2)=SP\,.\,SP'.$$

22. Let the coordinates of P, Q, R be (x_1, y_1), (x_2, y_2) and (x_3, y_3), respectively.

Then the ordinates of the points p, q, r will be $-\dfrac{4a^2}{y_1}$, $-\dfrac{4a^2}{y_2}$ and $-\dfrac{4a^2}{y_3}$ respectively.

Hence the equation of the diameter through p is $y=-\dfrac{4a^2}{y_1}$.

Also the equation of the chord QR is

$$y(y_2+y_3)-4ax-y_2y_3=0.$$

Hence the equation of any line through A is given by

$$y(y_2+y_3)-4ax-y_2y_3+\lambda(yy_1+4a^2)=0.$$

The last equation will be the equation of SA provided λ be so chosen that $$-4a^2-y_2y_3+4a^2\lambda=0.$$

Hence the equation of SA is

$$4a^2y(y_2+y_3)-16a^3x-4a^2y_2y_3+(4a^2+y_2y_3)(yy_1+4a^2)=0,$$

i.e. $$4a^2y(y_1+y_2+y_3)+yy_1y_2y_3-16a^3(x-a)=0,$$

and the symmetry of this last result shews that it is also the equation of SB and of SC.

23. Let PP' meet the diameter AV which bisects the parallel chords in V; then the equation of the parabola, referred to AV and the tangent at A as axes, will be $y^2=4ax$.

Also $$PO\,.\,OP'=PV^2-OV^2.$$

But $PO\,.\,OP'=\text{constant}=c^2$, and $PV^2=4a\,.\,AV$. Hence, if x, y be the coordinates of O, we have $4ax-y^2=c^2$. Thus the locus of O is a parabola.

24. Take for axes the diameter through O and the tangent at O; then the equation of the parabola will be of the form $y^2-4ax=0$, and every line whose equation is of the form $y=mx+\frac{a}{m}$ will be a tangent.

Let
$$y=m_1x+\frac{a}{m_1} \quad\text{(i)},$$
be the equation of one of the tangents from P; then, since the tangents meet on the axis of x, the equation of the other tangent will be
$$y=-m_1x-\frac{a}{m_1} \quad\text{(ii)}.$$

So also the two tangents from P' will be of the form
$$y=m_2x+\frac{a}{m_2} \quad\text{(iii)},$$
and
$$y=-m_2x-\frac{a}{m_2} \quad\text{(iv)}.$$

Now $OP \cdot OP'=\left(-\frac{a}{m_1^2}\right)\left(-\frac{a}{m_2^2}\right)=\text{constant}$; and $\therefore m_1m_2=\text{constant}$.

At the intersection of (i) and (iii), and also at the intersection of (ii) and (iv), we have
$$x-\frac{a}{m_1m_2}=0.$$

Again, at the intersection of (i) and (iv), and also at the intersection of (ii) and (iii), we have
$$x+\frac{a}{m_1m_2}=0.$$

Hence the four points of intersection are on the lines
$$x-a/m_1m_2=0,\ x+a/m_1m_2=0;$$
and, since $m_1m_2=\text{constant}$, these are fixed straight lines, and they are obviously parallel to and equidistant from the tangent at O.

25. Let the equations of the four tangents AB, BC, CD, DA be

$y=m_1x+\frac{a}{m_1}$, $y=m_2x+\frac{a}{m_2}$, $y=m_3x+\frac{a}{m_3}$ and $y=m_4x+\frac{a}{m_4}$ respectively.

Then the ordinate of B is $a\left(\frac{1}{m_1}+\frac{1}{m_2}\right)$, and the ordinate of D is $a\left(\frac{1}{m_3}+\frac{1}{m_4}\right)$; hence the ordinate of the middle point of BD is
$$\frac{a}{2}\left(\frac{1}{m_1}+\frac{1}{m_2}+\frac{1}{m_3}+\frac{1}{m_4}\right).$$

So also the ordinate of the middle point of AC is $\frac{a}{2}\left(\frac{1}{m_1}+\frac{1}{m_2}+\frac{1}{m_3}+\frac{1}{m_4}\right)$;

and, if Q be the point of intersection of BA and CD and R be the point of intersection of AD and BC, the middle point of QR will also be

$$\frac{a}{2}\left(\frac{1}{m_1}+\frac{1}{m_2}+\frac{1}{m_3}+\frac{1}{m_4}\right).$$

Hence the middle points of the three diagonals AC, BD and PQ are on a straight line parallel to the axis of the parabola.

26. Let the equations of the two tangents be $y=m_1x+\frac{a}{m_1}$, $y=m_2x+\frac{a}{m_2}$; then their point of intersection, T, is $\left(\frac{a}{m_1m_2}, \frac{a}{m_1}+\frac{a}{m_2}\right)$.

Let the tangent at *either* extremity of the focal chord be $y=mx+\frac{a}{m}$, then its point of contact, P, will be $\left(\frac{a}{m^2}, \frac{2a}{m}\right)$.

Since the three points S, P, T are on a straight line we have

$$\frac{\frac{a}{m_1m_2}-a}{a-\frac{a}{m^2}}=\frac{\frac{a}{m_1}+\frac{a}{m_2}}{-\frac{2a}{m}},$$

that is, $2m(1-m_1m_2)+(m^2-1)(m_1+m_2)=0$, which is equivalent to

$$\frac{m-m_1}{1+mm_1}=-\frac{m-m_2}{1+mm_2}.$$

Hence the tangents m_1, m_2 make equal angles with the tangent m.

27. Let the fixed straight line make an angle α with the axis of x, and let two tangents which satisfy the required condition make angles θ_1, θ_2 with the axis; then $2\alpha=\theta_1+\theta_2$. But, if y_1, y_2 be the ordinates of the points of contact, $\tan\theta_1=\frac{2a}{y_1}$ and $\tan\theta_2=\frac{2a}{y_2}$; also the equation of the chord of contact will be

$$y(y_1+y_2)-4ax-y_1y_2=0.$$

But
$$\tan 2\alpha=\frac{\frac{2a}{y_1}+\frac{2a}{y_2}}{1-\frac{4a^2}{y_1y_2}}=\frac{2a(y_1+y_2)}{y_1y_2-4a^2}.$$

Hence the equation of the chord of contact may be written

$$y(y_1y_2-4a^2)\tan 2\alpha-8a^2x-2ay_1y_2=0$$

or

$$(y\tan 2\alpha-2a)\,y_1y_2-4a^2(y\tan 2\alpha+2x)=0.$$

Hence, for all values of y_1y_2, the chord of contact passes through the fixed point given by $y\tan 2\alpha-2a=0$ and $y\tan 2\alpha+2x=0$, that is, through the point $(-a, 2a\cot 2\alpha)$.

28. Take the given focus for origin and the axis for axis of x; then the equations of the parabolas will be

$$y^2=4a\ (x+a)\dots\dots\dots\dots(\text{i}),\quad y^2=-4a'\ (x-a')\dots\dots\dots\dots(\text{ii}).$$

The equation of any tangent to (i) is $y=m\ (x+a)+\dfrac{a}{m}$; and this meets (ii) where

$$\left\{m\ (x+a)+\frac{a}{m}\right\}^2=-4a'\ (x-a'),$$

and
$$my^2=-4a'\left(y-ma-\frac{a}{m}-ma'\right).$$

Hence, if $(\xi,\ \eta)$ be the middle point of the intercept,

$$2\xi=-\{2\ (m^2+1)\ a+4a'\}/m^2,$$

and
$$2\eta=-\frac{4a'}{m}.$$

Hence, eliminating m, we have

$$4a'^2\ (\xi+a)+(a+2a')\ \eta^2=0.$$

Thus the locus of the middle point of the chords is the parabola

$$y^2=-\frac{4a'^2}{a+2a'}\ (x+a).$$

[It is not necessary that the two parabolas should be confocal, for the above would still be true if the equation (ii) were $y^2=-4a'\ (x-d)$.]

29. The equation of the chord whose middle point is $(\xi,\ \eta)$ is [see 15],

$$\eta\ (y-\eta)-2a\ (x-\xi)=0\dots\dots\dots\dots\dots\dots\dots\dots\dots\dots(\text{i}).$$

The equation of the lines joining the vertex to the two points where (i) cuts the parabola is [Art. 38],

$$y^2-4ax\ \frac{y\eta-2ax}{\eta^2-2a\xi}=0\dots\dots\dots\dots\dots\dots\dots\dots\dots\dots(\text{ii}).$$

Hence, if (ii) are at right angles, we have $1+8a^2/(\eta^2-2a\xi)=0$. Hence the required locus is the parabola whose equation is $y^2=2a\ (x-4a)$.

30. The normal $y=mx-2am-am^3$ meets the parabola $y^2-4ax=0$, where

$$my^2-4ay-8a^2m-4a^2m^3=0.$$

Hence, if $(\xi,\ \eta)$ be the middle point of the chord $\eta=\dfrac{2a}{m}$, and therefore

$$\eta=(\xi-2a)\ \frac{2a}{\eta}-a\ \frac{8a^3}{\eta^3},$$

or
$$\frac{\eta^2}{2a}+\frac{4a^3}{\eta^2}=\xi-2a.$$

31. Let PQ be

$$y-y_1=-\frac{y_1}{2a}(x-x_1);$$

then, if Q be (x_2, y_2),

$$y_2-y_1=-\frac{y_1}{2a}\left(\frac{y_2^2-y_1^2}{4a}\right);$$

therefore $$8a^2=-y_1(y_2+y_1)\ldots\ldots\text{(i)}.$$

Equation to AQ is $y=\frac{y_2}{x_2}\cdot x=\frac{4a}{y_2}\cdot x$;

therefore equation to PR is

$$y-y_1=\frac{4a}{y_2}(x-x_1).$$

Hence, putting $y=0$, we have

$$AR=x_1-\frac{y_1y_2}{4a},$$

$$=x_1+\frac{y_1^2+8a^2}{4a}, \text{ by (i)},$$

$$=2(x_1+a)=2SP.$$

32. Let y_1, y_2 be the ordinates at the extremities of any chord of a parabola which is parallel to the line $y=mx$.

Then the equation of the chord is $y(y_1+y_2)-4ax-y_1y_2=0$; and therefore $y_1+y_2=\frac{4a}{m}$.

Hence, from Art. 96, Ex. 1, the tangents at y_1, y_2 meet on the fixed line $y=\frac{2a}{m}$ ....(i).

Also, from Art. 106, the normals at y_1, y_2 meet on the normal at y_3, where $-y_3=y_1+y_2=\frac{4a}{m}$. Hence the locus of the intersection of the normals is the normal at the point whose ordinate is $-\frac{4a}{m}$, and whose equation is therefore $y+\frac{4a}{m}-\frac{2}{m}\left(x-\frac{4a}{m^2}\right)=0$ ....(ii).

The locus of the point of intersection of (i) and (ii) for different values of m is obtained by eliminating m from (i) and (ii). The result is $y^2=a(x-3a)$.

33. If the ordinates of P, Q, R be y_1, y_2, y_3; then $y_1+y_2+y_3=0$.

Let $x^2+y^2+2gx+2fy+c=0$ be the equation of the circle PQR. Then at the points common to the circle and the parabola $y^2-4ax=0$, we have

$$\left(\frac{y^2}{4a}\right)^2+y^2+2g\cdot\frac{y^2}{4a}+2fy+c=0.$$

Since the coefficient of y^3 is zero, we have

$$y_1+y_2+y_3+y_4=0,$$

where y_1, y_2, y_3, y_4 are the ordinates of the points of intersection. But we know that $y_1+y_2+y_3=0$. Hence $y_4=0$, so that the circle PQR goes through the vertex.

34. The directions of the three normals which meet in (x, y) are given by

$$y=mx-2am-am^3 \qquad \text{(i)}.$$

Hence $m_1m_2m_3=-\dfrac{y}{a}$.

If, therefore, $m_1m_2=2$, we have $m_3=-\dfrac{y}{2a}$.

Hence, as m_3 is a root of (i), we have

$$y=x\left(-\frac{y}{2a}\right)-2a\left(-\frac{y}{2a}\right)-a\left(-\frac{y}{2a}\right)^3,$$

whence $y^2=4ax$.

35. The directions of the normals which meet in (x, y) are given by

$$y=mx-2am-am^3 \qquad \text{(i)}.$$

Hence $m_1m_2m_3=-\dfrac{y}{a}$.

Now, if two of the normals make complementary angles with the axis, we have $m_1m_2=1$, and therefore $m_3=-\dfrac{y}{a}$.

Hence, as m_3 is a root of (i), we have

$$y=\left(-\frac{y}{a}\right)(x-2a)-a\left(-\frac{y}{a}\right)^3,$$

whence $y^2=a(x-a)$.

36. The directions of the three normals which meet in (x, y) are given by

$$y=mx-2am-am^3 \qquad \text{(i)}.$$

Since two of the normals make equal angles with a fixed line, $y=\tan\alpha \,.\, x$ suppose, we have a relation of the form $\tan^{-1}m_1+\tan^{-1}m_2=2\alpha$;

$$\therefore \frac{m_1+m_2}{1-m_1m_2}=\tan 2\alpha=k \text{ suppose}.$$

But

$$m_1+m_2+m_3=0,$$

$$m_2m_3+m_3m_1+m_1m_2=\frac{2a-x}{a},$$

and

$$m_1m_2m_3=-\frac{y}{a}.$$

Whence $$\frac{-m_3}{1+\frac{y}{am_3}}=k;$$

$$\therefore \quad am_3^2+akm_3+ky=0 \quad \text{(i)}.$$

Hence we have to eliminate m_3 between (i) and

$$am_3^3+(2a-x)\,m_3+y=0 \quad \text{(ii)}.$$

Multiply (i) by m_3 and subtract (ii); then

$$akm_3^2+(ky+x-2a)\,m_3-y=0 \quad \text{(iii)}.$$

Multiply (iii) by k and add to (i); then

$$a\,(1+k^2)\,m_3^2+k\,(ky+x-a)\,m_3=0,$$

whence $m_3=-k\,(ky+x-a)/a\,(1+k^2)$, since $m_3=0$;

$\therefore$ substituting in (i) we have

$$k\,(ky+x-a)^2+a\,(1+k^2)\,\{(1-k^2)\,y+kx-ka\}=0;$$

which is of the form $Y^2 \propto X$ and is therefore a parabola. [Art. 104.]

37. If x_1, y_1 be the coordinates of P; the coordinates of H will be x_1+3a and $-\frac{y_1}{2}$.

Now two of the normals which meet in (x, y) are at right angles provided $y^2=a\,(x-3a)$ [Art. 107, Ex. 4]; and this condition is satisfied by the coordinates of H.

38. If the normal at (x', y') pass through the point (h, k) we have [Art. 79]

$$(k-y')\,2a+y'\,(h-x')=0,$$

$$\therefore \quad 4a^2k^2=y'^2\,(2a-h+x')^2;$$

$$\therefore \quad 4a^2k^2=4ax'\,(2a-h+x')^2.$$

Hence if x_1, x_2, x_3 be the abscissae of the points P, Q, R the normals at which meet in (h, k) we have

$$x_1+x_2+x_3=2\,(h-2a);$$

$$\therefore \quad (x_1+a)+(x_2+a)+(x_3+a)+a=2h,$$

$$\text{i.e.} \quad SP+SQ+SR+SA=2OM.$$

39. Let the tangents be $y=m_1x+\frac{a}{m_1}$, $y=m_2x+\frac{a}{m_2}$, $y=m_3x+\frac{a}{m_3}$.

Let $\frac{1}{m_2}-\frac{1}{m_1}=\frac{1}{m_3}-\frac{1}{m_2}=$ given quantity $=c$ suppose.

Then the points of intersection of the tangents are

$$\left(\frac{a}{m_2m_3},\ \frac{a}{m_2}+\frac{a}{m_3}\right), \text{ \&c.}$$

Hence $2\Delta = x_1(y_2 - y_3) + \dots\dots$

$$= \frac{a}{m_2 m_3}\left\{\left(\frac{a}{m_3}+\frac{a}{m_1}\right)-\left(\frac{a}{m_1}+\frac{a}{m_2}\right)\right\}+\dots\dots$$

$$= \frac{a^2}{m_2 m_3}\left(\frac{1}{m_3}-\frac{1}{m_2}\right)+\frac{a^2}{m_3 m_1}\left(\frac{1}{m_1}-\frac{1}{m_3}\right)+\frac{a^2}{m_1 m_2}\left(\frac{1}{m_2}-\frac{1}{m_1}\right)$$

$$= a^2 c\left(\frac{1}{m_2 m_3}-\frac{2}{m_3 m_1}+\frac{1}{m_1 m_2}\right)$$

$$= a^2 c\left\{\frac{c}{m_3}-\frac{c}{m_1}\right\} = 2a^2c^3;$$

$\therefore\ \Delta = a^2 c^3.$

40. Let the normals be $y = m_1 x - 2am_1 - am_1^3$, &c.

Then, if (x_1, y_1), (x_2, y_2), (x_3, y_3) be the points of intersection, we have

$$x_1 = 2a + a(m_2^2 + m_3^2 + m_2 m_3),$$

and

$$y_1 = am_2 m_3(m_2 + m_3).$$

Hence $x_2 - x_3 = a(m_1 + m_2 + m_3)(m_3 - m_2)$.

Hence $2\Delta = \Sigma y_1(x_2 - x_3)$

$$= a^2(m_1 + m_2 + m_3)\,\Sigma m_2 m_3(m_3^2 - m_2^2)$$

$$= a^2(m_1 + m_2 + m_3)^2(m_2 - m_3)(m_3 - m_1)(m_1 - m_2).$$

Since $m_2 m_3(m_2^2 - m_3^2) + m_3 m_1(m_3^2 - m_1^2) + m_1 m_2(m_1^2 - m_2^2)$

$$= -(m_1 + m_2 + m_3)(m_2 - m_3)(m_3 - m_1)(m_1 - m_2).$$

41. Take the tangent parallel to the given parallel straight lines as axis of y, and the diameter through its point of contact for axis of x.

Then the equation of the parabola will be $y^2 - 4ax = 0$, and the equations of the lines on which P and Q lie will be $x = b$, $x = c$ respectively.

Let the equation of PQ be $y = mx + \dfrac{a}{m}$ ....(i)

and let the other tangents through P, Q be

$$y = m_1 x + \frac{a}{m_1} \quad \dots\dots\dots\dots\text{(ii)},$$

and

$$y = m_2 x + \frac{a}{m_2} \quad \dots\dots\dots\dots\text{(iii)}.$$

Then where (i) meets (ii) $x = \dfrac{a}{mm_1} = b$,

and where (i) meets (iii) $x = \dfrac{a}{mm_2} = c$;

$$\therefore\ bm_1 = cm_2 \quad \dots\dots\dots\dots\text{(iv)}.$$

Where (ii) meets (iii) we have $x=\frac{a}{m_1m_2}$ and $y=\frac{a}{m_1m_2}(m_1+m_2)$. Hence, from (iv), $bcy^2=(b+c)^2ax$.

42. If the three tangents $y=m_1x+\frac{a}{m_1}$, $y=m_2x+\frac{a}{m_2}$, $y=m_3x+\frac{a}{m_3}$ form an equilateral triangle it is easy to shew that

$$2\tan^{-1}m_1=\tan^{-1}m_2+\tan^{-1}m_3\pm\pi,$$

and two similar relations.

Hence $\frac{2m_1}{1-m_1^2}=\frac{m_2+m_3}{1-m_2m_3}$, and two similar relations.

The tangents (m_2, m_3) meet in the point $\left(\frac{a}{m_2m_3}, \frac{a}{m_2}+\frac{a}{m_3}\right)$. The point of contact of m_1 is $\left(\frac{a}{m_1^2}, \frac{2a}{m_1}\right)$.

Hence the equation of the line through $(a, 0)$ and the point of contact is

$$\frac{x-a}{a-\frac{a}{m_1^2}}=\frac{y}{-\frac{2a}{m_1}},$$

and this will go through the point of intersection of (m_2, m_3) if

$$\frac{\frac{1}{m_2m_3}-1}{1-\frac{1}{m_1^2}}=\frac{\frac{1}{m_2}+\frac{1}{m_3}}{-\frac{2}{m_1}},$$

that is, if $\frac{2m_1}{1-m_1^2}=\frac{m_2+m_3}{1-m_2m_3}$, which is known to be the case.

43. The ordinates of the feet of the normals which meet in (ξ, η) are given by the equation

$$8a^2(\eta-y)+y(4a\xi-y^2)=0. \quad \text{[Art. 106.]}$$

Hence, if η_1, η_2, η_3 be the roots we have

$$\eta_1+\eta_2+\eta_3=0 \text{.................................(i)},$$

$$\eta_2\eta_3+\eta_3\eta_1+\eta_1\eta_2=4a(2a-\xi) \text{.........................(ii)}.$$

Now the tangents at the points whose ordinates are η_1, η_2 meet in the point where $x=\frac{\eta_1\eta_2}{4a}$, $y=\frac{1}{2}(\eta_1+\eta_2)$.

Hence, if this point lie on $y^2=a(x+c)$, we have

$$(\eta_1+\eta_2)^2=\eta_1\eta_2+4ac,$$

or, from (i), $$(\eta_1+\eta_2)\eta_3+\eta_1\eta_2+4ac=0;$$

$\therefore$, from (ii), $$4a(2a-\xi)+4ac=0.$$

Hence the normals meet on the line $x=2a+c$.

44. The ordinates of the feet of the three normals which meet in (ξ, η) are given by the equation

$$8a^2(\eta - y) + y(4a\xi - y^2) = 0.$$

Hence, when the point is on the curve, so that $4a\xi - \eta^2 = 0$, we have

$$8a^2(\eta - y) + y(\eta^2 - y^2) = 0.$$

Hence $y = \eta$ is one ordinate, as is obvious, and the other two are given by

$$8a^2 + y(\eta + y) = 0.$$

Hence, if y_1, y_2 be the roots of the above, $y_1 y_2 = 8a^2$; and this shews that the chord joining the feet of these normals passes through the fixed point $x = -2a$, for the equation of the chord is

$$y(y_1 + y_2) - 4ax - y_1 y_2 = 0.$$

45. Let the ordinates of the extremities of the chord be y_1, y_2, and let (α, β) be the fixed point; then the equation of the chord is

$$y(y_1 + y_2) - 4ax - y_1 y_2 = 0.$$

Hence we have

$$\beta(y_1 + y_2) - 4a\alpha - y_1 y_2 = 0 \text{..........................(i)}.$$

The normals at y_1, y_2 are

$$8a^2(y - y_1) + y_1(4ax - y_1^2) = 0 \text{........................(ii)},$$

and

$$8a^2(y - y_2) + y_2(4ax - y_2^2) = 0 \text{.......................(iii)}.$$

Subtract (iii) from (ii), and divide by $y_2 - y_1$; then

$$8a^2 - 4ax + y_2^2 + y_1 y_2 + y_1^2 = 0 \text{.........................(iv)}.$$

Multiply (ii) by y_2 and (iii) by y_1 and divide their difference by $y_2 - y_1$; then

$$8a^2 y + y_1 y_2 (y_1 + y_2) = 0 \text{...............................(v)}.$$

We have now to eliminate $y_1 y_2$ and $y_1 + y_2$ from (i), (iv) and (v).

From (i) and (iv) we have

$$(y_1 + y_2)^2 - \beta(y_1 + y_2) = 4a(x - 2a - \alpha) \text{..................(vi)}.$$

Also, from (i) and (v) we have

$$\beta(y_1 + y_2)^2 - 4a\alpha(y_1 + y_2) + 8a^2 y = 0 \text{..................(vii)}.$$

Hence, from (vi) and (vii), we have

$$2\{\beta(x - 2a - \alpha) + 2ay\}^2 = (\beta^2 - 4a\alpha)\{2a(2a + \alpha - x) - \beta y\},$$

which is of the form $Y^2 \propto X$, and therefore represents a parabola.

46. As in 38, the abscissae of the three points the normals at which pass through (h, k) are the roots of the following cubic in x'

$$4a^2 k^2 = 4ax'(2a - h + x')^2 \text{..........................(i)}.$$

Now the normal at x' cuts the axis in the point $x'+2a$. Hence if x_1, x_2, x_3 be the three roots of (i) we have

$$x_1+2a+x_3+2a=2(x_2+2a);$$

$$\therefore\ x_1+x_2+x_3=3x_2.$$

Now $$x_1+x_2+x_3=2h-4a;$$

$$\therefore\ x_2=\frac{2}{3}(h-2a).$$

Hence, as x_2 is a root of (i), we have

$$ak^2=\frac{2}{3}(h-2a)\left\{2a-h+\frac{2}{3}(h-2a)\right\}^2,$$

or $$27ak^2=2(h-2a)^3.$$

Thus the point of intersection of the normals is on the curve whose equation is

$$27ay^2=2(x-2a)^3.$$

47. Let y_1, y_2, y_3, y_4 be the ordinates of the four points A, B, C, D on the parabola $y^2-4ax=0$; and let AB, BC and CD make given angles with the axis.

Then the equations of AB, BC, CD and DA are respectively

$$y(y_1+y_2)-4ax-y_1y_2=0,$$

$$y(y_2+y_3)-4ax-y_2y_3=0,$$

$$y(y_3+y_4)-4ax-y_3y_4=0,$$

and $$y(y_4+y_1)-4ax-y_4y_1=0.$$

Since AB, BC, CD make given angles with the axis it follows that

$$y_1+y_2,\quad y_2+y_3 \text{ and } y_3+y_4$$

are all constant, and therefore also $(y_1+y_2)+(y_3+y_4)-(y_2+y_3)$, that is y_1+y_4, is constant. Hence DA makes a fixed angle with the axis.

48. Let $\left(\frac{y_1^2}{4a}, y_1\right)$ and $\left(\frac{y_2^2}{4a}, y_2\right)$ be the extremities of one of the chords: then the equation of the corresponding circle is

$$\left(x-\frac{y_1^2}{4a}\right)\left(x-\frac{y_2^2}{4a}\right)+(y-y_1)(y-y_2)=0. \quad \text{[Art. 67, Ex. 4.]}$$

Similarly, if y_3, y_4 be the ordinates of the extremities of the other chord, the eqnation of the other circle will be

$$\left(x-\frac{y_3^2}{4a}\right)\left(x-\frac{y_4^2}{4a}\right)+(y-y_3)(y-y_4)=0.$$

The common chord of the two circles will pass through the origin provided

$$\frac{y_1^2 y_2^2}{16a^2}+y_1y_2-\left(\frac{y_3^2 y_4^2}{16a^2}+y_3y_4\right)=0,$$

or

$$(y_1y_2-y_3y_4)(y_1y_2+y_3y_4+16a^2)=0 \dots\dots\dots\dots\text{(i)}.$$

But since the chord whose equation is

$$y(y_1+y_2)-4ax-y_1y_2=0$$

passes through the focus, we have

$$y_1y_2=-4a^2.$$

Similarly $$y_3y_4=-4a^2.$$

Hence the condition (i) is satisfied.

[It is not necessary that the chords should be *focal* chords, for the relation $y_1y_2=y_3y_4$ holds good whenever the two chords cut the axis in the same point.]

49. Let the two tangents be

$$y=m_1x+\frac{a}{m_1} \text{ and } y=m_2x+\frac{a}{m_2}.$$

Let $$m_1=\frac{2\mu_1}{1-\mu_1^2}, \text{ and } m_2=\frac{2\mu_2}{1-\mu_2^2};$$

then $\mu_1\mu_2=\text{constant}=k$ suppose.

If (x, y) be the point of intersection of the tangents, we have

$$x=\frac{a}{m_1m_2}=a\frac{1-(\mu_1^2+\mu_2^2)+\mu_1^2\mu_2^2}{4\mu_1\mu_2}=\frac{a(1+k)^2}{4k}-a\frac{(\mu_1+\mu_2)^2}{4k},$$

and $$y=\frac{a}{m_1}+\frac{a}{m_2}=\frac{a}{2\mu_1\mu_2}(\mu_1+\mu_2)(1-\mu_1\mu_2)=\frac{a(1-k)}{2k}(\mu_1+\mu_2).$$

Hence $$y^2=-a\frac{(1-k)^2}{k}\left\{x-\frac{a(1+k)^2}{4k}\right\}.$$

The locus is therefore a parabola; moreover the focus of the parabola is at the point whose abscissa is $\frac{a(1+k)^2}{4k}-\frac{a(1-k)^2}{4k}=a.$

50. Let P, Q, R, S be (x_1, y_1), (x_2, y_2), (x_3, y_3) and (x_4, y_4) respectively. Then the equation of the circle whose diameter is PQ is

$$\left(x-\frac{y_1^2}{4a}\right)\left(x-\frac{y_2^2}{4a}\right)+(y-y_1)(y-y_2)=0.$$

Where the circle meets the parabola $y^2=4ax$, we have

$$\frac{1}{16a^2}(y^2-y_1^2)(y^2-y_2^2)+(y-y_1)(y-y_2)=0.$$

Hence y_3 and y_4 are the roots of

$$(y+y_1)(y+y_2)+16a^2=0;$$

and therefore $$y_3y_4=y_1y_2+16a^2 \text{ .... (i).}$$

But PQ and RS cut the axis in points whose abscissae are $-\frac{y_1y_2}{4a}$ and $-\frac{y_3y_4}{4a}$ respectively; and (i) shews that the difference of these abscissae is equal to $4a$.

51. The ordinates of the points the normals at which meet in the point (α, β) are given by the equation

$$8a^2(\beta-\eta)+\eta(4a\alpha-\eta^2)=0,$$

or $$\eta^3+4a(2a-\alpha)\eta-8a^2\beta=0 \text{ .... (i).}$$

Now, let η_1, η_2, η_3 be the roots of the above cubic; then the equations of PP', QQ', RR' will be

$$8a^2(y-\eta_1)-\eta_1(4ax-\eta_1^2)=0,$$

i.e. $$\eta_1^3+4a(-2a-x)\eta_1+8a^2y=0 \text{ .... (ii),}$$

and two similar equations.

Now the equation (ii), and each of the two similar equations, is clearly satisfied by the values $-2a-x=2a-\alpha$, $y=-\beta$, since η_1, η_2, η_3 are the roots of (i).

Hence PP', QQ', RR' all meet in the point $(\alpha-4a, -\beta)$.

The line OO' is $\frac{x-\alpha}{4a}=\frac{y-\beta}{2\beta}$, which is clearly perpendicular to the polar of O' whose equation is $$-\beta y=2a(x+\alpha-4a).$$

52. The equation of the normal at the point (x', y') is

$$\frac{y-y'}{y'}=\frac{x-x'}{-2a}=\frac{\sqrt{\{(y-y')^2+(x-x')^2\}}}{\sqrt{\{y'^2+4a^2\}}}.$$

Hence, if (x, y) be the coordinates of the point O, and (x', y') be either of the points P, Q, R, and if r be the distance between the points (x, y) and (x', y'), we have $$4a(x'+a)(x-x')^2=4a^2r^2.$$

Hence $a^3r_1^2r_2^2r_3^2=(x_1+a)(x_2+a)(x_3+a)(x-x_1)^2(x-x_2)^2(x-x_3^2)\text{......(i)}$

where x_1, x_2, x_3 are the three possible values of x'.

Now $\frac{y}{y'}=1-\frac{x-x'}{2a}$; $\therefore y^2=\frac{4ax'}{4a^2}(x'-x+2a)^2$; $\therefore x_1, x_2, x_3$ are the three roots of the following cubic in x'

$$x'(x'-x+2a)^2-ay^2=0 \text{ .... (ii).}$$

In (ii) put $a+x'=\lambda$; then we have $(\lambda-a)(\lambda-x+a)^2-ay^2=0$;

$$\therefore\ \lambda_1\lambda_2\lambda_3=a\{(x-a)^2+y^2\},$$

i.e.
$$(x_1+a)(x_2+a)(x_3+a)=a\{(x-a)^2+y^2\}.$$

Again, in (ii) put $x-x'=\mu$; then we have $(x-\mu)(2a-\mu)^2-ay^2=0$; $\therefore\ \mu_1\mu_2\mu_3=a(4ax-y^2)$, i.e. $(x-x_1)(x-x_2)(x-x_3)=a(4ax-y^2)$.

Hence from (i)

$$a^3r_1^2r_2^2r_3^2=a\{(x-a)^2+y^2\}\,.\,a^2(4ax-y^2)^2;$$

$$\therefore\ r_1^2r_2^2r_3^2=\{(x-a)^2+y^2\}(y^2-4ax)^2.$$

The point of contact of the tangent $y=mx+\frac{a}{m}$ is $\left(\frac{a}{m^2},\frac{2a}{m}\right)$. Hence if the two tangents which meet in (x, y) make angles $\tan^{-1}m_1$, $\tan^{-1}m_2$ with the axis, and t_1, t_2 be the lengths of the tangents, we have since $x=\frac{a}{m_1m_2}$ and $y=\frac{a}{m_1}+\frac{a}{m_2}$,

$$t_1^2t_2^2=\left\{\left(\frac{a}{m_1m_2}-\frac{a}{m_1^2}\right)^2+\left(\frac{a}{m_1}-\frac{a}{m_2}\right)^2\right\}\left\{\left(\frac{a}{m_1m_2}-\frac{a}{m_2^2}\right)^2+\left(\frac{a}{m_1}-\frac{a}{m_2}\right)^2\right\}$$

$$=a^4\,.\left(\frac{1}{m_1}-\frac{1}{m_2}\right)^4\left(\frac{1}{m_1^2}+1\right)\left(\frac{1}{m_2^2}+1\right)$$

$$=(y^2-4ax)^2\{(x-a)^2+y^2\}/a^2,$$

since m_1, m_2 are the roots of $m^2x-my+a=0$.

Hence $OP\,.\,OQ\,.\,OR=a\,.\,OL\,.\,OM$.

53. The ordinates of the feet of the normals to a parabola which meet in any point (ξ, η) are the three roots of

$$y^3-4a(\xi-2a)\,y-8a^2\eta=0.$$

Let these roots be y_1, y_2, y_3; then

$$\Sigma y_1=0,\ \Sigma y_2y_3=-4a(\xi-2a),\text{ and } y_1y_2y_3=8a^2\eta \ldots\ldots\ldots\ldots\text{(i)}.$$

The sum of the squares of the sides of the triangle formed by joining the feet of the normals will be

$$\Sigma(y_2-y_3)^2+\Sigma(x_2-x_3)^2=2\Sigma y_1^2-2\Sigma y_2y_3+\frac{1}{16a^2}\{2\Sigma y_1^4-2\Sigma y_2^2y_3^2\}\ldots\text{(ii)}.$$

Now
$$(\Sigma y_1)^2=\Sigma y_1^2+2\Sigma y_1y_2,$$

$$\Sigma y_1^2y_2^2+2y_1y_2y_3\Sigma y_1=(\Sigma y_1y_2)^2,$$

and
$$\Sigma y_1^4+2\Sigma y_1^2y_2^2=(\Sigma y_1^2)^2.$$

Hence from (i)
$$\Sigma y_1^2 = 8a(\xi - 2a)$$
$$\Sigma y_1^2 y_2^2 = 16a^2(\xi - 2a)^2$$
$$\Sigma y_1^4 = 32a^2(\xi - 2a)^2.$$

Hence, from (ii), the sum of the squares of the sides of the triangle is equal to

$$16a(\xi - 2a) + 8a(\xi - 2a) + 4(\xi - 2a)^2 - 2(\xi - 2a)^2 = 2(\xi - 2a)(\xi + 10a),$$

and is constant so long as ξ is constant.

54. Let A', B', C' be the points of contact of the sides BC, CA, AB respectively, and let the ordinates of A', B', C' be y_1, y_2, y_3 respectively.

Then D is the intersection of

$$y(y_2 + y_3) - 4ax - y_2y_3 = 0, \text{ and } y = y_1.$$

Hence D is the point $\left(\dfrac{-y_2y_3 + y_3y_1 + y_1y_2}{4a},\ y_1\right)$.

Similarly E is $\left(\dfrac{y_2y_3 - y_3y_1 + y_1y_2}{4a},\ y_2\right)$, and F is $\left(\dfrac{y_2y_3 + y_3y_1 - y_1y_2}{4a},\ y_3\right)$.

Hence the middle point of EF is $\left(\dfrac{y_2y_3}{4a}, \dfrac{y_2 + y_3}{2}\right)$, which is easily seen to be the intersection of the tangents at B' and C', that is A. Similarly B is the middle point of FD, and C of DE.

55. Let the sides $B'C'$, $C'A'$, $A'B'$ of $A'B'C'$ touch the parabola in D, E, F respectively. Then the diameter through D is midway between the diameters through B and C, and the diameter through E is midway between the diameters through C and A; hence the distance between the diameters through D and E is half the distance between the diameters through B and A.

Again, the diameter through B' is midway between the diameters through D and F, and the diameter through A' is midway between those through F and E; hence the distance between the diameters through B' and A' is half the distance between the diameters through D and E, and therefore one-fourth of the distance between the diameters through B and A. Hence, as $B'A'$ is parallel to BA, $B'A'$ must be one-fourth of AB, and so for the other sides.

56. Let the four tangents be $yy_1 = 2a(x + x_1)$, &c.

Then the tangents at (x_1, y_1), (x_2, y_2) meet where $x = \dfrac{y_1y_2}{4a}$; and the tangents at (x_3, y_3), (x_4, y_4) where $x = \dfrac{y_3y_4}{4a}$.

Hence the product of the squares of abscissae of these two points of intersection

$$=\frac{y_1^2 y_2^2 y_3^2 y_4^2}{(16a^2)^2}=x_1 x_2 x_3 x_4.$$

57. Let the equations of TP, TQ be respectively

$$y-m_1x-\frac{a}{m_1}=0 \text{ and } y-m_2x-\frac{a}{m_2}=0.$$

Then P, Q, T are respectively the points

$$\left(\frac{a}{m_1^2}, \frac{2a}{m_1}\right), \left(\frac{a}{m_2^2}, \frac{2a}{m_2}\right) \text{ and } \left(\frac{a}{m_1m_2}, \frac{a}{m_1}+\frac{a}{m_2}\right).$$

Hence, if $y=mx+\frac{a}{m}$ be the equation of any other tangent, we have

$$p_1=\left(\frac{2a}{m_1}-\frac{ma}{m_1^2}-\frac{a}{m}\right)\Big/\sqrt{(1+m^2)}$$

$$=-am\left(\frac{1}{m}-\frac{1}{m_1}\right)^2\Big/\sqrt{(1+m^2)}.$$

So $$p_3=-am\left(\frac{1}{m}-\frac{1}{m_2}\right)^2\Big/\sqrt{(1+m^2)}.$$

Also $$p_2=\left(\frac{a}{m_1}+\frac{a}{m_2}-\frac{ma}{m_1m_2}-\frac{a}{m}\right)\Big/\sqrt{(1+m^2)}$$

$$=-am\left(\frac{1}{m}-\frac{1}{m_1}\right)\left(\frac{1}{m}-\frac{1}{m_2}\right)\Big/\sqrt{(1+m^2)}.$$

Hence $$p_1p_3=p_2^2.$$

58. Let A, B be the points whose ordinates are y_1, y_2 respectively.

Let x, y be the co-ordinates of O, and ξ, η be the co-ordinates of P.

Then $$4ax=y_1y_2,\ 2y=y_1+y_2.$$

Also [see 45]

$$4a\xi=8a^2+(y_1+y_2)^2-y_1y_2,$$

and $$8a^2\eta=-y_1y_2(y_1+y_2).$$

Hence, if ξ = constant = c suppose,

$$4ac=8a^2+4y^2-4ax,$$

and therefore the locus of O is a parabola.

Again, if η = constant = d suppose,

$$8a^2d=-4ax\times 2y, \text{ or } xy+ad=0.$$

59. Let P be the point (x_1, y_1), and Q the point (ξ, η). Then, since $GP=PQ$, it follows that $\xi=x_1-2a$, and $\eta=2y_1$.

But $y_1^2 = 4ax_1$; $\therefore\ \eta^2 = 16a(\xi + 2a)$.

Hence the locus of Q is the parabola $y^2 = 16a(x + 2a)$.

The tangent to $y^2 - 4ax = 0$ at (x_1, y_1) is $yy_1 = 2a(x + x_1)$; and the tangent to $y^2 = 16a(x + 2a)$ at the corresponding point $(x_1 - 2a, 2y_1)$ is

$$2y_1y = 8a(x + x_1 - 2a) + 32a^2.$$

Hence the point of intersection of the corresponding tangents is given by

$$x = -x_1 - 4a,\ y = -\frac{8a^2}{y_1}.$$

$$\therefore \left(\frac{8a^2}{y}\right)^2 + 4a(x + 4a) = 0,$$

or

$$16a^3 + y^2(x + 4a) = 0.$$

CHAPTER VI.

Pages 119—120.

1. (i) We have $\dfrac{x^2}{\frac{1}{2}} + \dfrac{y^2}{\frac{1}{3}} = 1$; hence the squares of the axes are $\frac{1}{2}$ and $\frac{1}{3}$.

The eccentricity is given by $\frac{1}{3} = \frac{1}{2}(1-e^2)$, and therefore $e = \frac{1}{\sqrt{3}}$.

The foci are the points for which $x = \pm\sqrt{\left(\frac{1}{2} - \frac{1}{3}\right)}$, $y=0$.

(ii) Changing to the point $(1, -1)$ the equation becomes

$$8x^2 + 6y^2 = 1, \text{ or } \frac{x^2}{\frac{1}{8}} + \frac{y^2}{\frac{1}{6}} = 1.$$

Hence the equation represents an ellipse whose centre is the point $(1, -1)$, whose semi-major axis is $\frac{1}{6}\sqrt{6}$ and whose semi-minor axis is $\frac{1}{8}\sqrt{8}$.

The eccentricity is given by $\frac{1}{8} = \frac{1}{6}(1-e^2)$, and therefore $e = \frac{1}{2}$. The foci are at a distance $\pm\sqrt{\left(\frac{1}{6} - \frac{1}{8}\right)}$ from the centre and are on the new axis of y. Hence the foci are $\left(1, -1 \pm \frac{1}{12}\sqrt{6}\right)$.

2. The length of the latus rectum is $2\dfrac{b^2}{a}$; hence the latus rectum of (i) is $2 \cdot \frac{1}{3} \Big/ \frac{1}{\sqrt{2}} = \frac{2}{3}\sqrt{2}$, and of (ii) is $2 \cdot \frac{1}{8} \Big/ \frac{1}{\sqrt{6}} = \frac{1}{4}\sqrt{6}$.

3. The abscissae of the points of intersection are given by

$$2x^2+3\left(x+\sqrt{\frac{5}{6}}\right)^2=1,$$

that is
$$5x^2+x\sqrt{30}+\frac{3}{2}=0.$$

Hence the two values of x, and therefore also the two values of y, are equal.

4. The abscissae of the two points of intersection are given by

$$4x^2-\frac{1}{3}(x-3)^2-2x=0,$$

that is $\frac{11}{3}x^2-3=0$. The two values of x are thus equal and opposite, and therefore the points of intersection are equidistant from the axis of y.

5. Substituting the coordinates of (2, 1) in the equation of the ellipse, the result is equal to -1. Hence [Art. 112] the point is *within* the ellipse.

6. The line $y=\pm\sqrt{3}x+c$ makes an angle of 60^0 with the axis of x [the upper sign referring to lines for which the angle between the positive direction of the axis of x and the part of the line above the axis of x is 60^0].

Also, the line $y=\pm\sqrt{3}x+c$ will touch the ellipse $x^2/a^2+y^2/b^2=1$ provided $c^2=3a^2+b^2$ [Art. 113].

Hence the required equation is $y=\pm\sqrt{3}x\pm\sqrt{(3a^2+b^2)}$.

7. Writing the equation of the ellipse in the form $\frac{x^2}{3}+\frac{y^2}{2}=1$, we see that the squares of the semi-axes are 3 and 2. Hence the foci are at a distance ± 1 from the centre, and the length of the semi-latus rectum is $\frac{2}{\sqrt{3}}$. The co-ordinates of the extremities of the latera recta are therefore $\left(\pm 1, \pm\frac{2}{3}\sqrt{3}\right)$.

The equations of the tangents and of the normals are then found at once by substitution in the formulae of Articles 114 and 116.

8. If the intercepts made on the axes by any line be equal in absolute magnitude, the line must be parallel to $y=\pm x$. Hence from Art. 113, the tangents to $x^2/a^2+y^2/b^2=1$ which make equal intercepts are given by

$$y=\pm x\pm\sqrt{(a^2+b^2)}.$$

9. The equation may be written in the form

$$4\left(x-\frac{3}{4}\right)^2+2y^2=\frac{9}{4}, \text{ or } \frac{\left(x-\frac{3}{4}\right)^2}{\frac{9}{16}}+\frac{y^2}{\frac{9}{8}}=1.$$

Hence the point $\left(\frac{3}{4}, 0\right)$ is the centre of the ellipse, and the semi-axes are $\frac{3}{\sqrt{8}}$ and $\frac{3}{4}$, the major-axis lying along the new axis of y and the minor-axis along the axis of x. Since the minor-axis is along the axis of x, and the centre of the ellipse is at a distance from the original origin equal to the semi-minor-axis, it follows that the origin is at an extremity of the minor-axis of $4x^2+2y^2=6x$.

The eccentricity is given by $\frac{9}{16}=\frac{9}{8}(1-e^2)$; $\therefore e=\frac{1}{\sqrt{2}}$.

10. We have by definition

$$(x+1)^2+(y-1)^2=\frac{25}{36}\left(\frac{4x-3y}{5}\right)^2;$$

$$\therefore\ 20x^2+24xy+27y^2+72x-72y+72=0.$$

11. The equation of the normal at $\left\{ae, \frac{b^2}{a}\right\}$ is

$$\frac{x-ae}{\frac{e}{a}}=\frac{y-\frac{b^2}{a}}{\frac{1}{a}}.$$

This cuts $x=0$ where $y=\frac{b^2}{a}-a=-\frac{a^2e^2}{a}$.

Hence if the normal pass through an extremity of the minor-axis we must have $ae^2=b$; $\therefore\ a^2e^4=a^2(1-e^2)$, or $e^4+e^2-1=0$.

12. The equation of the tangent at $\left(ae, \frac{b^2}{a}\right)$ is $\frac{ex}{a}+\frac{y}{a}=1$.

This is met by $x=x'$ in the point whose ordinate $=a-ex'$. But, if $x=x'$ meet the curve in P, $SP=a-ex'$ [Art. 110]. Hence $MQ=SP$.

13. Take the lines OA, OB for axes, and let $AC=b$ and $CB=a$. Then, if the coordinates of C be x and y, we have

$$\frac{x^2}{a^2}+\frac{y^2}{b^2}=\left(\frac{OA}{AB}\right)^2+\left(\frac{OB}{AB}\right)^2=1.$$

Thus the locus of C is an ellipse whose semi-axes are BC and AC.

Pages 128—129.

1. The pole of $(ae, 0)$ with respect to $x^2/a^2+y^2/b^2=1$ is $aex/a^2=1$, i.e. $x=a/e$.

2. The necessary and sufficient condition that the line

$$x \cos \alpha + y \sin \alpha - p = 0$$

should touch the ellipse $x^2/a^2 + y^2/b^2 = 1$ is that $p^2 = a^2 \cos^2 \alpha + b^2 \sin^2 \alpha$. Now if r and θ are the polar coordinates of the foot of the perpendicular from the origin on the line $x \cos \alpha + y \sin \alpha - p = 0$, $r = p$ and $\theta = \alpha$. Hence the equation of the locus of the foot of the perpendicular from the centre of an ellipse on any tangent is $r^2 = a^2 \cos^2 \theta + b^2 \sin^2 \theta$.

3. Let r_1, r_2 be the lengths of the semi-diameters which make angles θ, $\theta + \frac{\pi}{2}$ respectively with the major-axis. Then [Art. 111]

$$\frac{1}{r_1^2} = \frac{\cos^2 \theta}{a^2} + \frac{\sin^2 \theta}{b^2} \text{ and } \frac{1}{r_2^2} = \frac{\sin^2 \theta}{a^2} + \frac{\cos^2 \theta}{b^2}.$$

Hence $\dfrac{1}{r_1^2} + \dfrac{1}{r_2^2} = \dfrac{1}{a^2} + \dfrac{1}{b^2}$.

4. Let one of the sides of the triangle make an angle θ with the axis of x; then the other sides will make angles $\theta + \frac{2\pi}{3}$ and $\theta + \frac{4\pi}{3}$ with the axis.

Hence, if r_1, r_2, r_3 be the lengths of the semi-diameters parallel to the sides of the triangle, we have

$$\frac{1}{r_1^2} + \frac{1}{r_2^2} + \frac{1}{r_3^2} = \left\{\frac{\cos^2 \theta}{a^2} + \frac{\sin^2 \theta}{b^2}\right\} + \left\{\frac{\cos^2\left(\theta + \frac{2\pi}{3}\right)}{a^2} + \frac{\sin^2\left(\theta + \frac{2\pi}{3}\right)}{b^2}\right\}$$

$$+ \left\{\frac{\cos^2\left(\theta + \frac{4\pi}{3}\right)}{a^2} + \frac{\sin^2\left(\theta + \frac{4\pi}{3}\right)}{b^2}\right\} = \frac{3}{2}\left(\frac{1}{a^2} + \frac{1}{b^2}\right),$$

since $\cos 2\theta + \cos 2\left(\theta + \frac{2\pi}{3}\right) + \cos 2\left(\theta + \frac{4\pi}{3}\right) = 0$.

5. The point of intersection of two perpendicular tangents to a fixed ellipse is at a constant distance from the centre; hence when an ellipse of given axes touches two fixed perpendicular lines its centre must be on a fixed circle having the intersection of the lines for centre.

6. The points S', H' are $(0, \pm ae)$. The equation of any tangent to the ellipse is

$$x \cos \alpha + y \sin \alpha - \sqrt{(a^2 \cos^2 \alpha + b^2 \sin^2 \alpha)} = 0;$$

and the sum of the squares of the perpendiculars from S', H' on this tangent

$$= \{ae \sin \alpha - \sqrt{(a^2 \cos^2 \alpha + b^2 \sin^2 \alpha)}\}^2 + \{-ae \sin \alpha - \sqrt{(a^2 \cos^2 \alpha + b^2 \sin^2 \alpha)}\}^2$$

$$= 2a^2.$$

7. Let the eccentric angles of the two points be $\phi + \alpha$ and $\phi - \alpha$, 2α being the given difference of the eccentric angles.

Then the equations of the tangents are

$$\frac{x}{a}\cos(\phi+a)+\frac{y}{b}\sin(\phi+a)=1,$$

and

$$\frac{x}{a}\cos(\phi-a)+\frac{y}{b}\sin(\phi-a)=1.$$

These meet in the point $\frac{x}{a}=\cos\phi\sec a$, $\frac{y}{b}=\sin\phi\sec a$. Hence the required locus is the ellipse whose equation is $\frac{x^2}{a^2}+\frac{y^2}{b^2}=\sec^2 a$.

8. Let (x', y') be the point P; then the equation of the polar is

$$\frac{xx'}{a^2}+\frac{yy'}{b^2}=1,$$

and the equation of POg is

$$\frac{x-x'}{\frac{x'}{a^2}}=\frac{y-y'}{\frac{y'}{b^2}}.$$

Hence $Ct=\frac{b^2}{y'}$, and $Cg=y'\left(1-\frac{a^2}{b^2}\right)$; $\therefore\ tC\,.\,Cg=a^2-b^2=SC\,.\,CS'$.

Hence t, S, g, S' are on a circle; and, since tg bisects SS' at right angles, tg is a diameter of the circle, and therefore the circle also passes through O.

9. If the line $lx+my+n=0$ be the normal at any point θ, the given equation represents the same straight line as the equation

$$ax\sec\theta-by\operatorname{cosec}\theta=a^2-b^2.$$

Hence we must have

$$\frac{l\cos\theta}{a}=\frac{m\sin\theta}{-b}=\frac{n}{b^2-a^2},$$

or

$$\frac{\cos\theta}{\frac{a}{l}}=\frac{\sin\theta}{-\frac{b}{m}}=\frac{1}{\frac{b^2-a^2}{n}};$$

$$\therefore\ \frac{a^2}{l^2}+\frac{b^2}{m^2}=\frac{(a^2-b^2)^2}{n^2}.$$

10. Let P be the point $(a\cos\theta, b\sin\theta)$. Then the equation of the tangent at P is $\frac{x}{a}\cos\theta+\frac{y}{b}\sin\theta=1$; hence the equation of the line through $(ae, 0)$ perpendicular to the tangent at P is

$$(x-ae)\,a\sec\theta-yb\operatorname{cosec}\theta=0 \qquad \text{(i)}.$$

The equation of CP is

$$\frac{x}{a\cos\theta}=\frac{y}{b\sin\theta} \qquad \text{(ii)}.$$

The lines (i) and (ii) meet where $(x-ae)\,a\sec\theta-\frac{b^2}{a}x\sec\theta=0$, or $x=\frac{a}{e}$.

Hence (i) and (ii) meet on the directrix.

11. Let P be the point $(a\cos\theta,\ b\sin\theta)$; then Q will be $(a\cos\theta,\ a\sin\theta)$. The equations of the normals at P and Q will be

$$ax/\cos\theta - by/\sin\theta = a^2 - b^2 \quad\text{.........................(i)},$$

and

$$x/a\cos\theta - y/a\sin\theta = 0 \quad\text{.............................(ii)}.$$

From (ii) $\tan\theta = \dfrac{y}{x}$ and therefore $\cos\theta = \dfrac{x}{\sqrt{(x^2+y^2)}}$ and $\sin\theta = \dfrac{y}{\sqrt{(x^2+y^2)}}$. Hence from (i) $x^2+y^2=(a+b)^2$.

12. Let P be $(a\cos\theta,\ b\sin\theta)$, then Q will be $(a\cos\theta,\ a\sin\theta)$, and the equation of the tangent at Q will be $x\cos\theta + y\sin\theta - a = 0$. Hence the perpendicular distances of the foci $(\mp ae,\ 0)$ from the tangent at Q are $\pm ae\cos\theta + a$ that is $a \pm ex$, where x is the abscissa of P. But, from Art. 110, the focal distances of P are also $a \pm ex$.

13. The area is $\dfrac{1}{2}\begin{vmatrix} a\cos\alpha, & b\sin\alpha, & 1 \\ a\cos\beta, & b\sin\beta, & 1 \\ a\cos\gamma, & b\sin\gamma, & 1 \end{vmatrix} = \dfrac{1}{2}ab\,\{\sin(\gamma-\beta)+\sin(\alpha-\gamma)$
$+\sin(\beta-\alpha)\}$.

Pages 138—145.

1. Let ϕ be the eccentric angle of P; then [Art. 110] $SP = a + ae\cos\phi$ and $S'P = a - ae\cos\phi$;

$$\therefore\ SP\,.\,S'P = a^2 - a^2e^2\cos^2\phi = a^2 - (a^2-b^2)\cos^2\phi = a^2\sin^2\phi + b^2\cos^2\phi = CD^2.$$
[Art. 130.]

2. Let ϕ be the eccentric angle of P, and let A be $(-a,\ 0)$ and therefore A' $(a,\ 0)$.

Thus Y is the point of intersection of

$$\frac{x}{a}\cos\phi + \frac{y}{b}\sin\phi = 1 \text{ and } x = -a;$$

$$\therefore\ CY \text{ is } \frac{x}{a}\cos\phi + \frac{y}{b}\sin\phi + \frac{x}{a} = 0.$$

The equation of $A'P$ is $\dfrac{x - a\cos\phi}{a\cos\phi - a} = \dfrac{y - b\sin\phi}{b\sin\phi}$; and CY and $A'P$ are parallel if $-\dfrac{b}{a}\,\dfrac{\cos\phi+1}{\sin\phi} = \dfrac{b}{a}\,\dfrac{\sin\phi}{\cos\phi-1}$, which is obvious.

3. Take for axes the bisectors of the angles between the given straight lines; and let their equations be

$$x\cos\alpha + y\sin\alpha = 0 \text{ and } x\cos\alpha - y\sin\alpha = 0.$$

Let $(x,\ y)$ be any point on the locus; then

$$(x\cos\alpha + y\sin\alpha)^2 + (x\cos\alpha - y\sin\alpha)^2 = \text{constant} = c^2;$$

$$\therefore\ \frac{x^2}{c^2\sec^2\alpha} + \frac{y^2}{c^2\operatorname{cosec}^2\alpha} = 1.$$

If 2α is the *acute* angle between the lines, so that $\operatorname{cosec}\alpha > \sec\alpha$, we have $c^2\sec^2\alpha = c^2\operatorname{cosec}^2\alpha\,(1-e^2)$, and therefore $e = \sqrt{(1-\tan^2\alpha)}$.

4. Let the eccentric angles of P, Q, R be α, β, θ respectively. Then the coordinates of V are $\frac{a}{2}(\cos\alpha+\cos\theta), \frac{b}{2}(\sin\alpha+\sin\theta)$; and the coordinates of V' are $\frac{a}{2}(\cos\beta+\cos\theta), \frac{b}{2}(\sin\beta+\sin\theta)$.

Hence the equation of VG is

$$\left\{x-\frac{a}{2}(\cos\alpha+\cos\theta)\right\}a\sin\frac{\alpha+\theta}{2}-\left\{y-\frac{b}{2}(\sin\alpha+\sin\theta)\right\}b\cos\frac{\alpha+\theta}{2}=0.$$

Hence, at G, $$x=\frac{a}{2}(\cos\alpha+\cos\theta)-\frac{b^2}{2a}(\sin\alpha+\sin\theta)\cot\frac{\alpha+\theta}{2}$$

$$=\frac{a^2-b^2}{2a}(\cos\alpha+\cos\theta).$$

Similarly at G' $$x=\frac{a^2-b^2}{2a}(\cos\beta+\cos\theta).$$

Hence GG' $$=\frac{a^2-b^2}{2a}(\cos\alpha-\cos\beta),$$

which is independent of θ.

5. Since $BC^2=a^2(1-e^2)$, $SC=ae$ and $XS=XC-SC=\frac{a}{e}-ae$, it follows that $BC^2=XS.SC.$. Hence the locus of B (or B') is a parabola whose vertex is S and whose latus rectum is equal to XS.

6. Let the eccentric angles of P, P', Q be α, $-\alpha$ and θ respectively. Then the equations of PQ, $P'Q$ will be

$$\frac{x}{a}\cos\frac{\alpha+\theta}{2}+\frac{y}{b}\sin\frac{\alpha+\theta}{2}=\cos\frac{\alpha-\theta}{2},$$

and $$\frac{x}{a}\cos\frac{-\alpha+\theta}{2}+\frac{y}{b}\sin\frac{-\alpha+\theta}{2}=\cos\frac{-\alpha-\theta}{2}.$$

Hence $$CM=a\cos\frac{\alpha-\theta}{2}\Big/\cos\frac{\alpha+\theta}{2},$$

and $$CM'=a\cos\frac{\alpha+\theta}{2}\Big/\cos\frac{\alpha-\theta}{2};$$

$$\therefore\ CM.CM'=a^2.$$

7. Let the conjugate diameters be $y=mx$ and $y=m'x$; then $mm'=-b^2/a^2$. The perpendicular lines through the foci are

$$y=-\frac{1}{m}(x-ae),\ y=-\frac{1}{m'}(x+ae);$$

hence, from the relation $mm'=-\frac{b^2}{a^2}$, we have

$$y^2=-\frac{a^2}{b^2}(x^2-a^2e^2),\text{ or }\frac{x^2}{b^2}+\frac{y^2}{a^2}=\frac{a^2e^2}{b^2}.$$

Thus the required locus is a concentric ellipse. The ellipse is similar to the original ellipse, but the major axis of one ellipse lies along the minor axis of the other.

8. Take the equi-conjugate diameters for axes; then the equation of the ellipse will be $x^2+y^2=c^2$.

Let P be (x', y'), then the equation of the tangent at P is $xx'+yy'=c^2$; and therefore $CT=\frac{c^2}{x'}$ and $CT'=\frac{c^2}{y'}$.

Now $\triangle\, TCP : \triangle\, T'CP = y'.CT : x'.CT' = \frac{CT}{CT'} : \frac{CT'}{CT} = CT^2 : CT'^2$.

9. Let Q be (x', y') and P be (x'', y'').

Then the equations of CQ, CP are $\frac{x}{x'}-\frac{y}{y'}=0$ and $\frac{x}{x''}-\frac{y}{y''}=0$. Also the equations of the normals at Q, P are

$$\frac{x-x'}{\frac{x'}{a^2}}=\frac{y-y'}{\frac{y'}{b^2}} \text{ and } \frac{x-x''}{\frac{x''}{a^2}}=\frac{y-y''}{\frac{y''}{b^2}}.$$

Since CQ is conjugate to the normal at P, we have [Art. 127]

$$-\frac{b^2}{a^2}=\frac{y'}{x'}\cdot\frac{a^2y''}{b^2x''}, \text{ or } \frac{x'x''}{a^4}+\frac{y'y''}{b^4}=0.$$

Also CP will be conjugate to the normal at Q, provided

$$\frac{y''}{x''}\cdot\frac{a^2y'}{b^2x'}=-\frac{b^2}{a^2}, \text{ or } \frac{x'x''}{a^4}+\frac{y'y''}{b^4}=0,$$

which is known to be true.

10. Let P be the point $(a\cos\phi,\ b\sin\phi)$ and D the point $(-a\sin\phi,\ b\cos\phi)$.

Then P' is $(-a\cos\phi,\ b\sin\phi)$ and D' is $(a\sin\phi,\ b\cos\phi)$.

Hence the equation of PD' is $\frac{x-a\cos\phi}{a\cos\phi-a\sin\phi}=\frac{y-b\sin\phi}{b\sin\phi-b\cos\phi}$,

that is $$b\,(x-a\cos\phi)+a\,(y-b\sin\phi)=0,$$
which is parallel to $bx+ay=0$.

Again the equation of $P'D$ is $\frac{x+a\cos\phi}{-a\cos\phi+a\sin\phi}=\frac{y-b\sin\phi}{b\sin\phi-b\cos\phi}$, which is parallel to $bx-ay=0$.

This proves the theorem, since the equi-conjugates are $bx\pm ay=0$.

11. Let θ be the eccentric angle of P; then $\theta\pm\frac{\pi}{2}$ will be the eccentric angle of D.

Then the equations of the tangents at P and D will be respectively

$$\frac{x}{a}\cos\theta+\frac{y}{b}\sin\theta=1,$$

and
$$\frac{x}{a}\sin\theta-\frac{y}{b}\cos\theta=\pm 1.$$

Hence T is $\left(\frac{a}{\cos\theta},\,0\right)$ and T' is $\left(0,\,\mp\frac{b}{\cos\theta}\right)$.

Hence TT' is
$$\frac{x}{\frac{a}{\cos\theta}}\mp\frac{y}{\frac{b}{\cos\theta}}=1,$$

which is parallel to $\frac{x}{a}\mp\frac{y}{b}=0$, that is parallel to one of the equi-conjugates.

12. Take the equi-conjugate diameters for axes; then the equation of the ellipse will be $x^2+y^2=c^2$. Hence, if CT meet QQ' in V and the curve in P, we have $QV^2+CV^2=c^2$.

But $CV\,.\,CT=CP^2=c^2$; $\therefore\ QV^2=CV\,(CT-CV)=CV\,.\,VT$, from which it follows that Q, T, Q', C lie on a circle.

13. From Art. 125, we have

$$KC=\frac{1}{\sqrt{\left(\frac{x'^2}{a^4}+\frac{y'^2}{b^4}\right)}},\ \text{and}\ PG=b^2\sqrt{\left(\frac{x'^2}{a^4}+\frac{y'^2}{b^4}\right)};$$

$$\therefore\ KC\,.\,PG=b^2=SZ\,.\,S'Z'.$$

Hence
$$KC:ZS::Z'S':PG.$$

14. Let the conjugate diameters be PCP', DCD', and O be the point.

Then
$$OP^2+OP'^2=2OC^2+2CP^2,$$
$$OD^2+OD'^2=2OC^2+2CD^2.$$

Hence the sum of the square of the distance of P, P', D, D' from O is

$$4OC^2+2\,(CP^2+CD^2),$$

which is constant, since OC is constant and CP^2+CD^2 is constant.

15. Let P be the point $(a\cos\theta,\ b\sin\theta)$, and P' the point $(a\cos\theta,\ -b\sin\theta)$; then O is the point of intersection of the lines whose equations are

$$\frac{ax}{\cos\theta}-\frac{by}{\sin\theta}=a^2-b^2\ \text{and}\ \frac{x}{a\cos\theta}=\frac{y}{-b\sin\theta}.$$

Hence at O we have

$$\frac{x}{a}=\frac{a^2-b^2}{a^2+b^2}\cos\theta,\ \text{and}\ \frac{y}{b}=-\frac{a^2-b^2}{a^2+b^2}\sin\theta.$$

Square and add; then we obtain the equation of the locus of O, namely

$$\frac{x^2}{a^2}+\frac{y^2}{b^2}=\left(\frac{a^2-b^2}{a^2+b^2}\right)^2.$$

16. The equation of the normal at P is $\frac{ax}{\cos\theta}-\frac{by}{\sin\theta}=a^2-b^2$. Hence G is the point $\left(\frac{a^2-b^2}{a}\cos\theta,\ 0\right)$. Hence, if (x, y) be the middle point of PQ, we have $2x=a\cos\theta+\frac{a^2-b^2}{a}\cos\theta$ and $2y=b\sin\theta$;

$$\therefore \left(\frac{2ax}{2a^2-b^2}\right)^2+\left(\frac{2y}{b}\right)^2=1;$$

the locus is therefore an ellipse whose semi-axes are $\frac{2a^2-b^2}{2a}$ and $\frac{b}{2}$.

17. Let θ be the eccentric angle of P; then the equations of AP, $A'P$ will be

$$\frac{x}{a}\cos\frac{\theta}{2}+\frac{y}{b}\sin\frac{\theta}{2}=\cos\frac{\theta}{2},$$

and
$$\frac{x}{a}\cos\frac{\theta+\pi}{2}+\frac{y}{b}\sin\frac{\theta+\pi}{2}=\cos\frac{\theta-\pi}{2}.$$

Hence the equations of PN, PM will be

$$(x-a\cos\theta)\,a\sin\frac{\theta}{2}-(y-b\sin\theta)\,b\cos\frac{\theta}{2}=0,$$

and
$$(x-a\cos\theta)\,a\cos\frac{\theta}{2}+(y-b\sin\theta)\,b\sin\frac{\theta}{2}=0.$$

Hence
$$MN=\left(a\cos\theta-\frac{b^2}{a}\sin\theta\cot\frac{\theta}{2}\right)$$
$$\sim\left(a\cos\theta+\frac{b^2}{a}\sin\theta\tan\frac{\theta}{2}\right)=\frac{2b^2}{a}.$$

18. The directions of the two tangents from (x, y) are given by the equation

$$y=mx+\sqrt{(a^2m^2+b^2)} \quad \text{[Art. 113.]}$$

Hence $\tan\theta_1$, $\tan\theta_2$ are the two roots of $(y-mx)^2=a^2m^2+b^2$.

Hence
$$\tan\theta_1+\tan\theta_2=\frac{2xy}{x^2-a^2},$$

and
$$\tan\theta_1\tan\theta_2=\frac{y^2-b^2}{x^2-a^2}.$$

Hence, (i) if $\tan\theta_1+\tan\theta_2=\text{constant}=k$ suppose, the locus of (x, y) is given by

$$2xy=k(x^2-a^2).$$

(ii) If $\cot\theta_1 + \cot\theta_2 = \text{constant} = l$ suppose, the locus of (x, y) is given by

$$2xy = l(y^2 - b^2).$$

(iii) If $\tan\theta_1 \tan\theta_2 = \text{constant} = m$ suppose, the locus of (x, y) is given by

$$y^2 - b^2 = m(x^2 - a^2).$$

19. Let PCP', DCD' be any two chords of an ellipse, and let the eccentric angles of P and D be θ, ϕ respectively.

Then the eccentric angles of P', D' will be $\theta + \pi$ and $\phi + \pi$ respectively.

Let pCp' and dCd' be the diameters conjugate respectively to PCP' and DCD'; then the eccentric angles of p, p' are $\theta \pm \frac{\pi}{2}$, and the eccentric angles of d, d' are $\phi \pm \frac{\pi}{2}$.

Hence the sum of the eccentric angles of p and d will be either $\theta + \phi$ or $\theta + \phi \pm \pi$ which shews [Art. 123 and 129] that pd is either parallel to PD or to the conjugate diameter of PD.

20. Let the eccentric angle of P be θ, then the eccentric angle of D will be $\theta \pm \frac{\pi}{2}$.

Hence the equations of the tangents at P and D will be respectively

$$\frac{x}{a}\cos\theta + \frac{y}{b}\sin\theta = 1,$$

$$\frac{x}{a}\cos\left(\theta \pm \frac{\pi}{2}\right) + \frac{y}{b}\sin\left(\theta \pm \frac{\pi}{2}\right) = 1.$$

Square and add; then θ is eliminated, and hence the locus of the point of intersection is the ellipse

$$\frac{x^2}{a^2} + \frac{y^2}{b^2} = 2.$$

Again, if (x, y) be the middle point of PD, we have

$$2x = a\cos\theta + a\cos\left(\theta \pm \frac{\pi}{2}\right),$$

$$2y = b\sin\theta + b\sin\left(\theta \pm \frac{\pi}{2}\right).$$

Hence $$\left(\frac{2x}{a}\right)^2 + \left(\frac{2y}{b}\right)^2 = \left\{\cos\theta + \cos\left(\theta \pm \frac{\pi}{2}\right)\right\}^2 + \left\{\sin\theta + \sin\left(\theta \pm \frac{\pi}{2}\right)\right\}^2 = 2.$$

Hence the locus of the middle point of PD is the ellipse

$$\frac{x^2}{a^2} + \frac{y^2}{b^2} = \frac{1}{2}.$$

21. The equation of the line parallel to the directrix $x - \frac{a}{e} = 0$ and midway between it and the corresponding focus $(ae, 0)$ is $x = \frac{1}{2}\left(ae + \frac{a}{e}\right)$.

The equation of any line through the focus is $y = m(x - ae)$. This line meets the ellipse where

$$\frac{x^2}{a^2} + \frac{m^2 (x - ae)^2}{b^2} = 1.$$

Now the distance, d, of the point (x, y) from the line $x = \frac{1}{2}\left(ae + \frac{a}{e}\right)$ is given by $d = x - \frac{a}{2e}(1 + e^2)$.

Hence d is given by

$$b^2 \left\{d + \frac{a}{2e}(1 + e^2)\right\}^2 + a^2 m^2 \left\{d + \frac{a}{2e}(1 - e^2)\right\}^2 = a^2 b^2;$$

$$\therefore\ d_1 d_2 = \left\{\frac{a^2 b^2}{4e^2}(1 + e^2)^2 + \frac{a^4 m^2}{4e^2}(1 - e^2)^2 - a^2 b^2\right\} \Big/ (b^2 + a^2 m^2)$$

$$= a^2 (1 - e^2)^2 / 4e^2.$$

22. The equation of the chord joining the points whose eccentric angles are α, β respectively is

$$\frac{x}{a}\cos\frac{\alpha + \beta}{2} + \frac{y}{b}\sin\frac{\alpha + \beta}{2} = \cos\frac{\alpha - \beta}{2}.$$

Hence
$$\frac{d}{a} = \frac{\cos\frac{\alpha - \beta}{2}}{\cos\frac{\alpha + \beta}{2}};$$

$$\therefore\ \frac{d - a}{d + a} = \frac{\cos\frac{\alpha - \beta}{2} - \cos\frac{\alpha + \beta}{2}}{\cos\frac{\alpha - \beta}{2} + \cos\frac{\alpha + \beta}{2}} = \tan\frac{\alpha}{2}\tan\frac{\beta}{2}.$$

23. From the previous question, if the chord (α, β) cut the axis in the point $(d, 0)$ and the chord (γ, δ) in the point $(-d, 0)$, we have

$$\tan\frac{\alpha}{2}\tan\frac{\beta}{2} = \frac{d - a}{d + a},$$

$$\text{and } \tan\frac{\gamma}{2}\tan\frac{\delta}{2} = \frac{-d - a}{-d + a} = \frac{d + a}{d - a};$$

$$\therefore\ \tan\frac{\alpha}{2}\tan\frac{\beta}{2}\tan\frac{\gamma}{2}\tan\frac{\delta}{2} = 1.$$

24. As in 22, we have

$$\tan\frac{\theta_1}{2}\tan\frac{\theta_2}{2}=\frac{ae-a}{ae+a},$$

$$\tan\frac{\theta_2}{2}\tan\frac{\theta_3}{2}=\frac{-ae-a}{-ae+a},$$

$$\tan\frac{\theta_3}{2}\tan\frac{\theta_4}{2}=\frac{ae-a}{ae+a},\ \&c.$$

Hence $$\tan\frac{\theta_1}{2}\tan\frac{\theta_2}{2}=\cot\frac{\theta_2}{2}\cot\frac{\theta_3}{2}=\tan\frac{\theta_3}{2}\tan\frac{\theta_4}{2}=\ldots.$$

25. The tangents at β, γ meet in the point

$$\left(a\frac{\cos\frac{1}{2}(\beta+\gamma)}{\cos\frac{1}{2}(\beta-\gamma)},\quad b\frac{\sin\frac{1}{2}(\beta+\gamma)}{\cos\frac{1}{2}(\beta-\gamma)}\right),$$

and so for the other points of intersection.

Hence the area of the triangle formed by the tangents

$$=\frac{1}{2}\begin{vmatrix} a\dfrac{\cos\frac{1}{2}(\beta+\gamma)}{\cos\frac{1}{2}(\beta-\gamma)}, & b\dfrac{\sin\frac{1}{2}(\beta+\gamma)}{\cos\frac{1}{2}(\beta-\gamma)}, & 1 \\ a\dfrac{\cos\frac{1}{2}(\gamma+\alpha)}{\cos\frac{1}{2}(\gamma-\alpha)}, & b\dfrac{\sin\frac{1}{2}(\gamma+\alpha)}{\cos\frac{1}{2}(\gamma-\alpha)}, & 1 \\ a\dfrac{\cos\frac{1}{2}(\alpha+\beta)}{\cos\frac{1}{2}(\alpha-\beta)}, & b\dfrac{\sin\frac{1}{2}(\alpha+\beta)}{\cos\frac{1}{2}(\alpha-\beta)}, & 1 \end{vmatrix}$$

$$=\frac{ab}{2\cos\frac{1}{2}(\beta-\gamma)\cos\frac{1}{2}(\gamma-\alpha)\cos\frac{1}{2}(\alpha-\beta)}\begin{vmatrix} \cos\frac{1}{2}(\beta+\gamma), & \sin\frac{1}{2}(\beta+\gamma), & \cos\frac{1}{2}(\beta-\gamma) \\ \cos\frac{1}{2}(\gamma+\alpha), & \sin\frac{1}{2}(\gamma+\alpha), & \cos\frac{1}{2}(\gamma-\alpha) \\ \cos\frac{1}{2}(\alpha+\beta), & \sin\frac{1}{2}(\alpha+\beta), & \cos\frac{1}{2}(\alpha-\beta) \end{vmatrix}$$

$$=\frac{ab}{2}\sec\frac{1}{2}(\beta-\gamma)\sec\frac{1}{2}(\gamma-\alpha)\sec\frac{1}{2}(\alpha-\beta)\,.\,\Sigma\cos\frac{1}{2}(\beta-\gamma)\sin\frac{1}{2}(\beta-\gamma)$$

$$=ab\tan\frac{1}{2}(\beta-\gamma)\tan\frac{1}{2}(\gamma-\alpha)\tan\frac{1}{2}(\alpha-\beta).$$

26. Let ABC be the triangle formed by the tangents; then

$$R \,.\, 4\Delta = BC \,.\, CA \,.\, AB.$$

Now
$$BC^2 = \left\{ a \frac{\cos\frac{1}{2}(\alpha+\gamma)}{\cos\frac{1}{2}(\alpha-\gamma)} - a \frac{\cos\frac{1}{2}(\alpha+\beta)}{\cos\frac{1}{2}(\alpha-\beta)} \right\}^2$$
$$+ \left\{ b \frac{\sin\frac{1}{2}(\alpha+\gamma)}{\cos\frac{1}{2}(\alpha-\gamma)} - b \frac{\sin\frac{1}{2}(\alpha+\beta)}{\cos\frac{1}{2}(\alpha-\beta)} \right\}^2;$$

$$\therefore BC^2 \cos^2\frac{1}{2}(\alpha-\gamma)\cos^2\frac{1}{2}(\alpha-\beta) = (a^2\sin^2\alpha + b^2\cos^2\alpha)\sin^2\frac{1}{2}(\gamma-\beta) \;\ldots\; \text{(i)}.$$

Also the diameter parallel to the tangent at α meets the curve in points whose eccentric angles are $\alpha \pm \frac{\pi}{2}$. Hence

$$\left(\frac{p}{2}\right)^2 = a^2\cos^2\left(\alpha+\frac{\pi}{2}\right) + b^2\sin^2\left(\alpha+\frac{\pi}{2}\right) = a^2\sin^2\alpha + b^2\cos^2\alpha.$$

Hence (i) may be written

$$BC^2\cos^2\frac{1}{2}(\alpha-\gamma)\cos^2\frac{1}{2}(\alpha-\beta) = \frac{p^2}{4}\sin^2\frac{1}{2}(\beta-\gamma),$$

and similarly for the other sides of the triangle ABC.

$$\therefore R \,.\, 4\Delta = \frac{pqr\sin\frac{1}{2}(\beta-\gamma)\sin\frac{1}{2}(\gamma-\alpha)\sin\frac{1}{2}(\alpha-\beta)}{8\cos^2\frac{1}{2}(\beta-\gamma)\cos^2\frac{1}{2}(\gamma-\alpha)\cos^2\frac{1}{2}(\alpha-\beta)}.$$

Also, $\Delta = ab\tan\frac{1}{2}(\beta-\gamma)\tan\frac{1}{2}(\gamma-\alpha)\tan\frac{1}{2}(\alpha-\beta)$. [See 25.]

Hence $R = \frac{pqr}{32ab}\sec\frac{1}{2}(\beta-\gamma)\sec\frac{1}{2}(\gamma-\alpha)\sec\frac{1}{2}(\alpha-\beta)$.

27. Let P be the point (x', y'). Then the equation of PS will be $\frac{x-x'}{x'-ae} = \frac{y-y'}{y'}$.

Hence, at Q, $x = \frac{a}{e}$, $y = \frac{ay'(1-e^2)}{ex'-ae^2}$.

Hence the equation of HQ will be found to be

$$ay'(1-e^2)(x+ae) = ay(1+e^2)(x'-ae).$$

The equation of SR will be found by changing the sign of e, and will therefore be

$$ay'(1-e^2)(x-ae) = ay(1+e^2)(x'+ae).$$

Hence, at the point of intersection of QH and RS, we have

$$ay'(1-e^2)x = ay(1+e^2)x', \text{ and}$$

$$ay'(1-e^2) = -ay(1+e^2).$$

Hence $$x=-x', \text{ and } y(1+e^2) = -y'(1-e^2);$$

$$\therefore 1 = \frac{x'^2}{a^2} + \frac{y'^2}{b^2} = \frac{x^2}{a^2} + \frac{y^2}{b^2}\left(\frac{1+e^2}{1-e^2}\right)^2.$$

Hence the required locus is the ellipse whose equation is

$$\frac{x^2}{a^2} + \frac{y^2}{b^2}\left(\frac{1+e^2}{1-e^2}\right)^2 = 1.$$

28. Let the eccentric angle of P be θ; then the equation of Cp will be

$$\frac{x}{\cos\theta} = \frac{y}{\sin\theta} \qquad \text{(i)}.$$

The eccentric angle of Q is

$$\tan^{-1}\frac{NP}{CN} = \tan^{-1}\frac{b\sin\theta}{a\cos\theta} = \phi \qquad \text{(ii)}.$$

The equation of the tangent at Q is

$$\frac{x}{a}\cos\phi + \frac{y}{b}\sin\phi = 1 \qquad \text{(iii)}.$$

Now (i) is perpendicular to (ii) since $a\tan\phi = b\tan\theta$, from (ii). Also, since the tangent at Q is perpendicular to Cp, it cuts off from Cp a length equal to the perpendicular from the centre on (iii); and this perpendicular

$$= \frac{ab}{\sqrt{(b^2\cos^2\phi + a^2\sin^2\phi)}} = CP,$$

for $\sin^2\phi = \dfrac{PN^2}{CP^2} = \dfrac{b^2\sin^2\theta}{CP^2}$ and $\cos^2\phi = \dfrac{CN^2}{CP^2} = \dfrac{a^2\cos^2\theta}{CP^2}$.
and therefore

$$b^2\cos^2\phi + a^2\sin^2\phi = a^2b^2(\sin^2\theta + \cos^2\theta) / CP^2 = a^2b^2 / CP^2.$$

29. Let the eccentric angles of P, Q be θ, ϕ respectively.

Then, since the tangents at P, Q are at right angles, we have

$$\frac{\cos\theta\cos\phi}{a^2} + \frac{\sin\theta\sin\phi}{b^2} = 0 \qquad \text{(i)}.$$

The equations of Cp, Cq are

$$\frac{x}{\cos\theta} = \frac{y}{\sin\theta} \text{ and } \frac{x}{\cos\phi} = \frac{y}{\sin\phi};$$

and the condition that these may be conjugate diameters is

$$\tan\theta\tan\phi = -b^2/a^2,$$

which is true from (i).

30. Take C for origin, and the fixed straight line to which CQ and Cq are equally inclined for axis of x, and let $2a$, $2b$ be the radii of the two circles.

Then, if CQ make an angle θ with the axis, Cq will make an angle $-\theta$; hence the coordinates of Q will be $2a\cos\theta$, $2a\sin\theta$, and those of q will be $2b\cos\theta, -2b\sin\theta$.

Hence, if (x, y) be the middle point of Qq, we have

$$2x=(2a+2b)\cos\theta,\quad 2y=(2a-2b)\sin\theta;$$

$$\therefore\ \frac{x^2}{(a+b)^2}+\frac{y^2}{(a-b)^2}=1.$$

Thus the locus of P is an ellipse whose semi-axes are $a\pm b$.

The eccentric angle of P is clearly θ, and therefore the equation of the normal at P to the ellipse it describes is

$$\frac{(a+b)x}{\cos\theta}-\frac{(a-b)y}{\sin\theta}=(a+b)^2-(a-b)^2,$$

and it is easily seen that the points

$$(2a\cos\theta,\ 2a\sin\theta)\text{ and }(2b\cos\theta,\ -2b\sin\theta)$$

are both on the normal, so that QPq is the normal at P.

The square of the semi-diameter conjugate to CP is equal to

$$(a+b)^2\sin^2\theta+(a-b)^2\cos^2\theta.$$

But $Qq^2=(2a\cos\theta-2b\cos\theta)^2+(2a\sin\theta+2b\sin\theta)^2$; and therefore Qq is equal to the diameter conjugate to CP.

31. Let θ, θ' be the eccentric angles of any two points P, P' on an ellipse; then the semi-diameters conjugate to CP, CP' are equal to

$$\sqrt{(a^2\sin^2\theta+b^2\cos^2\theta)}\text{ and }\sqrt{(a^2\sin^2\theta'+b^2\cos^2\theta')}$$

respectively.

Hence $\lambda^2\mu^2=(a^2\sin^2\theta+b^2\cos^2\theta)(a^2\sin^2\theta'+b^2\cos^2\theta')$..................(i).

But, since the tangents at θ, θ' are at right angles, we have

$$\frac{\cos\theta\cos\theta'}{a^2}+\frac{\sin\theta\sin\theta'}{b^2}=0;$$

$\therefore\ a^4\sin^2\theta\sin^2\theta'+b^4\cos^2\theta\cos^2\theta'+2a^2b^2\sin\theta\sin\theta'\cos\theta\cos\theta'=0$ ..(ii).

From (i) and (ii) we have

$$\lambda^2\mu^2=a^2b^2(\sin\theta\cos\theta'-\sin\theta'\cos\theta)^2$$
$$=a^2b^2\sin^2\omega.$$

32. Take the line of centres and the common tangent for axes; then the equations of the circles will be

$$x^2+y^2-2ax=0,\quad x^2+y^2+2ax=0.$$

Hence, if (x, y) be any point on the required locus, and $2c$ the constant sum of the tangents, we have

$$\sqrt{(x^2+y^2-2ax)}+\sqrt{(x^2+y^2+2ax)}=2c,$$

whence

$$x^2(c^2-a^2)+y^2c^2=c^4,$$

which is an ellipse whose semi-axes are $\dfrac{c^2}{\sqrt{(c^2-a^2)}}$ and c.

33. The eccentric angles of the two extremities of two conjugate diameters may be taken to be θ, $\theta+\pi$ and $\theta+\frac{\pi}{2}$, $\theta+\frac{3\pi}{2}$.

Hence, if $\frac{x}{a}\cos\alpha+\frac{y}{b}\sin\alpha-1=0$ be any tangent, we have to prove that, for all values of α and θ,

$$(\cos\theta\cos\alpha+\sin\theta\sin\alpha-1)(-\cos\theta\cos\alpha-\sin\theta\sin\alpha-1)$$
$$+(-\sin\theta\cos\alpha+\cos\theta\sin\alpha-1)(\sin\theta\cos\alpha-\cos\theta\sin\alpha-1)=1;$$

that is

$$1-(\cos\theta\cos\alpha+\sin\theta\sin\alpha)^2+1-(\sin\theta\cos\alpha-\cos\theta\sin\alpha)^2=1,$$

which is obvious.

34. Let θ be the eccentric angle of P; then the equation of the normal at P will be

$$\frac{ax}{\cos\theta}-\frac{by}{\sin\theta}=a^2-b^2.$$

The equation of CP is $\dfrac{x}{a\cos\theta}=\dfrac{y}{b\sin\theta}$; and therefore the equation of CQ is $\dfrac{x}{a\cos\theta}=-\dfrac{y}{b\sin\theta}$.

Hence, at Q $x=a\dfrac{a^2-b^2}{a^2+b^2}\cos\theta$, $y=b\dfrac{b^2-a^2}{b^2+a^2}\sin\theta$.

Hence $PQ^2=a^2\cos^2\theta\left(1-\dfrac{a^2-b^2}{a^2+b^2}\right)^2+b^2\sin^2\theta\left(1-\dfrac{b^2-a^2}{b^2+a^2}\right)^2$

$$=\frac{4a^2b^2}{(a^2+b^2)^2}(a^2\sin^2\theta+b^2\cos^2\theta)$$

$$=\frac{4a^2b^2}{(a^2+b^2)^2}CD^2,$$

where CD is the semi-diameter conjugate to CP.

35. Let (x', y') be the point of intersection of two perpendicular tangents; then the equation of the chord of contact is $xx'/a^2+yy'/b^2=1$, and we know that

$$x'^2+y'^2=a^2+b^2 \quad\text{.................................(i).}$$

The product of the perpendiculars from (0, 0) and (x', y') on the chord of contact is equal to

$$\frac{x'^2/a^2+y'^2/b^2-1}{\sqrt{(x'^2/a^4+y'^4/b^4)}}\times\frac{-1}{\sqrt{(x'^2/a^4+y'^2/b^4)}}$$

$$=-\frac{x'^2/a^2+y'^2/b^2-(x'^2+y'^2)/(a^2+b^2)}{x'^2/a^4+y'^2/b^4}, \text{ from (i)}$$

$$=-a^2b^2/(a^2+b^2).$$

36. Let (x', y') and (x'', y'') be the extremities of any chord of an ellipse, and let (ξ, η) be the middle point of the chord; then $2\xi=x'+x''$, $2\eta=y'+y''$, and the equation of the chord is [Art. 114]

$$\frac{x(x'+x'')}{a^2}+\frac{y(y'+y'')}{b^2}=1+\frac{x'x''}{a^2}+\frac{y'y''}{b^2},$$

that is

$$\frac{2x\xi}{a^2}+\frac{2y\eta}{b^2}=1+\frac{x'x''}{a^2}+\frac{y'y''}{b^2}.$$

But, since the point (ξ, η) is on the line, we have

$$\frac{2\xi^2}{a^2}+\frac{2\eta^2}{b^2}=1+\frac{x'x''}{a^2}+\frac{y'y''}{b^2}.$$

Hence, by subtraction, we have for the equation of the chord whose middle point is (ξ, η),

$$(x-\xi)\frac{\xi}{a^2}+(y-\eta)\frac{\eta}{b^2}=0 \text{..............................(i).}$$

If the chord whose equation is (i) pass through the fixed point (f, g) we have

$$(f-\xi)\frac{\xi}{a^2}+(g-\eta)\frac{\eta}{b^2}=0,$$

so that the locus of (ξ, η) is the ellipse $\frac{x}{a^2}(x-f)+\frac{y}{b^2}(y-g)=0$;

or thus:

The coordinates of the point at a distance r from (ξ, η) and which is on the line through (ξ, η) which makes an angle θ with the axis are given by

$$x=\xi+r\cos\theta,\ y=\eta+r\sin\theta \text{..........................(i).}$$

Hence the values of r which correspond to the two points in which the line meets the ellipse are given by

$$(\xi+r\cos\theta)^2/a^2+(\eta+r\sin\theta)^2/b^2=1.$$

If (ξ, η) be the middle point of the chord it is necessary and sufficient that the coefficient of r in the above quadratic should be zero, and therefore

$$\xi\cos\theta/a^2+\eta\sin\theta/b^2=0.$$

Hence, substituting for $\cos\theta$ and $\sin\theta$ from (i), we have the equation of the chord whose middle point is (ξ, η), namely

$$\xi(x-\xi)/a^2+\eta(y-\eta)/b^2=0.$$

The equation of the required locus is then found as above to be

$$x(x-f)/a^2+y(y-g)/b^2=0.$$

37. Let DCD' be the diameter parallel to PRQ, let V be the middle point of PQ.

Draw PM parallel to CV, and let the tangent at P cut $D'CD$ in t.

Then we know that $CM.Ct=CD^2$; hence as $PR=Ct$ and $PQ=2PV=2CM$, we have $PQ.PR=2CD^2=\frac{1}{2}(DD')^2$.

38. Let (ξ, η) be the middle point of the chord and let $2c$ be its length, and θ the angle it makes with the axis. Then the points $\xi \pm c\cos\theta$, $\eta \pm c\sin\theta$ are both on the ellipse, and therefore

$$(\xi+c\cos\theta)^2/a^2+(\eta+c\sin\theta)^2/b^2=1$$

and

$$(\xi-c\cos\theta)^2/a^2+(\eta-c\sin\theta)^2/b^2=1.$$

Hence

$$(\xi^2+c^2\cos^2\theta)/a^2+(\eta^2+c^2\sin^2\theta)/b^2=1 \quad\text{.................(i)},$$

and

$$\xi\cos\theta/a^2+\eta\sin\theta/b^2=0\text{.........................(ii)}.$$

From (ii),

$$\cos\theta=a^2\eta/\sqrt{(a^4\eta^2+b^4\xi^2)},$$

and

$$\sin\theta=-b^2\xi/\sqrt{(a^4\eta^2+b^4\xi^2)}.$$

Whence, from (i)

$$\{\xi^2/a^2+\eta^2/b^2-1\}(a^4\eta^2+b^4\xi^2)+c^2(a^2\eta^2+b^2\xi^2)=0.$$

39. The equation of the chord whose middle point is (ξ, η) is [as in 36]

$$(x-\xi)\xi/a^2+(y-\eta)\eta/b^2=0.$$

If this chord is the polar of (x', y') its equation is the same as

$$xx'/a^2+yy'/b^2=1;$$

whence

$$\frac{x'}{\xi}=\frac{y'}{\eta}=\frac{1}{\xi^2/a^2+\eta^2/b^2}.$$

But if the tangents from (x', y') are at right angles $x'^2+y'^2=a^2+b^2$, and therefore

$$\xi^2+\eta^2=(a^2+b^2)(\xi^2/a^2+\eta^2/b^2)^2.$$

Thus the locus of the middle point of the chord is the curve of the fourth degree

$$(x^2+y^2)/(a^2+b^2)=(x^2/a^2+y^2/b^2)^2.$$

40. Let $\alpha, \beta, \gamma, \delta$ be the eccentric angles of the points A, B, C, D respectively. Then, since AB, BC and CD are parallel to three fixed straight lines, $\alpha+\beta$, $\beta+\gamma$ and $\gamma+\delta$ are all constant; and therefore $\alpha+\delta$ is constant, from which it follows that AD is always parallel to a fixed straight line.

41. Let PCP', QCQ' be the two diameters, and let the eccentric angles of P, Q be α, β, respectively. Then the area of $PQP'Q'$ is $4\triangle PCQ$

$$=2\begin{vmatrix} a\cos\alpha, & b\sin\alpha \\ a\cos\beta, & b\sin\beta \end{vmatrix} = 2ab\sin(\alpha-\beta)\ldots\ldots\ldots\ldots\ldots\ldots\text{(i)}.$$

The equations of the pairs of tangents are

$$\frac{x}{a}\cos\alpha+\frac{y}{b}\sin\alpha=\pm1,\quad \frac{x}{a}\cos\beta+\frac{y}{b}\sin\beta=\pm1.$$

Now the area of a parallelogram is the product of the perpendicular distances between its pairs of sides multiplied by the cosecant of the angle between two intersecting sides.

The product of the perpendicular distances between the pairs of parallel sides is

$$4\left(\frac{\cos^2\alpha}{a^2}+\frac{\sin^2\alpha}{b^2}\right)^{-\frac{1}{2}}\left(\frac{\cos^2\beta}{a^2}+\frac{\sin^2\beta}{b^2}\right)^{-\frac{1}{2}},$$

and the cosecant of the angle between intersecting sides is

$$\sqrt{\left(\frac{\cos^2\alpha}{a^2}+\frac{\sin^2\alpha}{b^2}\right)}\sqrt{\left(\frac{\cos^2\beta}{a^2}+\frac{\sin^2\beta}{b^2}\right)}\Big/\left(\frac{\cos\alpha}{a}\frac{\sin\beta}{b}-\frac{\sin\alpha}{b}\frac{\cos\beta}{a}\right).$$

Hence the area of the parallelogram is

$$4ab/\sin(\alpha\sim\beta)\ldots\ldots\ldots\ldots\ldots\ldots\ldots\ldots\text{(ii)}.$$

From (i) and (ii) it follows that the area of one of the parallelograms varies inversely as the area of the other.

[The proposition can be very easily proved geometrically.]

42. Draw QN parallel to the tangent at P meeting CPq in N. Then, since $CN.Cq=CP^2$, and QN is parallel to Pp, we have

$$Cq : CP = CP : CN$$
$$= Cp : CQ.$$

Hence $Cq.CQ=Cp.CP$, so that the triangles PCp and QCq are equal, and therefore also the triangles TQp and TPq are equal.

43. If the eccentric angles of P, Q be θ_1, θ_2 respectively, the area $PCQ=$

$$\frac{1}{2}\begin{vmatrix} a\cos\theta_1, & b\sin\theta_1 \\ a\cos\theta_2, & b\sin\theta_2 \end{vmatrix} = \frac{1}{2}ab\sin(\theta_1-\theta_2).$$

Since the tangents at θ_1, θ_2 meet in (h, k), we have

$$\frac{h}{a}=\frac{\cos\frac{1}{2}(\theta_1+\theta_2)}{\cos\frac{1}{2}(\theta_1-\theta_2)},\quad \frac{k}{b}=\frac{\sin\frac{1}{2}(\theta_1+\theta_2)}{\cos\frac{1}{2}(\theta_1-\theta_2)};$$

$$\therefore \cos^2 \frac{1}{2}(\theta_1 - \theta_2) = 1 \Big/ \left(\frac{h^2}{a^2} + \frac{k^2}{b^2}\right),$$

and therefore $\sin(\theta_1 - \theta_2) = 2\sqrt{\left(\frac{h^2}{a^2} + \frac{k^2}{b^2} - 1\right)} \Big/ \left(\frac{h^2}{a^2} + \frac{k^2}{b}\right)$.

Hence $$\triangle PCQ = ab\sqrt{\left(\frac{h^2}{a^2} + \frac{k^2}{b^2} - 1\right)} \Big/ \left(\frac{h^2}{a^2} + \frac{k^2}{b^2}\right).$$

Again area $OPCQ : \triangle PCQ :: OC : VC$, where V is the point of intersection of CO and PQ.

The equation of PQ is $xh/a^2 + yk/b^2 = 1$, and $CV : OV$ is equal to the ratio of the perpendiculars from C and O on PQ; $\therefore CV : OV = -1 : h^2/a^2 + k^2/b^2 - 1$, and therefore $CV : CO = 1 : h^2/a^2 + k^2/b^2$.

Hence area $OPCQ = \triangle PCQ \times (h^2/a^2 + k^2/b^2)$

$$= ab\sqrt{\left(\frac{h^2}{a^2} + \frac{k^2}{b^2} - 1\right)}.$$

44. Since CT bisects PQ, area $CPTQ = 2\triangle CPT$. Now P is $(a\cos\phi, b\sin\phi)$ and T is $\left(a\dfrac{\cos\frac{1}{2}(\phi+\phi')}{\cos\frac{1}{2}(\phi-\phi')},\ b\dfrac{\sin\frac{1}{2}(\phi+\phi')}{\cos\frac{1}{2}(\phi-\phi')}\right)$.

Hence area required

$$= \begin{vmatrix} a\cos\phi & b\sin\phi \\ a\dfrac{\cos\frac{1}{2}(\phi+\phi')}{\cos\frac{1}{2}(\phi-\phi')}, & b\dfrac{\sin\frac{1}{2}(\phi+\phi')}{\cos\frac{1}{2}(\phi-\phi')} \end{vmatrix}$$

$$= ab\sec\frac{1}{2}(\phi-\phi')\begin{vmatrix} \cos\phi & \sin\phi \\ \cos\frac{1}{2}(\phi+\phi') & \sin\frac{1}{2}(\phi+\phi') \end{vmatrix}$$

$$= ab\tan\frac{1}{2}(\phi \sim \phi').$$

45. The tangents at P, Q meet on the axis at T the point $\left(\frac{a}{\cos\phi}, 0\right)$; and the tangents at P', Q' meet on the axis at T' the point $\left(-\frac{a}{\cos\phi}, 0\right)$.

Also the tangents at P, Q', whose equations are respectively

$$\frac{x}{a}\cos\phi + \frac{y}{b}\sin\phi = 1 \quad\text{and}\quad -\frac{x}{a}\cos\phi - \frac{y}{a}\sin\phi = 1,$$

meet in the point t, where $y = \dfrac{2ab}{a-b}\operatorname{cosec}\phi$.

Now parallelogram $=2\triangle TtT'=\frac{2a}{\cos\phi}\cdot\frac{2ab}{(a-b)\sin\phi}=\frac{8a^2b}{(a-b)\sin 2\phi}$.

46. Take the centre of the circle for origin and axes parallel and perpendicular to the given straight lines; and let the equation of the circle be $x^2+y^2=a^2$ and the equations of the fixed straight lines be $x=\pm c$.

Let α, β, $\pi+\alpha$, $\pi+\beta$ be the angular coordinates of the points on the circle; and let the tangents at α, β meet on the line $x=c$; then

$$c=a\frac{\cos\frac{1}{2}(\alpha+\beta)}{\cos\frac{1}{2}(\alpha-\beta)} \qquad \text{(i)}.$$

The tangents at α and $\pi+\beta$ will meet in the point given by

$$x=a\frac{\cos\frac{1}{2}(\alpha+\beta+\pi)}{\cos\frac{1}{2}(\alpha-\beta-\pi)},\quad y=a\frac{\sin\frac{1}{2}(\alpha+\beta+\pi)}{\cos\frac{1}{2}(\alpha-\beta-\pi)};$$

$$\therefore\ x\sin\frac{1}{2}(\alpha-\beta)=-a\sin\frac{1}{2}(\alpha+\beta) \qquad \text{(ii)},$$

$$y\sin\frac{1}{2}(\alpha-\beta)=a\cos\frac{1}{2}(\alpha+\beta) \qquad \text{(iii)};$$

$$\therefore\ \sin^2\frac{1}{2}(\alpha-\beta)=a^2/(x^2+y^2),$$

and therefore $$\cos^2\frac{1}{2}(\alpha-\beta)=(x^2+y^2-a^2)/(x^2+y^2).$$

Hence from (i) and (iii)

$$c^2(x^2+y^2-a^2)=y^2a^2.$$

Hence the locus required is the ellipse

$$x^2/a^2+y^2\Big/\frac{a^2c^2}{c^2-a^2}=1.$$

Since the minor-axis of the ellipse is equal to $2a$, the original circle is the minor auxiliary circle.

47. Take the fixed conjugate diameters for axes, and let (α, β) be the point O. Then the equations of OP, OQ will be

$$y-\beta=m_1(x-\alpha) \text{ and } y-\beta=m_2(x-\alpha),$$

where $$m_1m_2=-\frac{b^2}{a^2}.$$

The points P, Q are $\left(\alpha-\frac{\beta}{m_1},\ 0\right)$ and $(0,\ \beta-m_2\alpha)$. Hence, if (x, y) be the middle point of PQ we have

$$2x = a - \frac{\beta}{m_1},\ 2y = \beta - m_2 a\ ;$$

$$\therefore\ m_1 = \frac{\beta}{a - 2x} \quad \text{and} \quad m_2 = \frac{\beta - 2y}{a},$$

and therefore the locus of the middle point of PQ is given by

$$\frac{\beta}{a - 2x} \cdot \frac{\beta - 2y}{a} = -\frac{b^2}{a^2},$$

or

$$a^2\beta\left(\frac{\beta}{2} - y\right) + b^2 a\left(\frac{a}{2} - x\right) = 0,$$

the locus is therefore a straight line which is conjugate to CO and bisects CO.

48. Let O be the point (a, β).

The lines CM, CN are $bx - ay = 0$, $bx + ay = 0$.

Hence OM is $a(x - a) + b(y - \beta) = 0$, and ON is $a(x - a) - b(y - \beta) = 0$.

Hence M is the point $x = a(aa + b\beta) / (a^2 + b^2)$, $y = b(aa + b\beta) / (a^2 + b^2)$; also N is the point $x = a(aa - b\beta) / (a^2 + b^2)$, $y = -b(aa - b\beta) / (a^2 + b^2)$.

Hence the coordinates of V, the middle point of MN, are

$$a^2 a / (a^2 + b^2) \text{ and } b^2\beta / (a^2 + b^2).$$

The equation of OV, which clearly goes through P, is therefore $\frac{x - a}{ab^2} = \frac{y - \beta}{\beta a^2}$; and this is perpendicular to the line $\frac{xa}{a^2} + \frac{y\beta}{b^2} = 1$, which proves the theorem.

49. Let a, θ, β be the eccentric angles of A, P, B respectively.

Then the lines through P parallel to the tangents at A, B are respectively

$$(x - a\cos\theta)\frac{\cos a}{a} + (y - b\sin\theta)\frac{\sin a}{b} = 0 \quad \text{..............(i)},$$

$$(x - a\cos\theta)\frac{\cos\beta}{a} + (y - b\sin\theta)\frac{\sin\beta}{b} = 0 \quad \text{..............(ii)}.$$

Also the equations of CA, CB are respectively

$$\frac{x}{a\cos a} = \frac{y}{b\sin a} \quad \text{..............................(iii)},$$

and

$$\frac{x}{a\cos\beta} = \frac{y}{b\sin\beta} \quad \text{..............................(iv)}.$$

Then Q, the point of intersection of (i) and (iv), is

$$\{a\cos\beta\cos(\theta - a) / \cos(\beta - a),\ b\sin\beta\cos(\theta - a) / \cos(\beta - a)\};$$

also R, the point of intersection of (ii) and (iii) is

$$\{a\cos a\cos(\theta - \beta) / \cos(a - \beta),\ b\sin a\cos(\theta - \beta) / \cos(a - \beta)\}.$$

Hence the equation of QR is

$$\frac{x - a\cos\beta\cos(\theta-\alpha) / \cos(\beta-\alpha)}{a\cos\beta\cos(\theta-\alpha) / \cos(\beta-\alpha) - a\cos\alpha\cos(\theta-\beta) / \cos(\alpha-\beta)}$$
$$= \frac{y - b\sin\beta\cos(\theta-\alpha) / \cos(\beta-\alpha)}{b\sin\beta\cos(\theta-\alpha)/\cos(\beta-\alpha) - b\sin\alpha\cos(\theta-\beta) / \cos(\alpha-\beta)}.$$

Hence QR is parallel to

$$\frac{x}{a\{\cos\beta\cos(\theta-\alpha) - \cos\alpha\cos(\theta-\beta)\}} = \frac{y}{b\{\sin\beta\cos(\theta-\alpha) - \sin\alpha\cos(\theta-\beta)\}},$$

and therefore to $\frac{x}{a}\cos\theta + \frac{y}{b}\sin\theta = 1$: thus QR is parallel to the tangent at P.

50. The equations of the normals can be taken to be

$$\frac{ax}{\cos\theta} - \frac{by}{\sin\theta} = a^2 - b^2 = c^2 \text{..............................(i)},$$

and
$$\frac{ax}{\cos\left(\theta+\frac{\pi}{2}\right)} - \frac{by}{\sin\left(\theta+\frac{\pi}{2}\right)} = c^2,$$

that is
$$-\frac{by}{\cos\theta} - \frac{ax}{\sin\theta} = c^2 \text{..............................(ii)}.$$

From (i) and (ii) we have

$$\frac{\sec\theta}{by - ax} = \frac{\operatorname{cosec}\theta}{ax + by} = \frac{-c^2}{a^2x^2 + b^2y^2};$$

whence $\quad 1 / (ax - by)^2 + 1 / (ax + by)^2 = c^4 / (a^2x^2 + b^2y^2)^2,$

or $\quad 2(a^2x^2 + b^2y^2)^3 = c^4(a^2x^2 - b^2y^2)^2.$

51. If θ_1, θ_2 be the extremities of any chord parallel to one of the equi-conjugates; then

$$\frac{x}{a}\cos\frac{1}{2}(\theta_1+\theta_2) + \frac{y}{b}\sin\frac{1}{2}(\theta_1+\theta_2) = \cos\frac{1}{2}(\theta_1-\theta_2),$$

must be parallel to one or other of the lines $\frac{x}{a} \pm \frac{y}{b} = 0$. Hence

$$\theta_1 + \theta_2 = \frac{\pi}{2} \text{ or } \frac{3\pi}{2}.$$

The normals at θ_1, θ_2 are

$$\frac{ax}{\cos\theta_1} - \frac{by}{\sin\theta_1} = c^2 \text{ and } \frac{ax}{\cos\theta_2} - \frac{by}{\sin\theta_2} = c^2;$$

and hence the line through the centre and the point of intersection of the normals is

$$ax\left(\frac{1}{\cos\theta_1} - \frac{1}{\cos\theta_2}\right) - by\left(\frac{1}{\sin\theta_1} - \frac{1}{\sin\theta_2}\right) = 0 \text{.........(i)}.$$

Now, if $\theta_1+\theta_2=\frac{\pi}{2}$, (i) will become $ax+by=0$; and if $\theta_1+\theta_2=\frac{3\pi}{2}$, (i) will become $ax-by=0$.

Hence the normals at extremities of chords parallel to $\frac{x}{a}+\frac{y}{b}=0$ meet on the line $ax+by=0$ which is perpendicular to $\frac{x}{a}-\frac{y}{b}=0$; and similarly for the other equiconjugate.

52. Let PSP' be a focal chord of an ellipse, and let the normals at P, P' cut the axis in G, G' respectively and intersect in O. Draw OV parallel to the axis to meet PP' in V; then we have to shew that V is the middle point of PP'.

From Art. 125, $SG : SP = SG' : SP'$;

but $SG : SP :: VO : VP$,

and $SG' : SP' :: VO : VP'$. Hence $VP=VP'$.

53. Let CF be the perpendicular on the tangent at P; then

$$CQ^2=CP^2+PQ^2\pm 2PQ\,.\,CF \qquad\qquad \text{(i)},$$

the upper or lower sign being taken according as PQ is measured outwards or inwards.

Since $PQ=CD$, $CD\,.\,CF=ab$, and $CP^2+CD^2=a^2+b^2$,

(i) becomes $CQ^2=a^2+b^2\pm 2ab$.

Hence Q lies on one or other of two circles concentric with the ellipse.

Or thus:—

The equation of PQ is

$$\frac{x-a\cos\theta}{\frac{\cos\theta}{a}}=\frac{y-b\sin\theta}{\frac{\sin\theta}{b}}=\frac{r}{\sqrt{\{\cos^2\theta/a^2+\sin^2\theta/b^2\}}}=\pm\frac{abr}{CD},$$

since $CD^2=a^2\sin^2\theta+b^2\cos^2\theta$.

Hence, if $PQ=CD$, the coordinates of Q are given by $x=a\cos\theta\pm b\cos\theta$, and $y=b\sin\theta\pm a\sin\theta$; whence the locus of Q is one or other of the circles $x^2+y^2=(a\pm b)^2$.

54. The directions of the two tangents drawn from (x', y') to the ellipse are given by

$$y'=mx'+\sqrt{(a^2m^2+b^2)}, \text{ or}$$

$$m^2(x'^2-a^2)-2mx'y'+y'^2-b^2=0 \qquad\qquad \text{(i)}.$$

Hence, if ϕ be the angle between the two tangents, we have

$$\tan^2\phi=\frac{(m_1-m_2)^2}{(1+m_1m_2)^2}=\frac{(m_1+m_2)^2-4m_1m_2}{(1+m_1m_2)^2},$$

$$=4(b^2x'^2+a^2y'^2-a^2b^2)\,/\,(x'^2+y'^2-a^2-b^2)^2.$$

55. If T be the point (x', y') the equation of PQ will be

$$xx' / a^2 + yy' / b^2 = 1.$$

Hence the abscissae x_1, x_2 of P, Q are the roots of

$$\frac{x^2}{a^2} + \left(1 - \frac{xx'}{a^2}\right)^2 \frac{b^2}{y'^2} = 1,$$

$$\therefore\ x_1 + x_2 = 2x'a^2b^2 / (a^2y'^2 + b^2x'^2),$$

and $$x_1x_2 = a^4(b^2 - y'^2) / (a^2y'^2 + b^2x'^2).$$

Now $$SP \,.\, SQ = (a - ex_1)(a - ex_2)$$
$$= a^2 - 2a^3eb^2x' / (a^2y'^2 + b^2x'^2)$$
$$+ e^2a^4(b^2 - y'^2) / (a^2y'^2 + b^2x'^2),$$
$$= a^2b^2 \{y'^2 + (x' - ae)^2\} / (a^2y'^2 + b^2x'^2).$$

But $$ST^2 = y'^2 + (x' - ae)^2.$$

Hence $$\frac{ST^2}{SP \,.\, SQ} = \frac{x'^2}{a^2} + \frac{y'^2}{b^2}.$$

56. The directions of the tangents from (x, y) are given by the equation $y = mx + \sqrt{(a^2m^2 + b^2)}$, or $m^2(x^2 - a^2) - 2mxy + y^2 - b^2 = 0$.

Hence $$\tan^2\theta = \frac{4(b^2x^2 + a^2y^2 - a^2b^2)}{(x^2 + y^2 - a^2 - b^2)^2};$$

$$\therefore\quad \cos^2\theta = \frac{(x^2 + y^2 - a^2 - b^2)^2}{(x^2 + y^2 - a^2 - b^2)^2 + 4(b^2x^2 + a^2y^2 - a^2b^2)} \text{..............(i)}.$$

Now $$ST^2 \,.\, HT^2 = \{y^2 + (x - ae)^2\}\{y^2 + (x + ae)^2\}$$
$$= (x^2 + y^2 - a^2 - b^2)^2 + 4(b^2x^2 + a^2y^2 - a^2b^2).$$

Hence from (i), $$ST^2 \,.\, HT^2 \cos^2\theta = CT^2 - a^2 - b^2.$$

57. Let P be the point whose eccentric angle is θ.

Then the equation of CR will be

$$\frac{ax}{\cos\theta} - \frac{by}{\sin\theta} = 0 \text{..................................(i)}.$$

The equation of SP will be

$$\frac{x - ae}{ae - a\cos\theta} = \frac{y}{-b\sin\theta} \text{..............................(ii)}.$$

The locus of R will be found by eliminating θ between (i) and (ii), that is from

$$ax \sin\theta - by \cos\theta = 0,$$

and
$$b(x - ae)\sin\theta - ay\cos\theta + aey = 0.$$

We have
$$\frac{\sin\theta}{abey} = \frac{\cos\theta}{a^2ex} = \frac{1}{a^2x - b^2(x - ae)};$$

$$\therefore\ a^2b^2e^2y^2 + a^4e^2x^2 = a^2e^2(aex + b^2)^2;$$

$$\therefore\ (x - ae)^2 + y^2 = a^2.$$

[The proposition can easily be proved geometrically.]

58. Let S, S' be the foci of one ellipse, and H, H' the foci of the other, C being the common centre. Then $SHS'H'$ is clearly a parallelogram; and, since $SH + HS' = HS' + S'H$, the major axes of the two ellipses are equal to one another.

Hence $SC = CS' = ae$, and $HC = CH' = ae'$.

Hence
$$SH^2 = a^2e^2 + a^2e'^2 - 2a^2ee'\cos\theta,$$
$$HS'^2 = a^2e^2 + a^2e'^2 + 2a^2ee'\cos\theta,$$

and
$$SH^2 + HS'^2 = 2a^2e^2 + 2a^2e'^2;$$

$$\therefore\ 4a^2 = (SH + HS')^2 = 2a^2e^2 + 2a^2e'^2 + 2a^2\sqrt{(e^2 + e'^2 - 2ee'\cos\theta)}\sqrt{(e^2 + e'^2 + 2ee'\cos\theta)},$$

whence
$$\cos\theta = \sqrt{(e^2 + e'^2 - 1)}/ee'.$$

59. Let the conjugate diameters be PCP', DCD'. Then if P be θ, P' will be $\theta + \pi$, D will be $\theta + \frac{\pi}{2}$ and D' will be $\theta - \frac{\pi}{2}$.

Hence, if A be 0, and A' be π, and B be $\frac{\pi}{2}$ and B' be $\frac{3\pi}{2}$,

the sum of the eccentric angles of A and P is equal to the sum of the eccentric angles of B and D', and therefore AP is parallel to BD'.

Similarly BD is parallel to PA', and the proposition is then obvious.

60. Let P be the point α, then the equations of AP, $A'P$ are

$$\frac{x}{a}\cos\frac{1}{2}\alpha + \frac{y}{b}\sin\frac{1}{2}\alpha = \cos\frac{1}{2}\alpha,$$

and
$$\frac{x}{2}\cos\frac{1}{2}(\alpha + \pi) + \frac{y}{b}\sin\frac{1}{2}(\alpha + \pi) = \cos\frac{1}{2}(\alpha - \pi).$$

Hence
$$\cot^2 APA' = \left(\frac{a^2 - b^2}{2ab}\right)^2 \sin^2\alpha.$$

Similarly $$\cot^2 ADA' = \left(\frac{a^2-b^2}{2ab}\right)^2 \cos^2\alpha.$$

Hence $$\cot^2\theta + \cot^2\theta' = \left(\frac{a^2-b^2}{2ab}\right)^2.$$

61. Since the normal bisects the angle between the focal distances SP, $S'P$, $\tan\theta = \frac{p_2}{p_1}$, where p_1, p_2 are respectively the perpendiculars from S on the tangent and normal at P.

Hence, if α be the eccentric angle of P,

$$p_1{}^2 = \frac{(e\cos\alpha - 1)^2}{\frac{\cos^2\alpha}{a^2} + \frac{\sin^2\alpha}{b^2}} \text{ and } p_2{}^2 = \frac{\left(\frac{a^2 e}{\cos\alpha} - a^2e^2\right)^2}{\frac{a^2}{\cos^2\alpha} + \frac{b^2}{\sin^2\alpha}}.$$

Hence $$\tan^2\theta = \frac{p_2{}^2}{p_1{}^2} = \frac{a^2e^2}{b^2}\sin^2\alpha.$$

Similarly $$\tan^2\theta' = \frac{a^2e^2}{b^2}\cos^2\alpha,$$

and therefore $$\tan^2\theta + \tan^2\theta' = \frac{a^2e^2}{b^2}.$$

62. Let θ, $\pi+\theta$ be the eccentric angles of the extremities of the diameter PCP', and $\theta - \frac{\pi}{2}$, $\theta + \frac{\pi}{2}$ the extremities of the conjugate diameter DCD'. Then, if ϕ be the eccentric angle of any other point Q on the curve, the equations of QP, QP' are

$$\frac{x}{a}\cos\frac{1}{2}(\theta+\phi) + \frac{y}{b}\sin\frac{1}{2}(\theta+\phi) = \cos\frac{1}{2}(\theta-\phi),$$

$$\frac{x}{a}\cos\frac{1}{2}(\theta+\pi+\phi) + \frac{y}{b}\sin\frac{1}{2}(\theta+\pi+\phi) = \cos\frac{1}{2}(\theta+\pi-\phi).$$

Hence $$\cot^2\lambda = \left(\frac{a^2-b^2}{2ab}\right)^2 \sin^2(\theta+\phi).$$

Similarly $$\cot^2\lambda' = \left(\frac{a^2-b^2}{2ab}\right)^2 \sin^2\left(\theta - \frac{\pi}{2} + \phi\right).$$

Hence $$\cot^2\lambda + \cot^2\lambda' = \left(\frac{a^2-b^2}{2ab}\right)^2.$$

63. Take the diameter parallel to the cutting line and its conjugate for axes, and let the equation of the cutting line be $x = k$.

Then $y=m_1x$ and $y=m_2x$ are conjugate if $m_1m_2=-\frac{b^2}{a^2}$.

But $y=m_1x$ and $y=m_2x$ are cut by $x=k$ in points the product of whose distances from $(k, 0)$ is $m_1m_2k^2$, that is $-\frac{b^2}{a^2}k^2$, which is the same for all the different pairs of conjugate diameters. This proves the proposition [Art. 61].

64. Let ABC be any triangle whose sides touch an ellipse and enclose it, and let A', B', C' be the points of contact of the sides BC, CA, AB respectively.

A tangent line to the ellipse at any point P in the arc $C'A'B'$ near to C' will clearly be cut by AB, AC in points X, Y respectively so that $XP<PY$; and a tangent at any point P' near B' will be cut by AB, AC in points X', Y' respectively so that $X'P>PY'$. There must therefore be some tangent, MDN suppose, whose point of contact is the middle point of the intercept made by AB, AC; also, if $BA'<A'C$, B will be between A and M and therefore N between A and C.

Let NM and BC meet in O, and draw Nn parallel to AB to meet BC in n.

Then $ND=Dm$, and therefore $OM<ON$; hence also $BO<On<OC$; and since OM is less than ON and OB less than OC, the triangle BOM is less than the triangle NOC, and therefore $\triangle MAN<\triangle BAC$.

Thus if AB, AC be fixed tangents, and a third tangent be drawn so as to enclose the ellipse; then when the area of the triangle formed by the three tangents is the least possible, the moveable side of the triangle is bisected in its point of contact. Hence when all three tangents to the ellipse are moveable, and the area of the circumscribing triangle is least, each side of the triangle must be bisected in its point of contact. [It will be seen at once that the theorem is true for any polygon, of a given number of sides, whose sides touch and enclose *any oval* curve.]

Now let ABC be a triangle which circumscribes an ellipse so that the points of contact A', B', C', are the middle points of BC, CA, AB respectively.

Let O be the centre of the ellipse, and let AO produced cut the ellipse in a. Then we know that AO bisects $B'C'$, and that the tangent at a is parallel to $B'C'$; the portion of the tangent at a intercepted by AB, AC must therefore be bisected at a, whence it follows that a coincides with A'.

Thus the triangle $A'B'C'$ is such that the tangent to the ellipse at A' is parallel to $B'C'$, and similarly the tangents at B', C' are parallel to CA, AB respectively. Hence [Art. 138, (i)] $A'B'C'$ is a maximum triangle.

65. Let the eccentric angles of A, B, C, D be α, β, γ, δ; and let θ be the eccentric angle of any other point P on the curve.

Then the product of the perpendiculars from P on AB and CD is equal to

$$\frac{\cos\theta\cos\frac{1}{2}(\alpha+\beta)+\sin\theta\sin\frac{1}{2}(\alpha+\beta)-\cos\frac{1}{2}(\alpha-\beta)}{\sqrt{\left\{\cos^2\frac{1}{2}(\alpha+\beta)\Big/a^2+\sin^2\frac{1}{2}(\alpha+\beta)\Big/b^2\right\}}}$$
$$\times\frac{\cos\theta\cos\frac{1}{2}(\gamma+\delta)+\sin\theta\sin\frac{1}{2}(\gamma+\delta)-\cos\frac{1}{2}(\gamma-\delta)}{\sqrt{\left\{\cos^2\frac{1}{2}(\gamma+\delta)\Big/a^2+\sin^2\frac{1}{2}(\gamma+\delta)\Big/b^2\right\}}}$$

$$a^2b^2\,.\,\frac{2\sin\frac{1}{2}(\theta-\alpha)\sin\frac{1}{2}(\theta-\beta)}{\sqrt{\left\{b^2\cos^2\frac{1}{2}(\alpha+\beta)+a^2\sin^2\frac{1}{2}(\alpha+\beta)\right\}}}\,.\,\frac{2\sin\frac{1}{2}(\theta-\gamma)\sin\frac{1}{2}(\theta-\delta)}{\sqrt{\left\{b^2\cos^2\frac{1}{2}(\gamma+\delta)+a^2\sin^2\frac{1}{2}(\gamma+\delta)\right\}}}$$

Hence the ratio of the product of the perpendiculars on AB and CD to the product of the perpendiculars of BC and DA is independent of θ, and this ratio is equal to

$$\sqrt{\left\{b^2\cos^2\frac{1}{2}(\beta+\gamma)+a^2\sin^2\frac{1}{2}(\beta+\gamma)\right\}}\,.\,\sqrt{\left\{b^2\cos^2\frac{1}{2}(\delta+\alpha)+a^2\sin^2\frac{1}{2}(\delta+\alpha)\right\}}$$
$$:\sqrt{\left\{b^2\cos^2\frac{1}{2}(\alpha+\beta)+a^2\sin^2\frac{1}{2}(\alpha+\beta)\right\}}\,.\,\sqrt{\left\{b^2\cos^2\frac{1}{2}(\gamma+\delta)+a^2\sin^2\frac{1}{2}(\gamma+\delta)\right\}}$$
$$=d_3\,.\,d_4:d_1\,.\,d_2,$$

where d_1, d_2, d_3, d_4 are the semi-diameters conjugate to AB, BC, CD and DA respectively.

66. If the normals at P, P' are at right angles, the tangents at P, P' will be at right angles; and if the tangents intersect in T and the normals in N, $NPTP'$ will be a rectangle and NT will bisect PP'. Hence NT will pass through the centre C, of the ellipse.

Now $$NC=CT-2CV;$$

also $$CT=\sqrt{a^2+b^2}\quad\text{[Art. 120]},$$

and $$CV\,.\,CT=CQ^2,$$

where Q is on the ellipse.

Hence, if (r, θ) be the polar co-ordinates of N, we have

$$-r=\sqrt{(a^2+b^2)}-2\frac{CQ^2}{\sqrt{(a^2+b^2)}}$$
$$=\sqrt{(a^2+b^2)}-\frac{2}{\sqrt{(a^2+b^2)}}\left(\frac{\cos^2\theta}{a^2}+\frac{\sin^2\theta}{b^2}\right)^{-1},$$

since $$CQ^2\left(\frac{\cos^2\theta}{a^2}+\frac{\sin^2\theta}{b^2}\right)=1.\quad\text{[Art. 111]}$$

Hence $\sqrt{(a^2+b^2)}\,\{r+\sqrt{(a^2+b^2)}\}\left(\frac{\cos^2\theta}{a^2}+\frac{\sin^2\theta}{b^2}\right)=2;$

$$\therefore\ (a^2-b^2)(a^2\sin^2\theta-b^2\cos^2\theta)=-r\sqrt{(a^2+b^2)}\,(a^2\sin^2\theta+b^2\cos^2\theta),$$

whence $(a^2-b^2)^2(a^2y^2-b^2x^2)^2=(x^2+y^2)(a^2+b^2)(a^2y^2+b^2x^2)^2,$

the equation required.

Or thus:

Let θ, ϕ be the eccentric angles of P, P'; then N is the point of intersection of

$$\frac{ax}{\cos\theta}-\frac{by}{\sin\theta}=a^2-b^2 \quad\text{.............(i)},$$

$$\frac{ax}{\cos\phi}-\frac{by}{\sin\phi}=a^2-b^2 \quad\text{.............(ii)},$$

where θ, ϕ are subject to the condition

$$a^2\sin\theta\sin\phi+b^2\cos\theta\cos\phi=0 \quad\text{.............(iii)}.$$

From (i) and (ii) we have

$$\frac{ax(\cos\phi-\cos\theta)}{\cos\theta\cos\phi}=\frac{by(\sin\phi-\sin\theta)}{\sin\theta\sin\phi};$$

$\therefore$, from (iii), $\tan\frac{1}{2}(\theta+\phi)=\frac{ay}{bx}.$

so that $\sin\frac{1}{2}(\theta+\phi)=ay/\sqrt{(a^2y^2+b^2x^2)},$

and $\cos\frac{1}{2}(\theta+\phi)=bx/\sqrt{(a^2y^2+b^2x^2)}$ ....(iv).

Again, from (i) and (ii) by addition

$$\frac{ax(\cos\theta+\cos\phi)}{\cos\theta\cos\phi}-\frac{by(\sin\theta+\sin\phi)}{\sin\theta\sin\phi}=2(a^2-b^2),$$

or, from (iii),

$$ax\cos\frac{1}{2}(\theta+\phi)\cos\frac{1}{2}(\theta-\phi)+\frac{a^2}{b}y\sin\frac{1}{2}(\theta+\phi)\cos\frac{1}{2}(\theta-\phi)$$
$$=(a^2-b^2)\cos\theta\cos\phi$$
$$=(a^2-b^2)\left\{\cos^2\frac{1}{2}(\theta-\phi)-\sin^2\frac{1}{2}(\theta+\phi)\right\}.$$

Hence, from (iv)

$$\frac{a}{b}\sqrt{(a^2y^2+b^2x^2)}\cos\frac{1}{2}(\theta-\phi)=(a^2-b^2)\cos^2\frac{1}{2}(\theta-\phi)$$
$$-(a^2-b^2)\,a^2y^2/(a^2y^2+b^2x^2) \quad\text{......(v)}.$$

Now (iii) may be written

$$(a^2+b^2)\cos(\theta-\phi)=(a^2-b^2)\cos(\theta+\phi);$$

$\therefore$, using (iv),

$$(a^2+b^2)\cos^2\frac{1}{2}(\theta-\phi)=a^2b^2(x^2+y^2)/(a^2y^2+b^2x^2)\text{..........(vi)}.$$

We have now to eliminate $\cos\frac{1}{2}(\theta-\phi)$ between (v) and (vi), and we obtain as before the equation

$$(a^2+b^2)(x^2+y^2)(a^2y^2+b^2x^2)^2=(a^2-b^2)^2(b^2x^2-a^2y^2)^2.$$

67. Let θ, ϕ be the eccentric angles of the extremities of any focal chord; then the equations of the tangent at θ and the normal at ϕ are respectively

$$\frac{x}{a}\cos\theta+\frac{y}{b}\sin\theta=1,$$

and

$$\frac{ax}{\cos\phi}-\frac{by}{\sin\phi}=c^2=a^2-b^2.$$

The point of intersection is given by

$$\frac{x}{\frac{c^2}{b}\sin\theta+\frac{b}{\sin\phi}}=\frac{y}{\frac{a}{\cos\phi}-\frac{c^2}{a}\cos\theta}=\frac{-1}{-\frac{b}{a}\frac{\cos\theta}{\sin\phi}-\frac{a}{b}\frac{\sin\theta}{\cos\phi}},$$

or

$$\frac{xb\sin\phi}{b^2+c^2\sin\theta\sin\phi}=\frac{ya\cos\phi}{a^2-c^2\cos\theta\cos\phi}=\frac{ab\sin\phi\cos\phi}{a^2\sin\theta\sin\phi+b^2\cos\theta\cos\phi}$$

But, since the line joining θ, ϕ passes through a focus, we have

$$e\cos\frac{1}{2}(\theta+\phi)=\cos\frac{1}{2}(\theta-\phi);$$

$$\therefore\ a^2e^2\{1+\cos(\theta+\phi)\}=a^2\{1+\cos(\theta-\phi)\};$$

$$\therefore\ b^2+(a^2-b^2)\sin\theta\sin\phi=-\{b^2\cos\theta\cos\phi+a^2\sin\theta\sin\phi\}.$$

Hence

$$xb\sin\phi=-ab\sin\phi\cos\phi;$$

$$\therefore\ x=-a\cos\phi,$$

whence from the equation of the normal

$$y=-\frac{2a^2-b^2}{b}\sin\phi.$$

Eliminating ϕ, we have the required equation, namely

$$\left(\frac{x}{a}\right)^2+\left(\frac{by}{2a^2-b^2}\right)^2=1.$$

68. The equations of the two parallel lines are

$$y^2-\frac{a^2b^2}{a^2-b^2}=0\text{..............................(i)}.$$

The equation of the tangent at θ is

$$\frac{x}{a}\cos\theta+\frac{y}{b}\sin\theta=1 \quad\text{.............................(ii).}$$

The equation of the lines joining the centre to the points of intersection of (i) and (ii) is

$$y^2-\frac{a^2b^2}{a^2-b^2}\left(\frac{x}{a}\cos\theta+\frac{y}{b}\sin\theta\right)^2=0 \quad\text{.................. (iii).}$$

The equation of the lines bisecting the angles between the lines given by (iii) is

$$\frac{x^2-y^2}{\dfrac{b^2\cos^2\theta}{a^2-b^2}-\dfrac{a^2\sin^2\theta}{a^2-b^2}+1}=\frac{xy}{\dfrac{ab\sin\theta\cos\theta}{a^2-b^2}},$$

or $$ab\sin\theta\cos\theta\,(x^2-y^2)=(a^2\cos^2\theta-b^2\sin^2\theta)\,xy \quad\text{...........(iv).}$$

It is clear that the point of contact of the tangent, namely the point $(a\cos\theta,\ b\sin\theta)$, is on (iv): this proves the proposition.

69. Let θ be the eccentric angle of P; then the equation of PG is

$$ax\sec\theta-by\operatorname{cosec}\theta=a^2-b^2.$$

Hence G is the point $\left(\dfrac{a^2-b^2}{a}\cos\theta,\ 0\right)$.

If therefore Q be (x, y) we have

$$x=a\cos\theta+a\cos\theta-\frac{a^2-b^2}{a}\cos\theta=\frac{a^2+b^2}{a}\cos\theta,\quad y=2b\sin\theta.$$

Hence the locus of Q is the ellipse

$$\left(\frac{ax}{a^2+b^2}\right)^2+\left(\frac{y}{2b}\right)^2=1.$$

The eccentricity of the ellipse is given by

$$4b^2=\left(\frac{a^2+b^2}{a}\right)^2(1-E^2),$$

whence $$E=\frac{a^2-b^2}{a^2+b^2}.$$

The tangents at P and Q are

$$\frac{x}{a}\cos\theta+\frac{y}{b}\sin\theta=1,$$

$$\frac{ax}{a^2+b^2}\cos\theta+\frac{y}{2b}\sin\theta=1.$$

whence
$$\frac{x\cos\theta}{\frac{1}{b}-\frac{1}{2b}}=\frac{y\sin\theta}{\frac{a}{a^2+b^2}-\frac{1}{a}}=\frac{-1}{\frac{1}{2ab}-\frac{a}{b(a^2+b^2)}},$$

or
$$\frac{\cos\theta}{\frac{a(a^2+b^2)}{x}}=\frac{\sin\theta}{-\frac{2b^3}{y}}=\frac{1}{a^2-b^2}.$$

Eliminating θ we have the required equation, namely

$$\frac{a^2(a^2+b^2)^2}{x^2}+\frac{4b^6}{y^2}=(a^2-b^2)^2.$$

CHAPTER VII.

Pages 162—165.

1. Take OA, OC for axes of x and y respectively, and let the coordinates of A, B be $\pm a$, 0, and those of C, D be 0, $\pm b$. Then, if (x, y) be any point on the locus, we have

$$\sqrt{\{(x-a)^2+y^2\}} \cdot \sqrt{\{(x+a)^2+y^2\}} = \sqrt{\{x^2+(y-b)^2\}}\sqrt{\{x^2+(y+b)^2\}};$$

whence $x^2 - y^2 = \frac{1}{2}(a^2 - b^2)$.

The locus is therefore a rectangular hyperbola.

2. Taking the asymptotes for axes, the equation of the hyperbola will be $xy = c^2$. Now any line $y = mx + b$ will cut $xy - c^2 = 0$ in points whose ordinates are given by $x(mx+b) - c^2 = 0$; and the abscissa of the middle point of the chord is $-\frac{b}{2m}$, which is independent of c.

Hence the middle point of the chord of the hyperbola is the same as the middle point of the intercept by the asymptotes.

3. Take the fixed straight lines for axes; and let the equation of the chord in any one of its possible positions be $x/h + y/k = 1$. Then if the line always pass through the fixed point (α, β) we have

$$\alpha/h + \beta/k = 1 \text{..(i).}$$

And, if (x, y) be the middle point of the chord, $2x = h$ and $2y = k$.

Hence, from (i), $\frac{\alpha}{2x} + \frac{\beta}{2y} = 1$, or $(2x - \beta)(2y - \alpha) = \alpha\beta$.

Thus the locus is an hyperbola whose asymptotes are the lines $2x - \beta = 0$ and $2y - \alpha = 0$.

4. Take the fixed straight lines for axes; and let the equation of the line be $x/h + y/k = 1$.

Then, since the line cuts off a triangle of fixed area from the fixed straight lines, we have $hk = \text{constant} = c^2$ suppose.

If (x, y) be the middle point of the moving line, we have $2x = h$, $2y = k$. Hence $4xy = c^2$ is the equation of the required locus.

5. Take the bisectors of the angle AOB for axes; and suppose the equations of OA, OB to be respectively

$$x\cos\alpha + y\sin\alpha = 0 \text{ and } x\cos\alpha - y\sin\alpha = 0.$$

Let P be the point (ξ, η); then

$$PM = \xi\cos\alpha + \eta\sin\alpha \text{ and } PN = \xi\cos\alpha - \eta\sin\alpha.$$

The equations of PM, PN will be respectively $(\xi - x)\sin\alpha - (\eta - y)\cos\alpha = 0$ and $(\xi - x)\sin\alpha + (\eta - y)\cos\alpha = 0$.

Hence $OM = \xi\sin\alpha - \eta\cos\alpha$, and $ON = \xi\sin\alpha + \eta\cos\alpha$.

Now area $PMONP$

$$= \frac{1}{2}.OM.MP + \frac{1}{2}ON.NP$$

$$= \frac{1}{2}(\xi\sin\alpha - \eta\cos\alpha)(\xi\cos\alpha + \eta\sin\alpha)$$

$$+ \frac{1}{2}(\xi\sin\alpha + \eta\cos\alpha)(\xi\cos\alpha - \eta\sin\alpha)$$

$$= (\xi^2 - \eta^2)\sin\alpha\cos\alpha.$$

Hence the equation of the locus of P is $x^2 - y^2 =$ constant, which represents a rectangular hyperbola.

6. Let the equation of the hyperbola be $x^2 - y^2 = a^2$, and let (x', y') be any point P; then the distance of P from the centre is $\sqrt{(x'^2 + y'^2)}$. The equation of the polar of P is $xx' - yy' - a^2 = 0$, and the distance of the polar from the centre is $\dfrac{-a^2}{\sqrt{(x'^2 + y'^2)}}$.

Hence the distance of P from the centre varies inversely as the distance of the polar of P from the centre.

7. Let $x^2/a^2 - y^2/b^2 = 1$ be the equation of the hyperbola, then the asymptotes are given by $x/a \pm y/b = 0$.

Let P be the point (x', y'); then the normal at P is

$$\frac{x - x'}{\dfrac{x'}{a^2}} = \frac{y - y'}{-\dfrac{y'}{b^2}}.$$

Hence G is the point $\left\{x'\left(1 + \dfrac{b^2}{a^2}\right), 0\right\}$; also Q is the point $\left(x', \mp\dfrac{x'b}{a}\right)$.

The equation of QG is therefore

$$\frac{x - x'}{-\dfrac{b^2}{a^2}x'} = \frac{y \pm \dfrac{x'b}{a}}{\mp\dfrac{x'b}{a}},$$

and hence QG is perpendicular to $\dfrac{x}{a} \pm \dfrac{y}{b} = 0$.

8. Let the equations of the hyperbolas be

$$x^2/a^2 - y^2/b^2 = 1 \text{ and } -x^2/a^2 + y^2/b^2 = 1.$$

Then $b^2 = a^2(e^2 - 1)$ and $a^2 = b^2(e'^2 - 1)$; $\therefore 1 = (e^2 - 1)(e'^2 - 1)$, or $\dfrac{1}{e^2} + \dfrac{1}{e'^2} = 1$.

9. Take the asymptotes as axes, and let (x', y') and (x'', y'') be the points of contact P, P' of the tangents.

Then the tangents meet the asymptotes in the points q, r and q', r' whose coordinates are respectively $(2x', 0)$, $(0, 2y')$ and $(2x'', 0)$, $(0, 2y'')$.

Hence the equations of the lines qr', $q'r$ are

$$\frac{x}{2x'} + \frac{y}{2y''} = 1 \text{ and } \frac{x}{2x''} + \frac{y}{2y'} = 1,$$

and these lines are parallel since $x'y' = x''y''$.

Since qr' is parallel to $q'r$, and P, P' are the middle points qr, $q'r'$ respectively, it follows that PP' is midway between the parallel lines qr' and $q'r$.

10. If the tangent at $P(x', y')$ meet the axis in T, then T is the point $\left(\dfrac{a^2}{x'}, 0\right)$.

We have to shew that the perpendiculars from S, T, S' on the normal at P are in harmonical progression.

The equation of the normal at P is

$$\frac{a^2x}{x'} + \frac{b^2y}{y'} = a^2 + b^2.$$

Hence we have to shew that

$$\frac{2}{a^2 + b^2 - \dfrac{a^4}{x'^2}} = \frac{1}{a^2 + b^2 - \dfrac{a^3e}{x'}} + \frac{1}{a^2 + b^2 + \dfrac{a^3e}{x'}},$$

or

$$\frac{2x'^2}{(a^2 + b^2)x'^2 - a^4} = \frac{2x'^2(a^2 + b^2)}{(a^2 + b^2)^2 x'^2 - a^6e^2},$$

which is obvious since $a^2 + b^2 = a^2e^2$.

11. Take the asymptotes for axes, and let the equation of the curve be $2xy = c^2$. Then the equation of the tangent at (x', y') is $\dfrac{x}{x'} + \dfrac{y}{y'} = 2$ [Art. 153], or $xy' + x'y = c^2$.

Hence, as in Art. 118, the equation of the polar of (x', y') is $xy' + x'y = c^2$.

The line $y = y'$ which is parallel to one of the asymptotes meets the polar of (x', y') in the point Q whose abscissa is $x = \dfrac{c^2}{y'} - x'$. Hence the abscissa of the middle point of OQ is $\dfrac{c^2}{2y'}$, which is on the curve since $2y' \cdot \dfrac{c^2}{2y'} = c^2$.

12. Let (x', y') and (x'', y'') be the extremities of the chord. Then the other angular points of the parallelogram are (x', y'') and (x'', y'). Hence the equation of the diagonal is

$$\frac{x-x'}{x'-x''}=\frac{y-y''}{y''-y'}, \text{ or } x(y''-y')+y(x''-x')+x'y'-x''y''=0,$$

which passes through the centre since $x'y'=x''y''$.

13. Let P be the point (x', y'); then A, A' being $(-a, 0)$, $(a, 0)$ respectively, the equation of PA and PA' is

$$(xy'-x'y)^2-a^2(y-y')^2=0.$$

Hence the bisectors of the angle APA' are parallel to

$$\frac{x^2-y^2}{y'^2-x'^2+a^2}=\frac{xy}{-x'y'}, \qquad \text{[Art. 39]}$$

or

$$x^2-y^2=0, \text{ since } x'^2-y'^2=a^2.$$

Thus the bisectors are parallel to the asymptotes $x^2-y^2=0$.

14. Let $(-a, 0)$, $(a, 0)$ be the two points A, A'; and let $(a\cos\theta, a\sin\theta)$, $(a\cos\theta, -a\sin\theta)$ be the coordinates of P, P'. Then the equations of AP, $A'P'$

are

$$(x+a)\sin\theta-y\cos\theta=y,$$

and

$$(x-a)\sin\theta+y\cos\theta=y.$$

Hence at the intersection of AP, $A'P'$ we have

$$\frac{\sin\theta}{2y^2}=\frac{\cos\theta}{2ay}=\frac{1}{2yx};$$

$$\therefore y^2-x^2+a^2=0,$$

so that the required locus is a rectangular hyperbola.

15. The equation of the tangents at (x', y') (x'', y'') to the hyperbola $2xy=c^2$ are $xy'+x'y=c^2$ and $xy''+x''y=c^2$.

These meet in the point given by

$$\frac{x}{x'-x''}=\frac{y}{y''-y'}=\frac{-c^2}{y'x''-x'y''};$$

$$\therefore x=\frac{-c^2(x'-x'')}{y'x''-x'y''}=\frac{-2c^2(x'-x'')(x'x'')}{c^2(x''^2-x'^2)}=\frac{2x'x''}{x'+x''};$$

$$\therefore \frac{2}{x}=\frac{1}{x'}+\frac{1}{x''},$$

and similarly

$$\frac{2}{y}=\frac{1}{y'}+\frac{1}{y''}.$$

16. Let the point (x', y') be on the hyperbola $2xy=c^2$. Then the equation of the chord of contact of the tangents from (x', y') to the hyperbola

$2xy = c'^2$ will be [from 11]

$$xy' + yx' = c'^2.$$

Hence the area of the triangle cut off from the asymptotes by the polar

$$= \frac{1}{2}\frac{c'^4}{x'y'}\sin\omega = \frac{c'^4}{c^2}\sin\omega.$$

17. Let (x', y') $(-x', -y')$ be the extremities A, A' of any diameter of the hyperbola $2xy = c^2$; and let (ξ, η) be any other point P on the curve.

Then the equations of PA, PA' are respectively

$$x(y' - \eta) - y(x' - \xi) + x'\eta - y'\xi = 0,$$

and

$$x(-y' - \eta) - y(-x' - \xi) - x'\eta + y'\xi = 0.$$

These will make equal angles with the axes provided

$$\frac{\eta - y'}{\eta + y'} = -\frac{\xi - x'}{\xi + x'},$$

or

$$\xi\eta = x'y',$$

which is true since (x', y') and (ξ, η) are both on $2xy = c^2$.

18. The equation of the normal at (x', y') is

$$\frac{x - x'}{x'} = \frac{y - y'}{-y'}.$$

Put each fraction equal to λ; then, where the normal meets the curve, we have

$$(x' + \lambda x')^2 - (y' - \lambda y')^2 = a^2,$$

or

$$\lambda^2(x'^2 - y'^2) + 2\lambda(x'^2 + y'^2) = 0;$$

hence at the extremity of the chord $\lambda = 2\frac{x'^2 + y'^2}{y'^2 - x'^2}$. If therefore (ξ, η) be the middle point of the normal chord

$$2\xi = x'(2 + \lambda) = \frac{4x'y'^2}{x'^2 - y'^2},$$

and

$$2\eta = y'(2 - \lambda) = \frac{4x'^2y'}{x'^2 - y'^2}.$$

Hence $$(\eta^2 - \xi^2)^3 = 64x'^6y'^6a^{-6} = 4a^2\xi^2\eta^2.$$

19. We have $x(3x + 2y + 4) = 9$; hence the asymptotes are

$$x = 0 \text{ and } 3x + 2y + 4 = 0.$$

20. We have $(x - 2)(y - 3) = 6$; hence the asymptotes are

$$x - 2 = 0 \text{ and } y - 3 = 0.$$

The conjugate hyperbola is

$$(x - 2)(y - 3) = -6,$$

or

$$xy - 3x - 2y + 12 = 0.$$

21. It is well known that tangents to a conic subtend equal angles at a focus [Art. 165 (i) and Art. 228, Cor. 2]. Let then S, S' be the foci of the hyperbola, and P any point on the curve; and let the tangent at the vertex A, on the same branch as P, meet the tangent at P in T. Then

$$\tan\frac{1}{2}S'SP : \tan\frac{1}{2}SS'P' = \tan AST : \tan AS'T = S'A : SA.$$

22. Take the asymptotes for axes, and let the equation of the hyperbola be $xy=\lambda$. Then equation of any circle is

$$x^2+y^2+2xy\cos\omega+2gx+2fy+c=0.$$

The abscissae of the points of intersection of the circle and hyperbola are given by

$$x^2+\frac{\lambda^2}{x^2}+2\lambda\cos\omega+2gx+2f\cdot\frac{\lambda}{x}+c=0,$$

or
$$x^4+2gx^3+(c+2\lambda\cos\omega)x^2+2f\lambda x+\lambda^2=0.$$

Hence
$$x_1x_2x_3x_4=\lambda^2,$$
and similarly
$$y_1y_2y_3y_4=\lambda^2.$$

Therefore the product of the distances of the four points from the asymptotes, namely, $x_1x_2x_3x_4\sin^4\omega$ and $y_1y_2y_3y_4\sin^4\omega$, are equal.

23. Let the equation of the hyperbola be $xy=\lambda$, and the equation of the circle

$$x^2+y^2+2gx+2fy+c=0.$$

Then at the points of intersection of the two curves

$$x^4+\lambda^2+2gx^3+2f\lambda x+cx^2=0;$$

$$\therefore \frac{1}{4}(x_1+x_2+x_3+x_4)=-\frac{g}{2}\quad\text{.........................(i).}$$

Similarly
$$\frac{1}{4}(y_1+y_2+y_3+y_4)=-\frac{f}{2}\quad\text{.........................(ii).}$$

Now the centre of the circle is $(-g, -f)$ and the centre of mean position of the four points of intersection is the point

$$\left\{\frac{1}{4}(x_1+x_2+x_3+x_4),\ \frac{1}{4}(y_1+y_2+y_3+y_4)\right\}.$$

Hence, from (i) and (ii), the centre of mean position of the points of intersection is midway between the centres of the hyperbola and circle.

24. Let the four points be (x_1, y_1) &c., the equation of the hyperbola being $xy=\lambda$. Then, if the join of (x_1, y_1) and (x_2, y_2) be perpendicular to the join of (x_3, y_3) and (x_4, y_4), we have

$$(x_1-x_2)(x_3-x_4)+(y_1-y_2)(y_3-y_4)=0;$$

$$\therefore (x_1-x_2)(x_3-x_4)+\lambda^2\left(\frac{1}{x_1}-\frac{1}{x_2}\right)\left(\frac{1}{x_3}-\frac{1}{x_4}\right)=0.$$

Hence $$x_1x_2x_3x_4+\lambda^2=0.$$

Similarly $$y_1y_2y_3y_4+\lambda^2=0.$$

Hence $$\frac{y_1}{x_1}\cdot\frac{y_2}{x_2}\cdot\frac{y_3}{x_3}\cdot\frac{y_4}{x_4}=1,$$

that is $$\tan\alpha\tan\beta\tan\gamma\tan\delta=1.$$

25. The polar of (x', y') is

$$xx'/a^2-yy'/b^2=1.$$

If this touch the circle

$$x^2+y^2=a^2+b^2,$$

we have $$x'^2/a^4+y'^2/b^4=1/(a^2+b^2).$$

Hence the locus of the pole of chords which satisfy the given condition is the ellipse

$$x^2/a^4+y^2/b^4=1/(a^2+b^2).$$

26. The lines joining (x', y') to the two fixed points $(\pm a, 0)$ are

$$(yx'-xy')^2-a^2(y-y')^2=0.$$

These are parallel to

$$y^2(x'^2-a^2)-2x'y'xy+y'^2x^2=0,$$

the bisectors of which are $\dfrac{x^2-y^2}{y'^2-x'^2+a^2}+\dfrac{xy}{x'y'}=0.$

Since these bisectors are fixed lines, we have $\dfrac{y'^2-x'^2+a^2}{x'y'}=\text{constant}=k.$ Hence the equation of the locus of (x', y') is

$$x^2-y^2+kxy=a^2,$$

which represents a rectangular hyperbola since the straight lines

$$x^2-y^2+kxy=0$$

are at right angles.

27. See solution of 63, page 145.

28. If P be any one of the points of intersection; then, since the circles are equal, PA subtends equal angles at their circumferences; so that the angles PBA and PCA are equal.

Hence the difference of the angles CBP and BCP is constant.

Then see solution of question 15, page 44.

CHAPTER VIII.

Pages 176—178.

1. The tangents at α, β are

$$\frac{l}{r}=\cos\theta+\cos(\theta-\alpha)=2\cos\frac{\alpha}{2}\cos\left(\theta-\frac{\alpha}{2}\right),$$

$$\frac{l}{r}=\cos\theta+\cos(\theta-\beta)=2\cos\frac{\beta}{2}\cos\left(\theta-\frac{\beta}{2}\right).$$

Hence, comparing with $\frac{p}{r}=\cos(\theta-A)$, we see that the perpendiculars on the tangents make angles $\frac{\alpha}{2}$, $\frac{\beta}{2}$ respectively with the initial line; and therefore the angle between the tangents is $\frac{1}{2}(\beta-\alpha)$.

2. The angle between two tangents to a parabola being α, the difference of the vectorial angles of the points of contact will, from question 1, be equal to 2α.

Hence, if the equation of one tangent be

$$\frac{l}{r}=\cos\theta+\cos(\theta-\beta) \quad\text{..............................}\quad \text{(i)},$$

the equation of the other will be

$$\frac{l}{r}=\cos\theta+\cos(\theta-\beta-2\alpha) \quad\text{........................}\quad \text{(ii)}.$$

Now (i) and (ii) meet where

$$\theta-\beta=-(\theta-\beta-2\alpha), \text{ or } \theta=\alpha+\beta.$$

Hence, substituting for β in (i), we have

$$\frac{l}{r}=\cos\theta+\cos\alpha, \text{ or } \frac{l\sec\alpha}{r}=1+\sec\alpha\cos\theta.$$

Thus the locus required is a conic having the focus of the parabola for one focus, and the directrix of the parabola for corresponding directrix; also the latus-rectum is $l\sec\alpha$ and the eccentricity $\sec\alpha$.

3. Let the equation of the conic be $\frac{l}{r}=1+e\cos\theta$, and let the vectorial angle of P be α.

Then
$$\frac{l}{SP}=1+e\cos\alpha,\quad \frac{l}{SP'}=1+e\cos(\alpha+\pi),$$
$$\frac{l}{SQ}=1+e\cos\left(\alpha+\frac{\pi}{2}\right),\quad \frac{l}{SQ'}=1+e\cos\left(\alpha+\frac{3\pi}{2}\right).$$

Hence
$$\frac{l^2}{SP\,.\,SP'}+\frac{l^2}{SQ\,.\,SQ'}=1-e^2\cos^2\alpha$$
$$+1-e^2\sin^2\alpha=2-e^2.$$

4. Let α, β, γ be the vectorial angles of A, B, C respectively. Then the equations of the tangents are
$$\frac{l}{r}=\cos\theta+\cos(\theta-\alpha),\ \frac{l}{r}=\cos\theta+\cos(\theta-\beta),$$
and
$$\frac{l}{r}=\cos\theta+\cos(\theta-\gamma).$$

Hence at A',
$$\theta=\frac{1}{2}(\beta+\gamma),\ r=l/2\cos\frac{\beta}{2}\cos\frac{\gamma}{2};$$
and so for B', C'.

Hence
$$SA'\,.\,SB'\,.\,SC'=l^3/8\cos^2\frac{\alpha}{2}\cos^2\frac{\beta}{2}\cos^2\frac{\gamma}{2}.$$

Now A is the point $\theta=\alpha$, $r=l/2\cos^2\frac{\alpha}{2}$; and so for B', C. Hence
$$SA\,.\,SB\,.\,SC=l^3/8\cos^2\frac{\alpha}{2}\cos^2\frac{\beta}{2}\cos^2\frac{\gamma}{2}=SA'\,.\,SB'\,.\,SC'.$$

5. The tangent at α is
$$\frac{l}{r}=e\cos\theta+\cos(\theta-\alpha)$$
$$=(e+\cos\alpha)\cos\theta+\sin\alpha\sin\theta$$
$$=A\cos(\theta-B),$$
where
$$\tan B=\frac{\sin\alpha}{e+\cos\alpha}.$$

Hence the perpendicular from the focus on the tangent at α makes with the axis an angle $\tan^{-1}\frac{\sin\alpha}{e+\cos\alpha}$. So also the perpendicular from the focus on the tangent at $\alpha+\pi$ makes with the axis an angle $\tan^{-1}\frac{-\sin\alpha}{e-\cos\alpha}$.

Hence the angle between the tangents is
$$\tan^{-1}\frac{\sin\alpha}{e+\cos\alpha}-\tan^{-1}\frac{-\sin\alpha}{e-\cos\alpha}=\tan^{-1}\frac{2e\sin\alpha}{1-e^2}.$$

6. Let the point S' on the axis be at a distance c from S; let P be any point (r, θ) on the curve, and let $S'P=r'$.

Then
$$r'^2=r^2+c^2+2cr\cos\theta$$
$$=r^2+c^2+\frac{2c}{e}(l-r)$$
$$=\left(r\sim\frac{c}{e}\right)^2 \quad\text{..................... (i),}$$

provided
$$c^2+\frac{2cl}{e}=\frac{c^2}{e^2},$$

or
$$c=2le/(1-e^2).$$

Now when r is greatest it is equal to $\dfrac{l}{1-e}$, which is less than $\dfrac{2l}{1-e^2}$, that is less than $\dfrac{c}{e}$. Hence, from (i), we have
$$r'=\frac{c}{e}-r,$$

or
$$r+r'=\frac{c}{e}=\frac{2l}{1-e^2}=\text{constant.}$$

7. As in Art. 165 (3), the locus of the pole of the chord is the conic
$$\frac{l\sec\alpha}{r}=1+e\sec\alpha\,.\,\cos\theta \quad\text{..................... (i).}$$

The eccentricity of the conic (i) is $e\sec\alpha$; and therefore the conic is an ellipse, parabola or hyperbola according as $\cos\alpha >=< e$.

8. As in Art. 165 (3), the locus of the pole of PQ is a conic whose semi-latus-rectum is $l\sec 45^0=l\sqrt{2}$; also the envelope of PQ is a conic whose semi-latus-rectum is $l\cos 45^0=\dfrac{l}{\sqrt{2}}$.

9. Let P, Q be any two points on a conic, and let PQ cut the directrix in K and let T be its pole.

Then, if the vectorial angles of P, Q be $\alpha+\beta$ and $\alpha-\beta$ respectively, the equation of PQ will be
$$\frac{l}{r}=e\cos\theta+\sec\beta\cos(\theta-\alpha).$$

Hence PQ will meet the directrix
$$\frac{l}{r}=e\cos\theta \text{ where } \theta=\alpha+\frac{\pi}{2}.$$

Again, the tangents at P, Q will be
$$\frac{l}{r}=e\cos\theta+\cos(\theta-\alpha-\beta) \text{ and } \frac{l}{r}=e\cos\theta+\cos(\theta-\alpha-\beta).$$

Hence the tangents meet where $\theta = a$.

Thus the angle KST is a right angle.

Now suppose that the focus and directrix are given, and that T is a fixed point. Then, if SK be drawn perpendicular to ST to meet the directrix in K, the polar of T will pass through the *fixed point* K.

10. Let the equations of the two conics be

$$\frac{l}{r} = 1 + e\cos\theta \text{ and } \frac{l'}{r'} = 1 + e'\cos(\theta - a).$$

Transformed to Cartesian coordinates the equations become

$$(l - ex)^2 - x^2 - y^2 = 0 \quad\text{.......(i)},$$

and $$\{l' - e'(x\cos a + y\sin a)\}^2 - x^2 - y^2 = 0 \quad\text{.......(ii)}.$$

Now $$(l - ex)^2 - \{l' - e'(x\cos a + y\sin a)\}^2 = 0 \quad\text{.......(iii)}$$

clearly represents some curve through the intersection of the conics (i) and (ii), for (iii) must be true whenever (i) and (ii) are simultaneously true. But (iii) is a pair of straight lines whose equations are

$$(l - ex) \mp \{l' - e'(x\cos a + y\sin a)\} = 0,$$

or in polars

$$\frac{l}{r} - e\cos\theta \mp \left\{\frac{l'}{r} - e'\cos(\theta - a)\right\} = 0 \quad\text{.......(iv)}.$$

Hence two of the chords of intersection of the conics are represented by the equations (iv), and these lines clearly pass through the intersection of the directrices whose equations are

$$\frac{l}{r} - e\cos\theta = 0 \text{ and } \frac{l'}{r} = e'\cos(\theta - a).$$

11. Let the equations of the conics be

$$\frac{l}{r} = 1 + e\cos\theta \text{ and } \frac{l'}{r} = 1 + e'\cos(\theta - a).$$

Let $\theta = \beta$ be the straight line which cuts the parabolas; then the tangents at the four points P, P', Q, Q' are

$$\frac{l}{r} = e\cos\theta + \cos(\theta - \beta) \quad\text{.......(i)},$$

$$\frac{l}{r} = e\cos\theta + \cos(\theta - \beta - \pi) \quad\text{.......(ii)},$$

$$\frac{l'}{r} = e'\cos(\theta - a) + \cos(\theta - \beta) \quad\text{.......(iii)},$$

and $$\frac{l'}{r} = e'\cos(\theta - a) + \cos(\theta - \beta - \pi) \quad\text{.......(iv)}.$$

Now the tangents at P and Q, and also the tangents at P', Q', intersect on the line

$$\frac{l-l'}{r}=e\cos\theta-e'\cos(\theta-\alpha)\text{.......................(v)}$$

for all values of β.

And the tangents at P and Q', and also the tangents at P' and Q, intersect on the line

$$\frac{l+l'}{r}=e\cos\theta+e'\cos(\theta-\alpha)\text{.....................(vi)}$$

for all values of β.

The lines (v) and (vi) clearly pass through the intersection of the directrices.

If $e=e'$, the equations (v) and (vi) may be written

$$\frac{l-l'}{r}=2e\sin\frac{\alpha}{2}\sin\left(\theta-\frac{\alpha}{2}\right)\text{ and }\frac{l+l'}{r}=2e\cos\frac{\alpha}{2}\cos\left(\theta-\frac{\alpha}{2}\right):$$

the lines are therefore at right angles.

12. Let the equation of the parabola be $\frac{l}{r}=1+\cos\theta$. Let the vectorial angles of L, L', M, M' be α, $\pi+\alpha$, β, $\pi+\beta$ respectively.

Then the equations of the tangents at L, L' M, M' respectively are

$$l/r=\cos\theta+\cos(\theta-\alpha)\text{...........................(i)},$$

$$l/r=\cos\theta+\cos(\theta-\alpha-\pi)\text{.....................(ii)},$$

$$l/r=\cos\theta+\cos(\theta-\beta)\text{ ....(iii)},$$

and

$$l/r=\cos\theta+\cos(\theta-\beta-\pi)\text{ ....(iv)}.$$

Hence $\theta=\frac{1}{2}(\alpha+\beta)$ at N, $\theta=\frac{1}{2}(\alpha+\beta+\pi)$ at N',

$$\theta=\frac{1}{2}(\alpha+\beta+\pi)\text{ at }K'\text{ and }\theta=\frac{1}{2}(\alpha+\beta+2\pi)\text{ at }K.$$

Hence NK and $N'K'$ pass through the focus and are at right angles.

13. Let the equation of the fixed conic be $\frac{l}{r}=1+e\cos\theta$; and let the equation of the moving conic in any one of its possible positions be

$$\frac{l'}{r}=1+e'\cos(\theta-\alpha).$$

Then, from 10, two of the common chords are given by the equations

$$\frac{l\pm l'}{r}=e\cos\theta\pm e'\cos(\theta-\alpha),$$

or $$\frac{\frac{1}{e'}(l+l')}{r}=\frac{e}{e'}\cos\theta+\cos(\theta-\alpha)\quad\text{.........(i)},$$

and $$\frac{\frac{1}{e'}(l-l')}{r}=\frac{e}{e'}\cos\theta-\cos(\theta-\alpha)\quad\text{.........(ii)}.$$

Now (i) is the tangent at α to the fixed conic

$$\frac{\frac{1}{e'}(l+l')}{r}=1+\frac{e}{e'}\cos\theta,$$

and (ii) is the tangent at $\pi+\alpha$ to the fixed conic

$$\frac{\frac{1}{e'}(l-l')}{r}=1+\frac{e}{e'}\cos\theta.$$

14. Let the two tangents be

$$\frac{l}{r}=e\cos\theta+\cos(\theta-\alpha)\text{ and }\frac{l}{r}=e\cos\theta+\cos(\theta-\beta).$$

The tangents will meet where

$$\theta=\frac{1}{2}(\alpha+\beta)\text{ and }\frac{l}{r}=e\cos\theta+\cos\frac{1}{2}(\alpha-\beta)\quad\text{.........(i)}.$$

Writing the equations of the tangents in the forms

$$\frac{l}{r}=\cos\theta(e+\cos\alpha)+\sin\theta\sin\alpha,$$

and $$\frac{l}{r}=\cos\theta(e+\cos\beta)+\sin\theta\sin\beta,$$

we see that the condition of perpendicularity is

$$(e+\cos\alpha)(e+\cos\beta)+\sin\alpha\sin\beta=0,$$

or $$e^2+2e\cos\frac{1}{2}(\alpha+\beta)\cos\frac{1}{2}(\alpha-\beta)+2\cos^2\frac{1}{2}(\alpha-\beta)-1=0.$$

Hence, from (i),

$$e^2-1+2e\cos\theta\left(\frac{l}{r}-e\cos\theta\right)+2\left(\frac{l}{r}-e\cos\theta\right)^2=0,$$

or $$r^2(1-e^2)+2elr\cos\theta-2l^2=0.$$

15. We know that

$$\frac{1}{PS}+\frac{1}{QS}=\frac{2}{l},$$

and $$\frac{1}{PH}+\frac{1}{HR}=\frac{2}{l}.$$

Multiply by PS, PH respectively and add; then

$$\frac{PS}{SQ}+\frac{PH}{HR}=\frac{2}{l}(PS+PH)-2$$

= constant, since $PS+PH$ is constant.

16. Let d be the distance of the focus from the directrix; then the equations of the conics may be supposed to be

$$\frac{ed}{r}=1+e\cos\theta \text{ and } \frac{e'd}{r}=1+e'\cos(\theta-\alpha).$$

If the conics touch one another at some point whose vectorial angle is β, the equations

$$\frac{ed}{r}=e\cos\theta+\cos(\theta-\beta),$$

and

$$\frac{e'd}{r}=e'\cos(\theta-\alpha)+\cos(\theta-\beta)$$

will represent the same straight line.

Hence, by writing the equations in the forms

$$\frac{d}{r}=\cos\theta\left(1+\frac{\cos\beta}{e}\right)+\sin\theta\frac{\sin\beta}{e}.$$

and

$$\frac{d}{r}=\cos\theta\left(\cos\alpha+\frac{\cos\beta}{e'}\right)+\sin\theta\left(\sin\alpha+\frac{\sin\beta}{e'}\right),$$

we see that

$$1+\frac{\cos\beta}{e}=\cos\alpha+\frac{\cos\beta}{e'},$$

and

$$\frac{\sin\beta}{e}=\sin\alpha+\frac{\sin\beta}{e'}.$$

Hence $\cos\beta\left(\frac{1}{e}-\frac{1}{e'}\right)=\cos\alpha-1$ and $\sin\beta\left(\frac{1}{e}-\frac{1}{e'}\right)=\sin\alpha$;

and by eliminating β we have the required result, namely

$$2\sin\frac{\alpha}{2}=\frac{1}{e}\sim\frac{1}{e'}.$$

17 and 18. Let the equation of the conic be $\frac{l}{r}=1+e\cos\theta$. The equation of any circle of radius a which passes through the focus is $r=2a\cos(\theta-\alpha)$.

In the circle

$$(r-2a\cos\theta\cos\alpha)^2=4a^2\sin^2\alpha\sin^2\theta=4a^2\sin^2\alpha-4a^2\sin^2\alpha\cos^2\theta.$$

Hence, substituting for $\cos\theta$ from the equation of the conic, we have the following equation, which gives the focal distances of the points of intersection;

$$\{er^2-2a\cos\alpha\,(l-r)\}^2-4a^2e^2r^2\sin^2\alpha+4a^2(l-r)^2\sin^2\alpha=0\ldots\ldots\text{(i)}.$$

Hence, if r_1, r_2, r_3, r_4 be the four distances we have

$$r_1 \cdot r_2 \cdot r_3 \cdot r_4 = \text{coefficient of } r^0/\text{coefficient of } r^4 = 4a^2l^2/e^2,$$

which is constant if a be constant.

Also $$\frac{1}{r_1}+\frac{1}{r_2}+\frac{1}{r_3}+\frac{1}{r_4} = -\text{coefficient of } r/\text{coefficient of } r^0$$

$$= -(-8a^2l)/4a^2l^2$$

$$=\frac{2}{l}.$$

19. Let the parabola be

$$a/r = 1+\cos\theta,$$

and the given circle $$r = c\cos\theta.$$

Let the equation of any one of the conics be

$$\frac{l}{r} = 1 + e\cos(\theta-\beta).$$

The directrix of this conic is

$$\frac{l}{r} = e\cos(\theta-\beta),$$

and this is a tangent to the parabola. Hence, comparing $\frac{l}{r} = e\cos(\theta-\beta)$ with the general equation of the tangent to the parabola, namely

$$\frac{a}{r} = \cos\theta + \cos(\theta-\gamma),$$

or $$\frac{a}{r} = 2\cos\frac{\gamma}{2}\cos\left(\theta-\frac{\gamma}{2}\right),$$

we see that $\beta = \frac{\gamma}{2}$, and that $$ae = 2l\cos\beta \qquad \text{(i)}.$$

Now the focal distances of the points in which the circle is cut by the conic are given by eliminating θ between the two equations; we then have

$$\{c(l-r) - er^2\cos\beta\}^2 = c^2e^2r^2\sin^2\beta - e^2r^4\sin^2\beta.$$

Hence $$r_1+r_2+r_3+r_4 = -2ec\cos\beta/e^2$$

$$= -\frac{ac}{l}, \text{ from (i)}.$$

20. Let the equations of the conics be

$$\frac{l}{r} = 1 + e\cos\theta \text{ and } \frac{l'}{r} = 1 + e'\cos\theta.$$

Let the vectorial angles of P and Q be α and $\alpha \pm \frac{\pi}{2}$. Then the equations of the tangents at P and Q will be

$$\frac{l}{r} = e \cos \theta + \cos (\theta - \alpha),$$

and
$$\frac{l'}{r} = e' \cos \theta + \cos \left(\theta - \alpha \mp \frac{\pi}{2}\right).$$

Hence at their point of intersection

$$\left(\frac{l}{r} - e \cos \theta\right)^2 + \left(\frac{l'}{r} - e' \cos \theta\right)^2 = 1,$$

or in Cartesian co-ordinates

$$(l - ex)^2 + (l' - e'x)^2 = x^2 + y^2,$$

or
$$y^2 + (1 - e^2 - e'^2) \left\{x + \frac{el + e'l'}{1 - e^2 - e'^2}\right\}^2 = l^2 + l'^2 + \frac{(el + e'l')^2}{1 - e^2 - e'^2}.$$

Hence the locus is a conic whose centre is $\left(-\frac{el + e'l'}{1 - e^2 - e'^2}, 0\right)$, and the squares of whose axes are in the ratio $1 : 1 - e^2 - e'^2$; hence, if E be the eccentricity of the conic

$$1 - E^2 = 1 - e^2 - e'^2, \text{ or } E^2 = e^2 + e'^2.$$

21. Let one of the conics be $\frac{l}{r} = 1 + e \cos \theta$.

The tangent at α to this conic is

$$\frac{l}{r} = e \cos \theta + \cos (\theta - \alpha) = \cos \theta (\cos \alpha + e) + \sin \theta \sin \alpha.$$

Comparing this equation with the equation

$$\frac{p}{r} = \cos (\theta - A) = \cos \theta \cos A + \sin \theta \sin A,$$

we have, when $p = l$,

$$\cos A = \cos \alpha + e \text{ and } \sin A = \sin \alpha;$$

$$\therefore e + 2 \cos \alpha = 0.$$

The required locus will be found by eliminating e between the equations

$$\frac{l}{r} = 1 + e \cos \theta \text{ and } e + 2 \cos \theta = 0;$$

and the result is
$$l = -r \cos 2\theta.$$

22. Let the equation of the conic, referred to the focus S as pole, be

$$\frac{l}{r} = 1 + e \cos \theta.$$

Let $\alpha-\beta$ and $\alpha+\beta$ be the vectorial angles of P and P' respectively, and let O be (r_1, θ_1).

Then the equation of POP' is

$$\frac{l}{r}=e\cos\theta+\sec\beta\cos(\theta-\alpha).$$

But, since O is on the chord,

$$\frac{l}{r_1}=e\cos\theta_1+\sec\beta\cos(\theta_1-\alpha).$$

Hence $$\frac{\cos(\theta_1-\alpha)}{\cos\beta}=\frac{l}{r_1}-e\cos\theta_1=\text{constant}.$$

Hence also $$\frac{\cos(\theta_1-\alpha)-\cos\beta}{\cos(\theta_1-\alpha)+\cos\beta}=\text{constant},$$

and therefore $\tan\frac{1}{2}(\theta_1-\overline{\alpha-\beta})\,.\,\tan\frac{1}{2}(\theta_1-\overline{\alpha+\beta})$ is constant,

that is $$\tan\frac{1}{2}PSO\,.\,\tan\frac{1}{2}P'SO \text{ is constant.}$$

23. Let any one of the conics be

$$\frac{l}{r}=1+e\cos(\theta-\alpha)\quad\text{(i)},$$

where l is the given semi-latus-rectum.

Let the fixed confocal conic be

$$\frac{l'}{r}=1+e'\cos\theta\quad\text{(ii)}.$$

The directrix of (i) is

$$\frac{l}{r}=e\cos(\theta-\alpha)\quad\text{(iii)};$$

and, since the directrix of (i) touches the conic (ii), the equation (iii) will for some value of β represent the same line as

$$\frac{l'}{r}=e'\cos\theta+\cos(\theta-\beta).$$

Hence $$\frac{l'}{l}=\frac{e'+\cos\beta}{e\cos\alpha}=\frac{\sin\beta}{e\sin\alpha};$$

$$\therefore\ l\cos\beta=l'e\cos\alpha-le' \text{ and } l\sin\beta=l'e\sin\alpha;$$

$$\therefore\ l^2=l'^2e^2-2ll'ee'\cos\alpha+l^2e'^2\quad\text{(iv)}.$$

Now the conic (i) will touch the conic

$$\frac{L}{r}=1+E\cos\theta$$

at the point whose vectorial angle is ϕ provided

$$\frac{l}{r}=e\cos(\theta-\alpha)+\cos(\theta-\phi)$$

and
$$\frac{L}{r}=E\cos\theta+\cos(\theta-\phi)$$

represent the same straight line.

Hence we must have, for some value of ϕ,

$$\frac{L}{l}=\frac{E+\cos\phi}{e\cos\alpha+\cos\phi}=\frac{\sin\phi}{e\sin\alpha+\sin\phi},$$

or
$$(L-l)\cos\phi=El-eL\cos\alpha,$$

and
$$(L-l)\sin\phi=-eL\sin\alpha;$$

$$\therefore\ (L-l)^2=E^2l^2+e^2L^2-2el\,EL\cos\alpha \ldots\ldots\ldots\ldots\ldots\ldots(\text{v}).$$

Now comparing (iv) and (v) we see that the condition (v) will always be satisfied provided

$$\frac{(L-l)^2}{l^2}=\frac{L^2}{l'^2}=\frac{EL}{e'l'}=\frac{E^2}{e'^2};$$

$$\text{or } \frac{1}{L}=\frac{1}{l}\mp\frac{1}{l'},\ \text{and } E=\frac{e'L}{l'}.$$

[The question becomes obvious when reciprocated with respect to the common focus; for the reciprocal theorem is:—"Circles are described with constant radius and with their centres on a fixed circle; shew that the circles all touch two other circles whose radii are respectively the sum and the difference of the fixed and the moving circles."]

CHAPTER IX.

Pages 190—191.

1. (i) The centre is given by the equations

$$3x-\frac{5}{2}y+\frac{11}{2}=0, \quad \text{and} \quad -\frac{5}{2}x+6y-\frac{17}{2}=0,$$

whence $$x=-1, \quad y=1.$$

The equation when referred to parallel axes through the centre will be

$$3x^2-5xy+6y^2+\frac{11}{2}(-1)-\frac{17}{2}(1)+13=0,$$

or $$3x^2-5xy+6y^2-1=0.$$

(ii) The equations for finding the centre are

$$\frac{1}{2}y+\frac{3}{2}a=0, \quad \frac{1}{2}x-\frac{3}{2}a=0,$$

the centre is therefore $$(3a, \ -3a).$$

The equation when referred to parallel axes through the centre will be

$$xy+\frac{3}{2}a(3a)-\frac{3}{2}a(-3a)=0,$$

or $$xy+9a^2=0.$$

(iii) The centre is given by the equations

$$3x-\frac{7}{2}y+\frac{3}{2}=0 \quad \text{and} \quad -\frac{7}{2}x-6y-9=0,$$

whence $$x=-\frac{9}{11}, \quad y=-\frac{3}{11}.$$

The equation referred to parallel axes through the centre will therefore be

$$3x^2-7xy-6y^2+\frac{3}{2}\left(-\frac{9}{11}\right)-\frac{9}{2}\left(-\frac{3}{11}\right)+5=0,$$

or $$3x^2-7xy-6y^2+5=0.$$

2. (i) The equation may be written $(x+1)(y-2)=0$, and therefore represents the two straight lines $x+1=0$, $y-2=0$.

(ii) Writing the equation in the form

$$(y-a)^2=-4a\left(x-\frac{1}{4}a\right),$$

it is obvious that the equation represents a parabola of which the axis is

$$y-a=0,$$

the tangent at the vertex

$$x-\frac{1}{4}a=0,$$

and whose latus-rectum $=4a$, the parabola lying wholly on the *negative* side of the line

$$x-\frac{1}{4}a=0.$$

(iii) Writing the equation in the form

$$\left(y+\frac{a}{2}\right)^2=-a\left(x+\frac{3}{4}a\right),$$

it is obvious that the given equation represents a parabola of which the axis is

$$y+\frac{a}{2}=0,$$

the tangent at the vertex

$$x+\frac{3}{4}a=0,$$

and whose latus-rectum $=a$, the parabola lying entirely on the *negative* side of

$$x+\frac{3}{4}a=0.$$

(iv) Writing the equation in the form

$$\left(\frac{x+y}{\sqrt{2}}\right)^2=\frac{a}{\sqrt{2}}\left(\frac{x-y}{\sqrt{2}}\right),$$

we see that the given equation represents a parabola of which the axis is

$$x+y=0,$$

the tangent at the vertex $\quad x-y=0,$

and whose latus-rectum $=\frac{a}{\sqrt{2}}$, the parabola lying wholly on the positive side of the line

$$x-y=0.$$

(v) Writing the equation in the form

$$4\left(\frac{x+2y}{\sqrt{5}}\right)^2+\left(\frac{y-2x}{\sqrt{5}}\right)^2=a^2,$$

we see that the equation when referred to the two perpendicular lines

$$x+2y=0, \quad y-2x=0$$

as axes of y and x respectively would take the form

$$\frac{x^2}{\frac{a^2}{4}}+\frac{y^2}{a^2}=1.$$

Hence the given equation represents an ellipse whose axes are along the lines $x+2y=0$, $y-2x=0$, the lengths of the semi-axes being respectively $\frac{a}{2}$ and a.

(vi) The equation may be written

$$y^2-(x+a)^2=-a^2,$$

and therefore represents a rectangular hyperbola, whose centre is $(a, 0)$, and whose imaginary axis is along the axis of x.

3. (1) We have $\qquad (x-2a)(y+a)=-2a^2.$

Hence the equation represents a rectangular hyperbola whose asymptotes are $x-2a=0$, $y+a=0$, and which passes through the origin.

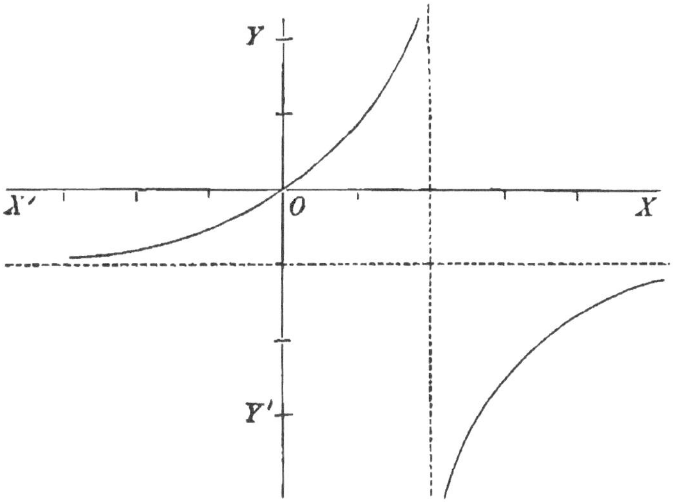

(2) We have $(x+y)^2=2x+1$, which may be written

$$(x+y+k)^2=2(1+k)x+2ky+k^2+1.$$

The lines $\qquad x+y+k=0$ and $2(1+k)x+2ky+k^2+1=0$

are at right angles, if $2+2k+2k=0$, or $k=-\frac{1}{2}$. Give k this value; then we have

$$\left(x+y-\frac{1}{2}\right)^2=x-y+\frac{5}{4};$$

$$\therefore \quad \left(\frac{x+y-\frac{1}{2}}{\sqrt{2}}\right)^2=\frac{1}{\sqrt{2}}\,\frac{x-y+\frac{5}{4}}{\sqrt{2}}.$$

Hence the given equation represents a parabola of which the axis is

$$x+y-\frac{1}{2}=0,$$

the tangent at the vertex

$$x-y+\frac{5}{4}=0,$$

and whose latus-rectum is $\frac{1}{\sqrt{2}}$, the curve lying altogether on the positive side of the line

$$x-y+\frac{5}{4}=0.$$

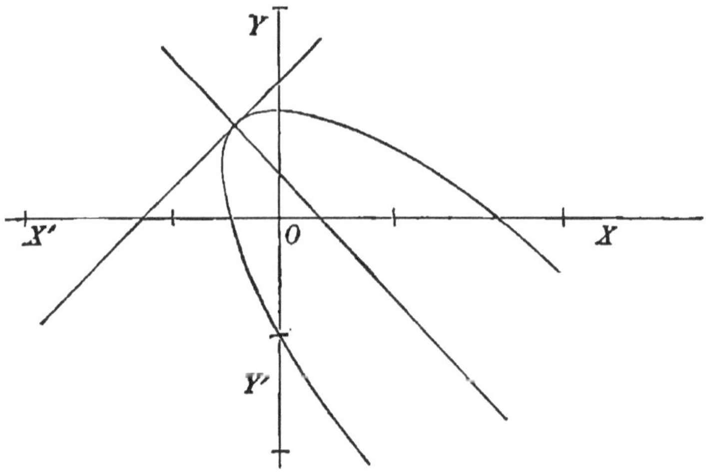

(3) The equations giving the centre are

$$2x+\frac{5}{2}y=0 \text{ and } \frac{5}{2}x+2y+\frac{3}{2}=0;$$

hence the centre is $\left(-\frac{5}{3}, \frac{4}{3}\right)$.

The equation when referred to parallel axes through the centre will be

$$2x^2+5xy+2y^2+\frac{3}{2}\left(\frac{4}{3}\right)-2=0,$$

that is

$$2x^2+5xy+2y^2=0.$$

Hence the given equation represents a pair of straight lines. These lines intersect in the point $\left(-\frac{5}{3}, \frac{4}{3}\right)$, and they cut the old axis of x where $x=\pm 1$; they can therefore be at once drawn.

(4) From Art. 171, the axes are the roots of the equation

$$\frac{1}{r^4}-\frac{2}{11}\frac{1}{r^2}+\frac{1}{121}-\frac{4}{121}=0;$$

$$\therefore r^2=\frac{11}{3} \text{ or } -11.$$

The equation of the real axis is given by the equation

$$\left(\frac{1}{11}-\frac{3}{11}\right)x+\frac{2}{11}y=0, \text{ or } x-y=0.$$

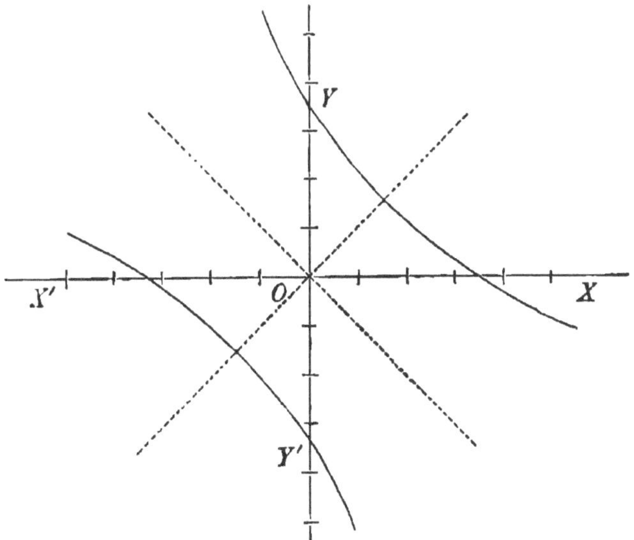

(5) The equation may be written

$$(2x+3y+\lambda)^2=(4\lambda-2)x+(6\lambda-2)y+\lambda^2-2.$$

The lines

$$2x+3y+\lambda=0 \text{ and } (4\lambda-2)x+(6\lambda-2)y+\lambda^2-2=0$$

are at right angles, if

$$8\lambda - 4 + 18\lambda - 6 = 0, \text{ or } \lambda = \frac{5}{13}.$$

The given equation is therefore equivalent to

$$\left(\frac{2x + 3y + \frac{5}{13}}{\sqrt{13}}\right)^2 = \frac{2}{13\sqrt{13}}\left\{\frac{-3x + 2y - \frac{313}{26}}{\sqrt{13}}\right\};$$

hence the equation represents a parabola of which the axis is

$$2x + 3y + \frac{5}{13} = 0,$$

the tangent at the vertex

$$-3x + 2y - \frac{313}{26} = 0,$$

and whose latus-rectum $= \frac{2}{13\sqrt{13}}$, the parabola lying wholly on the positive side of the line

$$-3x + 2y - \frac{313}{26} = 0.$$

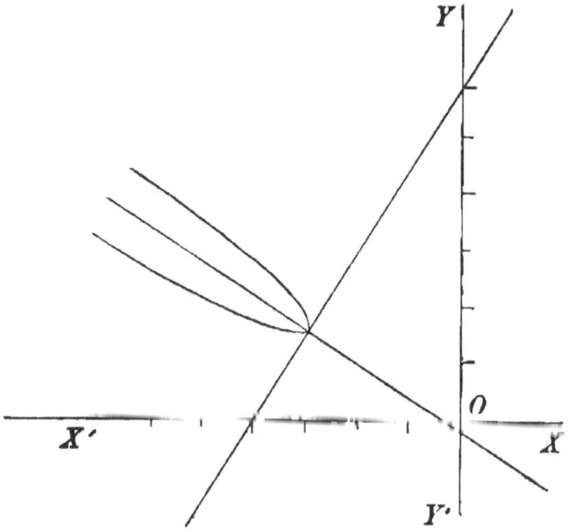

(6) The equations for finding the centre are

$$x - 2y + 5 = 0 \text{ and } -2x - 2y + 2 = 0.$$

Hence the centre is $(-1, 2)$, and the equation referred to parallel axes through the centre is

$$x^2 - 4xy - 2y^2 + 5(-1) + 2(2) = 0,$$

or

$$x^2 - 4xy - 2y^2 = 1.$$

The semi-axes are given by the equation

$$\frac{1}{r^4} + \frac{1}{r^2} - 2 - 4 = 0;$$

$$\therefore\ r^2 = \frac{1}{2} \text{ or } r^2 = -\frac{1}{3}.$$

Hence the equation represents an hyperbola whose real semi-axis $= \frac{1}{2}\sqrt{2}$ and is along the line whose equation is

$$(1-2)x - 2y = 0 \text{ or } x + 2y = 0.$$

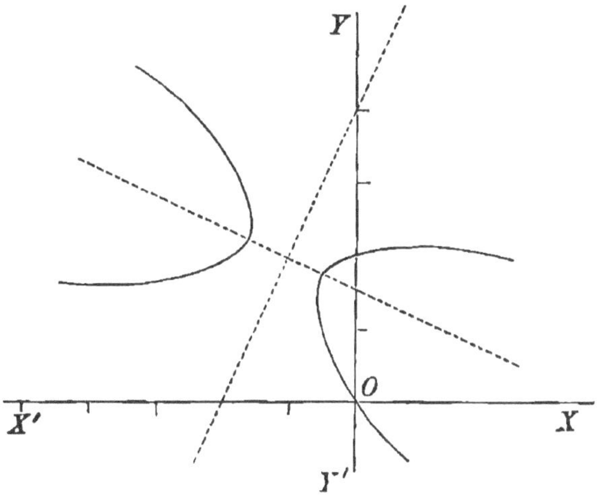

(7) The equations for the centre are

$$41x + 12y - 65a = 0 \text{ and } 12x + 9y - 30a = 0.$$

Hence the centre is $(a, 2a)$; and the equation referred to parallel axes through the centre will be

$$41x^2 + 24xy + 9y^2 = 9a^2.$$

The semi-axes are the roots of

$$\frac{1}{r^4} - \frac{50}{9a^2}\frac{1}{r^2} + \frac{9 \cdot 41 - 12^2}{81a^4} = 0;$$

$\therefore$ the squares of the semi-axes are $\frac{a^2}{5}$ and $\frac{9a^2}{5}$. Hence the curve represents an ellipse, whose major-axis is along the line whose equation is

$$\left(\frac{41}{9a^2}-\frac{5}{9a^2}\right)x+\frac{12}{9a^2}y=0 \quad \text{or} \quad 3x+y=0.$$

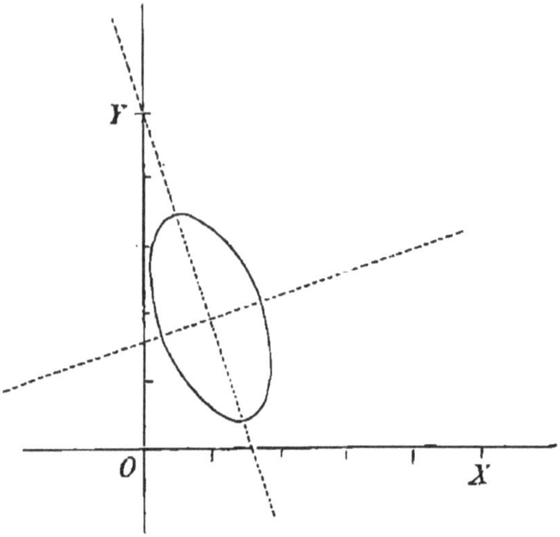

4. Take the two straight lines for axes; then if the equation be

$$ax^2+2hxy+by^2+2gx+2fy+c=0,$$

the conic will cut the axis of x where

$$ax^2+2gx+c=0\,;$$

and, since the two points are equidistant from the origin, we have $g=0$. Similarly, since the two points in which the conic cuts the axis of y are equidistant from the origin, we have $f=0$.

Hence the equation of the conic is of the form

$$ax^2+2hxy+by^2+c=0\,;$$

the origin is therefore the centre of the curve.

5. The equation may be written in the form

$$\frac{\left(\dfrac{x-2y+1}{\sqrt{5}}\right)^2}{2}+\frac{\left(\dfrac{2x+y-\frac{3}{2}}{\sqrt{5}}\right)^2}{\frac{1}{2}}=1.$$

Since the lines

$$x - 2y + 1 = 0 \text{ and } 2x + y - \frac{3}{2} = 0$$

are at right angles, the equation represents an ellipse whose semi-axes are $\sqrt{2}$ and $\frac{1}{2}\sqrt{2}$. Hence the product of the semi-axes $= 1$.

6. The centre will be found to be (2, 2).

The equation referred to parallel axes through the centre will be

$$x^2 - xy + 2y^2 = 1.$$

The product of the squares of the semi-axes is therefore [Art. 171 (iii)] $= 1 \Big/ \left(2 - \frac{1}{4}\right)$.

Hence the required product of the semi-axes $= \frac{2}{\sqrt{7}}$.

Referred to the centre, the polar equation of the axes is [Art. 167 (iii)]

$$\tan 2\theta = \frac{-1}{1 - 2} = 1;$$

$$\therefore \frac{2\tan\theta}{1 - \tan^2\theta} = \frac{2\frac{y}{x}}{1 - \frac{y^2}{x^2}} = 1.$$

Hence the equation of the axes when referred to the centre as origin is

$$x^2 - y^2 - 2xy = 0.$$

Hence the equation when referred to the original axes is

$$(x - 2)^2 - (y - 2)^2 - 2(x - 2)(y - 2) = 0,$$

or

$$x^2 - 2xy - y^2 + 8y - 8 = 0.$$

7. The condition is $\begin{vmatrix} 4, & \lambda, & -3, \\ \lambda, & -2, & 6, \\ -3, & 6, & -18, \end{vmatrix} = 0,$

whence $\lambda = 1$.

8. The conic is

$$(2x + 3y - 5)(5x + 3y - 8) = \lambda,$$

where λ has such a value that $(1, -1)$ is on the curve.

Hence $\lambda = (2 - 3 - 5)(5 - 3 - 8) = 36.$

Hence the required equation is

$$(2x + 3y - 5)(5x + 3y - 8) = 36.$$

9. The equation of the asymptotes is

$$3x^2 - 2xy - 5y^2 + 7x - 9y + c = 0,$$

where c is found from

$$\begin{vmatrix} 3, & -1, & \frac{7}{2} \\ -1, & -5, & -\frac{9}{2} \\ \frac{7}{2}, & -\frac{9}{2}, & c \end{vmatrix} = 0; \ \therefore c=2.$$

Thus the equation of the asymptotes of the given conic is

$$3x^2-2xy-5y^2+7x-9y+2=0.$$

The equation of any conic which has the given asymptotes is

$$3x^2-2xy-5y^2+7x-9y+2+\lambda=0.$$

If the conic pass through (2, 2) we have

$$12-8-20+14-18+2+\lambda=0, \text{ or } \lambda=18.$$

Hence the required equation is

$$3x^2-2xy-5y^2+7x-9y+20=0.$$

10. The equation of the asymptotes is

$$6x^2-7xy-3y^2-2x-8y-6+c=0,$$

where

$$\begin{vmatrix} 6, & -\frac{7}{2}, & -1 \\ -\frac{7}{2}, & -3, & -4 \\ -1, & -4, & -6+c \end{vmatrix} = 0, \text{ whence } c=2.$$

Hence the equation of the asymptotes is

$$6x^2-7xy-3y^2-2x-8y-4=0.$$

Since the equations of two conjugate hyperbolas differ from the equation of the asymptotes by constants which are equal and opposite to one another, the equation of the conjugate hyperbola must be

$$6x^2-7xy-3y^2-2x-8y-2=0.$$

11. From Art. 52 we have

$$a+b=a'+b',$$

$$ab-h^2=a'b'-h'^2.$$

Hence $$(a+b)^2-4(ab-h^2)=(a'+b')^2-4(a'b'-h'^2),$$

i.e. $$(a-b)^2+4h^2=(a'-b')^2+4h'^2.$$

12. If the axes be turned through an angle θ, we have [see Art. 167 (ii)]

$$g' = g\cos\theta + f\sin\theta \quad \text{and} \quad f' = -g\sin\theta + f\cos\theta;$$

$$\therefore \quad g'^2 + f'^2 = g^2 + f^2.$$

13. Take the line joining the centres of the circles for axis of x and the radical axis for axis of y, and let the equations of the circles be

$$x^2 + y^2 + 2ax + b = 0,$$

$$x^2 + y^2 + 2a'x + b = 0.$$

Let $lx + my = 1$ be the given straight line, and let (x', y') be any point on the line. Then the polars of (x', y') with respect to the two circles are

$$xx' + yy' + a(x + x') + b = 0$$

and
$$xx' + yy' + a'(x + x') + b = 0.$$

Hence the polars intersect where

$$x + x' = 0 \text{..(i)},$$

and
$$xx' + yy' + b = 0 \text{....................................(ii)}.$$

But (x', y') is on the given line, and therefore

$$lx' + my' = 1 \text{....................................(iii)}.$$

Eliminating x' and y' from the equations (i), (ii) and (iii) we have for the equation of the required locus

$$lxy - mx^2 + y + mb = 0.$$

The asymptotes of the locus are parallel to the lines

$$lxy - mx^2 = 0.$$

Hence one asymptote is parallel to $x = 0$ and is therefore perpendicular to the line joining the centres of the given circles; also the other asymptote is parallel to $ly - mx = 0$, and is therefore perpendicular to the given line.

14. Take the fixed point O for origin, and the diameter of the circle for the axis of x; then the equation of the circle will be

$$r = d\cos\theta \text{....................................(i)}.$$

Let the equation of the conic be

$$ax^2 + 2hxy + by^2 + 2gx + 2fy + c = 0,$$

or
$$r^2(a\cos^2\theta + 2h\sin\theta\cos\theta + b\sin^2\theta) + 2r(g\cos\theta + f\sin\theta) + c = 0 \text{...(ii)}.$$

Eliminating θ between (i) and (ii) we have

$$\{(a-b)r^4 + bd^2r^2 + 2gdr^2 + cd^2\}^2 = 4r^2(hr^2 + fd)^2(d^2 - r^2).$$

Hence
$$OP^2 \,.\, OQ^2 \,.\, OR^2 \,.\, OS^2 = \frac{c^2d^4}{(a-b)^2 + 4h^2}.$$

Now, since the origin is unaltered, c must be constant, and also

$$(a-b)^2+4h^2$$

by question 11. Hence $OP \,.\, OQ \,.\, OR \,.\, OS$ varies as the square of the radius.

15. The values of λ for which the expression

$$ax^2+2hxy+by^2+\lambda\,(Ax^2+2Hxy+By^2)$$

is a perfect square will be unaltered by any change of axes; and these values of λ are the roots of

$$(a+\lambda A)\,(b+\lambda B)-(h+\lambda H)^2=0,$$

that is, of $$ab-h^2+\lambda\,(aB+Ab-2hH)+\lambda^2\,(AB-H^2)=0.$$

Now [Art 52] $ab-h^2$ and $AB-H^2$ are unaltered by a change of rectangular axes [Art 52]; and hence

$$aB+bA-2hH$$

is unaltered by a change of rectangular axes.

Hence also $$(a+b)\,(A+B)-(aB+bA-2hH),$$

that is, $$aA+bB+2hH$$

is unaltered by any change of rectangular axes.

CHAPTER X.

Page 204.

1. Proceeding as in the previous chapter, it will be found that the centre of the conic is the point $(-1, 1)$; that the squares of the axes are 4 and -1; and that the *real* axis of the hyperbola lies along the line whose equation is

$$4x+3y=0.$$

The equation of the pair of tangents from (x', y') to the curve is

$$(11x^2+24xy+4y^2-2x+16y+11)\,(11x'^2+24x'y'+4y'^2-2x'+16y'+11)$$
$$=\{x\,(11x'+12y'-1)+y\,(12x'+4y'+8)-x'+8y'+11\}^2.$$

The necessary and sufficient condition that the lines may be at right angles is that the sum of the coefficients of x^2 and y^2 should be zero. Hence

$$15\,(11x'^2+24x'y'+4y'^2-2x'+16y'+11)$$
$$=(11x'+12y'-1)^2+(12x'+4y'+8)^2,$$

or $$100x'^2+100y'^2+200x'-200y'-100=0.$$

Hence (x', y') must be on the circle whose equation is

$$x^2+y^2+2x-2y-1=0.$$

2. The directrix of a parabola is the locus of the point of intersection of perpendicular tangents. Now the tangents from (x', y') to the curve are given by the equation

$$(x^2+2xy+y^2-4x+8y-6)\,(x'^2+2x'y'+y'^2-4x'+8y'-6)$$
$$-\{x\,(x'+y'-2)+y\,(x'+y'+4)-2x'+4y'-6\}^2=0.$$

These lines are at right angles if

$$2\,(x'^2+2x'y'+y'^2-4x'+8y'-6)-(x'+y'-2)^2-(x'+y'+4)^2=0,$$

or if $$-12x'+12y'-32=0.$$

Hence the locus of the point of intersection of perpendicular tangents is

$$3x-3y+8=0.$$

Pages 210—211.

1. It is easy to see that the sum of the coefficients of x^2 and y^2 is zero in either of the equations

$$h\{(ax+hy+g)^2-(hx+by+f)^2\}-(a-b)(ax+hy+g)(hx+by+f)=0,$$

and $$(ax+hy+g)^2-(hx+by+f)^2-(a-b)\,\phi(x,y)=0.$$

Hence the two conics represented by these equations are rectangular hyperbolas; but, from Art. 192 or Art. 193, these conics go through the four foci of the conic $\phi(x,y)=0$. Thus *two* conics through the foci of $\phi(x,y)=0$ are rectangular hyperbolas; and hence, from Art. 187, Ex. 1, *all* conics through the foci are rectangular hyperbolas.

Or thus:

It is easy to shew that the most general equation of a conic through the four points $\{\pm\sqrt{(a^2-b^2)},\,0\}$ and $\{0,\,\pm\sqrt{(b^2-a^2)}\}$ is

$$x^2+2hxy-y^2-a^2+b^2=0,$$

and this equation represents a rectangular hyperbola for all values of h.

2. The foci are given by

$$\frac{(ax+hy)^2-(hx+by)^2}{a-b}=\frac{(ax+hy)(hx+by)}{h}$$
$$=ax^2+2hxy+by^2-1.$$

Now $$(ax+hy)^2-(hx+by)^2=(a-b)(ax^2+2hxy+by^2-1)$$
is equivalent to
$$(h^2-ab)(x^2-y^2)=a-b.$$

Also $$(ax+hy)(hx+by)=h(ax^2+2hxy+by^2-1)$$
is equivalent to
$$(h^2-ab)\,xy=h.$$

Hence we have
$$\frac{x^2-y^2}{a-b}=\frac{xy}{h}=\frac{1}{h^2-ab}.$$

3. The foci are given by

$$\frac{(x-3y-1)^2-(-3x+y-1)^2}{1-1}=\frac{(x-3y-1)(-3x+y-1)}{-3}$$
$$=x^2-6xy+y^2-2x-2y+5.$$

Hence the equation of one conic on which the foci lie is

$$(x-3y-1)^2-(-3x+y-1)^2=0,\ \text{or}\ (x-y)(x+y+1)=0.$$

Another conic on which the foci lie is given by

$$(x-3y-1)(-3x+y-1)+3(x^2-6xy+y^2-2x-2y+5)=0\ldots\ldots\text{(i)}.$$

Hence the foci are the points where $x-y=0$ and $x+y+1=0$ cut the conic (i).

It will be found that $x-y=0$ cuts (i) in the points $(1, 1)$, $(-2, -2)$; and these points are the *real* foci. The two imaginary foci are the points of intersection of $x+y+1=0$ and (i).

4. The foci are the points of intersection of the conics given by the equations

$$\frac{(2x-4y)^2-(-4x-4y-2)^2}{2+4}=\frac{(2x-4y)(-4x-4y-2)}{-4}$$

$$=2x^2-8xy-4y^2-4y+1.$$

One conic through the four foci is given by

$$2(x-2y)^2-2(2x+2y+1)^2-3(x-2y)(2x+2y+1)=0,$$

or $$\{2(x-2y)+2x+2y+1\}\{(x-2y)-2(2x+2y+1)\}=0.$$

Hence the equations of the two axes are

$$4x-2y+1=0 \text{ and } 3x+6y+2=0.$$

The foci are the points where the axes cut the conic whose equation is

$$(x-2y)(2x+2y+1)=2x^2-8xy-4y^2-4y+1,$$

or $$6xy+x+2y-1=0 \quad\text{.............................(i)}.$$

Now it will be found that $4x-2y+1=0$ cuts (i) in the real points $\left(0, \frac{1}{2}\right)$ and $\left(-\frac{2}{3}, -\frac{5}{6}\right)$.

5. The foci of the conic are given by

$$\frac{(x+y-2)^2-(x+y+4)^2}{1-1}=\frac{(x+y-2)(x+y+4)}{1}$$

$$=x^2+2xy+y^2-4x+8y-6.$$

Hence $$x+y+1=0 \quad\text{.................................(i)}.$$

Also $$(x+y)^2+2(x+y)-8=(x+y)^2-4x+8y-6;$$

$$\therefore\ 6x-6y-2=0 \quad\text{.............................(ii)}.$$

The focus is therefore the point of intersection of (i) and (ii), namely, the point $\left(-\frac{1}{3}, -\frac{2}{3}\right)$.

6. The equation of any tangent to the ellipse is

$$x\cos\alpha+y\sin\alpha-\sqrt{(a^2\cos^2\alpha+b^2\sin^2\alpha)}=0.$$

The product of the perpendiculars from $\{0, \pm\sqrt{(b^2-a^2)}\}$ on the above tangent is

$$\{\sqrt{(b^2-a^2)}\sin\alpha-\sqrt{(a^2\cos^2\alpha+b^2\sin^2\alpha)}\}\{-\sqrt{(b^2-a^2)}\sin\alpha-\sqrt{(a^2\cos^2\alpha+b^2\sin^2\alpha)}\}$$

$$=a^2\cos^2\alpha+b^2\sin^2\alpha-(b^2-a^2)\sin^2\alpha=a^2.$$

7. The equation of any tangent is

$$x \cos \alpha + y \sin \alpha = \sqrt{(a^2 \cos^2 \alpha + b^2 \sin^2 \alpha)}.$$

The equation of the line through an imaginary focus perpendicular to the tangent is

$$x \sin \alpha - y \cos \alpha = \pm \sqrt{(b^2 - a^2)} \cos \alpha.$$

Square and add: then we have

$$x^2 + y^2 = b^2.$$

8. Let $S=0$ be the equation of the circle, and let $x - a = 0$ be the equation of the chord of contact of the ellipse and the circle; then the equation of the ellipse will be of the form

$$S - \lambda (x - a)^2 = 0.$$

Hence for any point (x, y) on the ellipse we have

$$S = \lambda (x - a)^2.$$

But S is equal to the square of the tangent from (x, y) to the circle, and $x - a$ is equal to the perpendicular distance of (x, y) from the chord of contact. Hence the tangent to the circle drawn from any point on the ellipse varies as the perpendicular distance of that point from the chord of contact.

Pages 219—230.

1. Let P be (x', y') and Q be (x'', y''), and let the equation of the conic be

$$ax^2 + by^2 + c = 0.$$

Then the equation of the polar of P is

$$ax'x + by'y + c = 0.$$

Hence the ratio of the perpendiculars from Q and C on the polar of P is

$$ax'x'' + by'y'' + c : c,$$

which is obviously equal to the ratio of the perpendiculars from P and C on the polar of Q.

2. Let TP, TP' be any two tangents to a conic, and let the normals at P, P' meet the axis in G, G' respectively: then we have to prove that

$$TP : TP' = PG : P'G'.$$

From Articles 125 and 131 it follows that

$$PG : P'G' = CD : CD',$$

where CD, CD' are the semi-diameters parallel to the tangents at P, P' respectively.

Also, from Art. 186, Cor. II.,

$$TP : TP' = CD : CD'.$$

Hence $$TP : TP' = PG : P'G'.$$

3. Draw through O the chords OQ, OQ' parallel to the axes of the conic; then QQ' is one of the chords which subtend a right angle at O, and it is obvious that QQ' is a diameter of the conic. Hence the fixed point, through which all chords which subtend a right angle at the point O pass, is the point where the normal at O is met by the diameter CQ, where OQ is parallel to an axis of the conic. The locus can now be easily found [see solution of question 15, page 139].

Again, to find the fixed point on the tangent at O through which any chord PQ passes which is such that OP, OQ are equally inclined to the normal at O: take OP, OQ indefinitely nearly coincident with the normal at O; then PQ will ultimately be the tangent at the other extremities of the normal chord, and therefore the locus required is the locus of the poles of normal chords of the conic which is found in Art. 138 (4).

4. Let the equation of the ellipse be $ax^2+by^2=1$, and let the chords make an angle θ with the axis.

Then, if O be (α, β), the equation of POQ will be

$$\frac{x-\alpha}{\cos\theta}=\frac{y-\beta}{\sin\theta}=r.$$

Hence OP, OQ are the roots of

$$a(\alpha+r\cos\theta)^2+b(\beta+r\sin\theta)^2=1;$$

$$\therefore\ r_1^2+r_2^2=\frac{4(a\alpha\cos\theta+b\beta\sin\theta)^2}{(a\cos^2\theta+b\sin^2\theta)^2}-2\frac{a\alpha^2+b\beta^2-1}{a\cos^2\theta+b\sin^2\theta}.$$

Hence if $OP^2+OQ^2=k^2$, the equation of the locus of O is

$$4(ax\cos\theta+by\sin\theta)^2-2(ax^2+by^2-1)(a\cos^2\theta+b\sin^2\theta)$$
$$=k^2(a\cos^2\theta+b\sin^2\theta)^2.$$

5. Take the fixed point for origin, and let the equation of the conic be

$$ax^2+2hxy+by^2+2gx+2fy+c=0.$$

Then, if OPP' make an angle θ with the axis of x, OP and OP' will be the roots of

$$r^2(a\cos^2\theta+2h\sin\theta\cos\theta+b\sin^2\theta)+2r(g\cos\theta+f\sin\theta)+c=0;$$

$$\therefore\ \frac{1}{OD^2}=\frac{1}{r_1^2}+\frac{1}{r_2^2}$$

$$=\frac{4}{c^2}(g\cos\theta+f\sin\theta)^2$$

$$-\frac{2}{c}(a\cos^2\theta+2h\sin\theta\cos\theta+b\sin^2\theta).$$

Hence the equation of the locus of D is

$$4(gx+fy)^2-2c(ax^2+2hxy+by^2)=c^2:$$

the locus is therefore a conic whose centre is O.

6. Let the equations of the conics be

$$\phi_1(x, y)=0 \text{ and } \phi_2(x, y)=0,$$

and let the parallel chords make an angle a with the axis of x.

Then, if O be (x', y'), we have from Art. 186

$$\frac{\phi_1(x', y')}{a_1\cos^2 a+2h_1\sin a\cos a+b_1\sin^2 a} : \frac{\phi_2(x', y')}{a_2\cos^2 a+2h_2\sin a\cos a+b_2\sin^2 a}=\text{const.}$$
$$=1:k.$$

Hence the equation of the locus of O is

$$k(a_2\cos^2 a+2h_2\sin a\cos a+b_2\sin^2 a)\,\phi_1(x, y)$$
$$=(a_1\cos^2 a+2h_1\sin a\cos a+b_1\sin^2 a)\,\phi_2(x, y),$$

so that the locus of O is a conic through the four points of intersection of the given conics.

7. Take any two perpendicular lines through O for axes, and let the equation of the conic be

$$ax^2+2hxy+by^2+2gx+2fy+c=0.$$

Let the lines POP', QOQ' make angles θ and $\theta+\frac{\pi}{2}$ respectively with the axis of x.

Then OP, OP' are the two values of r given by

$$r^2(a\cos^2\theta+2h\sin\theta\cos\theta+b\sin^2\theta)+2r(g\cos\theta+f\sin\theta)+c=0;$$

$$\therefore \frac{1}{OP\,.\,OP'}=\frac{1}{c}(a\cos^2\theta+2h\cos\theta\sin\theta+b\sin^2\theta).$$

So also $$\frac{1}{OQ\,.\,OQ'}=\frac{1}{c}(a\sin^2\theta-2h\cos\theta\sin\theta+b\cos^2\theta).$$

Hence $$\frac{1}{OP\,.\,OP'}+\frac{1}{OQ\,.\,OQ'}=\frac{a+b}{c},$$

which is independent of θ.

8. The equation of any line through the point $(ac, 0)$ is

$$\frac{x-ac}{\cos\theta}=\frac{y}{\sin\theta}=r.$$

This line will cut the ellipse $x^2/a^2+y^2/b^2=1$ where

$$(ac+r\cos\theta)^2/a^2+r^2\sin^2\theta/b^2=1.$$

Hence $$\frac{1}{r_1}+\frac{1}{r_2}=-\frac{2c\cos\theta}{a(c^2-1)},$$

and $$\frac{1}{r_1r_2}=\frac{\cos^2\theta/a^2+\sin^2\theta/b^2}{c^2-1}.$$

Hence $$\frac{1}{r_1^2}+\frac{1}{r_2^2}=\frac{\frac{4c^2}{a^2}\cos^2\theta-2(c^2-1)\left(\frac{\cos^2\theta}{a^2}+\frac{\sin^2\theta}{b^2}\right)}{(c^2-1)^2}$$

$$=\frac{2}{a^2(c^2-1)^2}\left\{(c^2+1)\cos^2\theta-\frac{c^2-1}{1-e^2}\sin^2\theta\right\}.$$

Hence $\frac{1}{r_1^2}+\frac{1}{r_2^2}$ will be constant for all values of θ provided $c^2+1=\frac{1-c^2}{1-e^2}$, or $a^2c^2=\frac{a^2e^2}{2-e^2}=a^2\frac{a^2-b^2}{a^2+b^2}$.

9. Since the conic is a rectangular hyperbola, and PP' is perpendicular to AA', it follows from Art. 187, Ex. 1 that PA is perpendicular to $A'P'$, and therefore PA and $A'P'$ will meet on the fixed circle whose diameter is AA'.

The second part is question 14, page 164.

10. Let $x\cos\alpha+y\sin\alpha-p=0$ be the equation of the fixed straight line.

Let (x_1, y_1) and (x_2, y_2) be the extremities of any focal chord; then we know that

$$x_1x_2=a^2 \text{ and } y_1y_2=-4a^2\text{.........................(i).}$$

Then $$\frac{PM}{PS}+\frac{P'M'}{P'S}=\frac{x_1\cos\alpha+y_1\sin\alpha-p}{x_1+a}+\frac{x_2\cos\alpha+y_2\sin\alpha-p}{x_2+a}$$

$$=\frac{x_1\cos\alpha+y_1\sin\alpha-p}{x_1+a}+\frac{a^2\cos\alpha-ay_1\sin\alpha-px_1}{a^2+ax_1}, \text{ from (i)}$$

$$=(a\cos\alpha-p)/a=\text{constant}.$$

11. Take the fixed point for origin, and the axis of x through the centre of the given circle, and let the equation of the circle be

$$x^2+y^2+2gx+c=0.$$

Let the line $y=mx$ meet the circle in the points (x_1, y_1), (x_2, y_2); then it is easily seen that

$$x_1+x_2=-\frac{2g}{1+m^2},\ x_1x_2=\frac{c}{1+m^2}$$

and $$y_1+y_2=-\frac{2mg}{1+m^2},\ y_1y_2=\frac{cm^2}{1+m^2}.$$

Hence the equation of the circle of which (x_1, y_1), (x_2, y_2) are extremities of a diameter is

$$x^2+y^2+\frac{2g}{1+m^2}x+\frac{2mg}{1+m^2}y+c=0.$$

The polar of the origin with reference to the circle is therefore

$$gx+mgy+c(1+m^2)=0,$$

which may be written $$y=-\frac{1}{m}\left(x+\frac{c}{g}\right)-\frac{mc}{g},$$
from which it is obvious that the polar always touches the parabola
$$y^2=\frac{4c}{g}\left(x+\frac{c}{g}\right).$$

12. Take the fixed diameter and its conjugate for axes, and let the equation of the conic be $ax^2+by^2=1$.

Let the fixed point P be (x', y'), and let PQ, PR be the straight lines through P; then, if (x'', y'') be the pole of QR, the equation of QR will be
$$ax''x+by''y-1=0,$$
and the equation of *any conic* touching the given conic at P and passing through Q, R will be
$$ax^2+by^2-1+\lambda(ax'x+by'y-1)(ax''x+by''y-1)=0\text{.........} \quad \text{(i)}.$$
The equation (i) will represent the straight lines PQ, PR if λ be properly chosen.

If the conic (i) cuts $y=0$ in two points equidistant from the centre the coefficient of x must be zero, and therefore $x'+x''=0$, for all values of λ. Thus the pole of QR is on the fixed straight line whose equation is $x+x'=0$.

13. The equation of any chord which passes through the *fixed point* $(c, 0)$ is $y=m(x-c)$; and where the chord cuts the ellipse we have
$$\frac{1}{a^2}\left(\frac{y+mc}{m}\right)^2+\frac{y^2}{b^2}=1;$$
$$\therefore\ y'y''=\frac{\dfrac{c^2}{a^2}-1}{\dfrac{1}{b^2}+\dfrac{1}{a^2m^2}}=\frac{b^2(c^2-a^2)m^2}{b^2+a^2m^2}\text{....................} \quad \text{(i)}.$$
Also $$\frac{x^2}{a^2}+\frac{m^2(x-c)^2}{b^2}=1;$$
$$\therefore\ \bar{x}=\frac{1}{2}(x'+x'')=\frac{\dfrac{m^2c}{b^2}}{\dfrac{1}{a^2}+\dfrac{m^2}{b^2}}=\frac{a^2cm^2}{b^2+a^2m^2}\text{....................} \quad \text{(ii)}.$$
From (i) and (ii) it is obvious that $y'y''\propto\bar{x}$.

The centre of a parabola is at infinity and therefore $\bar{x}$ is constant, so that in the parabola $y'y''$ is constant.

14. Let d be the distance of the points S, H from the centre; then, if α, θ_1, θ_2 be the eccentric angles of P, Q, Q' respectively, we have (as in example 22, page 140)
$$\tan\frac{\alpha}{2}\tan\frac{\theta_1}{2}=\frac{d-a}{d+a} \text{ and } \tan\frac{\alpha}{2}\tan\frac{\theta_2}{2}=\frac{d+a}{d-a}.$$

$$\therefore \frac{1-\cos\theta_1}{1+\cos\theta_1}\,\frac{1+\cos\theta_2}{1-\cos\theta_2}=\left(\frac{d-a}{d+a}\right)^4.$$

Hence, if R be (x, y), we have

$$\frac{a-x}{a+x}\cdot\frac{a+y}{a-y}=\left(\frac{d-a}{d+a}\right)^4,$$

which is of the form

$$Axy+Bx+Cy+D=0,$$

the locus of R is therefore a rectangular hyperbola whose asymptotes are parallel to the axes of the conic.

15. Let $\alpha, \theta_1, \theta_2$ be the eccentric angles of P, Q, Q' respectively; then we have to shew that $\tan\alpha \propto \tan\frac{1}{2}(\theta_1+\theta_2)$.

Now, as in example 14, we have

$$\tan\frac{\alpha}{2}\tan\frac{\theta_1}{2}=\frac{d-a}{d+a} \text{ and } \tan\frac{\alpha}{2}\tan\frac{\theta_2}{2}=\frac{d+a}{d-a}.$$

Hence
$$\frac{\tan\frac{1}{2}\theta_1+\tan\frac{1}{2}\theta_2}{1-\tan\frac{1}{2}\theta_1\,.\,\tan\frac{1}{2}\theta_2}=\frac{\frac{2(d^2+a^2)}{d^2-a^2}\cot\frac{\alpha}{2}}{1-\cot^2\frac{\alpha}{2}}=\tan\alpha\,.\,\frac{d^2+a^2}{a^2-d^2}:$$

$$\therefore \tan\frac{1}{2}(\theta_1+\theta_2)\propto\tan\alpha, \text{ since } d \text{ is constant.}$$

16. Let the eccentric angles of P, P' be θ, θ'; then the equation of PP' will be

$$\frac{x}{a}\cos\frac{1}{2}(\theta+\theta')+\frac{y}{b}\sin\frac{1}{2}(\theta+\theta')=\cos\frac{1}{2}(\theta-\theta').$$

The equations of PS, $P'S'$ are respectively

$$\frac{x-ae}{ae-a\cos\theta}=\frac{y}{b\sin\theta} \text{ and } \frac{x+ae}{-ae-a\cos\theta'}=\frac{y}{b\sin\theta'}.$$

Hence
$$\frac{\cos\theta-e}{\cos\theta'+e}=\frac{\sin\theta}{\sin\theta'};$$

$$\therefore \sin\frac{1}{2}(\theta-\theta')+e\sin\frac{1}{2}(\theta+\theta')=0 \text{.......................} \text{(i)}.$$

Now
$$\frac{AC^4}{CU^2}+\frac{BC^4}{CV^2}=\frac{a^2\cos^2\frac{1}{2}(\theta+\theta')+b^2\sin^2\frac{1}{2}(\theta+\theta')}{\cos^2\frac{1}{2}(\theta-\theta')}$$

$$=\frac{a^2-a^2e^2\sin^2\frac{1}{2}(\theta+\theta')}{1-e^2\sin^2\frac{1}{2}(\theta+\theta')}, \text{ from (i)}$$

$$=a^2.$$

17. Let the equation of the ellipse be $ax^2+by^2=1$; then the equation of the two tangents through the point (x', y') is

$$(ax^2+by^2-1)(ax'^2+by'^2-1)-(ax'x+by'y-1)^2=0.$$

The equation of any conic through the four points where the tangents cut the axes is given by

$$(ax^2+by^2-1)(ax'^2+by'^2-1)-(ax'x+by'y-1)^2+\lambda xy=0 \text{.........(i).}$$

Now the condition that (i) should represent a circle for some value of λ is

$$a(ax'^2+by'^2-1)-a^2x'^2=b(ax'^2+by'^2-1)-b^2y'^2,$$

or

$$x'^2-y'^2=\frac{1}{a}-\frac{1}{b}.$$

Hence the locus of (x', y') is a rectangular hyperbola.

18. The directions of the two tangents to $x^2/a^2+y^2/b^2=1$ through the point (x, y) are given by the equation

$$y=mx+\sqrt{a^2m^2+b^2},$$

or

$$m^2(x^2-a^2)-2mxy+y^2-b^2=0 \text{........................(i).}$$

If one tangent makes the same angle with the major axis that the other tangent makes with the minor, we must have $m_1m_2=1$, where m_1, m_2 are the roots of (i).

Hence $(y^2-b^2)/(x^2-a^2)=1$, or $x^2-y^2=a^2-b^2$.

Hence the required locus is a rectangular hyperbola which passes through the foci.

19. Take the given diameter and its conjugate for axes and let the equation of the conic be $ax^2+by^2=1$.

The equation of the tangents from (x', y') is

$$(ax^2+by^2-1)(ax'^2+by'^2-1)-(ax'x+by'y-1)^2=0.$$

The points where the tangents meet the axis of x are therefore given by

$$(ax^2-1)(ax'^2+by'^2-1)-(ax'x-1)^2=0.$$

Now (i) if the *sum* of the distances of the points of intersection from the centre be constant and equal to $2k$, we have

$$2k=-\frac{2ax'}{a(by'^2-1)}; \quad \therefore kby'^2+x'-k=0,$$

so that the locus of (x', y') is a parabola.

Again, (ii), if the *product* of the distances from the centre is constant and equal to c^2, we have

$$c^2=-\frac{ax'^2+by'^2}{a(by'^2-1)},$$

and therefore the locus of (x', y') is a co-axial conic.

And (iii), if the *sum of the reciprocals* of the distances from the centre be constant and equal to $\frac{2}{d}$, we have

$$\frac{2}{d}=\frac{2ax'}{ax'^2+by'^2},$$

the locus of (x', y') is therefore a conic.

20. Take O for origin, and the chord and its conjugate for axes; then the equation of the conic will be

$$ax^2+by^2+2fy+c=0.$$

The equation of the tangents from (x', y') is

$$(ax^2+by^2+2fy+c)(ax'^2+by'^2+2fy'+c)-\{axx'+byy'+f(y+y')+c\}^2=0.$$

The tangents meet AB in points whose abscissae are given by

$$(ax^2+c)(ax'^2+by'^2+2fy'+c)-(axx'+fy'+c)^2=0 \text{............(i)}.$$

In order that AS may be equal to BT it is necessary and sufficient that S, T should be equidistant from O, the condition for which is, from (i),

$$fy'+c=0.$$

But this condition is satisfied if (x', y') be on the polar of O.

Or thus:—Let the tangents at P and Q meet in K; then K is on the polar of O, and if KL be the polar of O, KL is parallel to AB, since O is the middle point of AB. If QOP meet the polar of O in L, $K\{QOPL\}$ is harmonic [Art. 181], and therefore $\{TOS\infty\}$ is harmonic, whence $TO=OS$, and therefore $AS=BT$.

21. The equation of the tangents to $ax^2+\beta y^2-1=0$ from (x', y') is

$$(ax^2+\beta y^2-1)(ax'^2+\beta y'^2-1)-(ax'x+\beta y'y-1)^2=0.$$

These are parallel to

$$a(\beta y'^2-1)x^2+\beta(ax'^2-1)y^2-2a\beta x'y'xy=0 \text{ .... (i)}.$$

The lines given by (i) are conjugate diameters of the conic

$$ax^2+2hxy+by^2=1,$$

if $$a\beta(ax'^2-1)+ba(\beta y'^2-1)+2ha\beta x'y'=0.$$

Hence the locus of (x', y') is the conic

$$ax^2+by^2+2hxy=\frac{a}{a}+\frac{b}{\beta}.$$

22. Take the tangent and normal at Q for axes, and let the equation of the conic be

$$ax^2+2hxy+by^2+2fy=0.$$

Let P be the point (x', y'); then the equation of PT, PT' is

$$(ax^2+2hxy+by^2+2fy)(ax'^2+2hx'y'+by'^2+2fy') - \{ax'x+h(xy'+x'y)+by'y+f(y+y')\}^2=0.$$

Hence QT, QT' are the roots of

$$ax^2(ax'^2+2hx'y'+by'^2+2fy')-(ax'x+hy'x+fy')^2=0.$$

Hence

$$QT \cdot QT' = \frac{-f^2y'}{(ab-h^2)y'+2af},$$

and
$$QT^2+QT'^2=\frac{4f^2(ax'+hy')^2}{\{(ab-h^2)y'+2af\}^2}+\frac{2f^2y'}{(ab-h^2)y'+2af}.$$

Hence (i) when $QT^2+QT'^2$ is constant, the locus of (x', y') is a conic.

Also (ii) when $QT \cdot QT'$ is constant the locus is a straight line.

23. Let the equation of the ellipse be

$$x^2/a^2+y^2/b^2=1,$$

and let T, (x', y'), be the point of intersection of the tangents which pass through P, P'. Then the equations of the tangents from T is

$$\{x^2/a^2+y^2/b^2-1\}\{x'^2/a^2+y'^2/b^2-1\}-\{x'x/a^2+y'y/b^2-1\}^2=0.$$

The ordinates of the two points where these tangents cut $x=a$ are given by the equation

$$\frac{y^2}{b^2}\left(\frac{x'^2}{a^2}-1\right)-2\frac{yy'}{b^2}\left(\frac{x'}{a}-1\right)-\left(\frac{x'}{a}-1\right)^2=0,$$

or
$$y^2\left(\frac{x'}{a}+1\right)-2yy'-\left(\frac{x'}{a}-1\right)b^2=0.$$

Since P, P' are equidistant from O, the sum of the ordinates of P and P' is equal to twice the ordinate of $O=2c$ suppose.

Hence we have

$$2c=2y' \Big/ \left(\frac{x'}{a}+1\right).$$

Hence the locus of T is the straight line whose equation is

$$c(x+a)-ay=0.$$

24. Take the intersection of the diagonals of the square for origin, and axes parallel to the sides of the square; then the equations of the circles are

$$x^2+y^2=a^2 \text{ and } x^2+y^2=2a^2.$$

The tangents from (x', y') to the inscribed circle are given by

$$(x^2+y^2-a^2)(x'^2+y'^2-a^2)-(xx'+yy'-a^2)^2=0.$$

Now the sum of the coefficients of x^2 and y^2 in the above equation is equal to

$$x'^2+y'^2-2a^2,$$

which is zero if (x', y') be on the outer circle.

Hence the tangents to the inner circle from any point on the outer are at right angles; also the diagonals of the square are at right angles; and hence, from Art. 187, Ex. 1, *all conics* through the four points in which the diagonals are cut by the tangents are rectangular hyperbolas.

25. Take the diameter parallel to the fixed straight line for axis of x, and let the equation of the conic be

$$ax^2+by^2-1=0.$$

Let the fixed straight line be $y=k$, and let $2l$ be the length of the intercept on $y=k$ made by a pair of tangents.

The tangents from (x', y') to the conic are given by

$$(ax^2+by^2-1)\,(ax'^2+by'^2-1)-(ax'x+by'y-1)^2=0.$$

These tangents cut $y=k$, where

$$(ax^2+bk^2-1)\,(ax'^2+by'^2-1)-(ax'x+by'k-1)^2=0;$$

and since the difference of the abscissae is $2l$, we have

$$l^2=\{x'(by'k-1)/(by'^2-1)\}^2-\{(bk^2-1)(ax'^2+by'^2-1)-(by'k-1)^2\}/a\,(by'^2-1).$$

Hence the locus of (x', y') is the curve of the fourth degree whose equation is

$$al^2(by^2-1)^2=ax^2(bky-1)^2-(bk^2-1)(by^2-1)(ax^2+by^2-1)+(by^2-1)(bky-1)^2.$$

26. Take the fixed point O for origin, and axes parallel and perpendicular to MN. Let the equation of MN be $y=k$, and let the equation of the conic be

$$ax^2+2hxy+by^2+2gx+2fg+c=0.$$

Then the equation of the tangents from (x', y') to the conic is

$$(ax^2+2hxy+by^2+2gx+2fy+c)\,\phi\,(x', y')-\{ax'x+h\,(xy'+x'y) \\ +byy'+g\,(x+x')+f\,(y+y')+c\}^2=0.$$

Hence the equation of the lines OP, OQ is

$$(ax^2+2hxy+by^2+2gxy/k+2fy^2/k+cy^2/k^2)\,\phi\,(x', y')-\{ax'x+h\,(xy'+x'y) \\ +by'y+gx+fy+(gx'+fy'+c)\,y/k\}^2=0.$$

Hence, if the lines OP, OQ are at right angles, we have

$$\{k^2\,(a+b)+2fk+c\}\,\phi\,(x', y')-k^2\,(ax'+hy'+g)^2-\{k\,(hx'+by'+f) \\ +(gx'+fy'+c)\}^2=0,$$

the locus of (x', y') is therefore a conic.

27. Take the fixed diameter for the axis of x, and let (x', y') be the point from which the tangents are drawn. Then the equation of the tangents is

$$(x^2+y^2-a^2)(x'^2+y'^2-a^2)-(xx'+yy'-a^2)^2=0.$$

Then meet $x=0$ in points given by

$$(y^2-a^2)(x'^2+y'^2-a^2)-(yy'-a^2)=0.$$

Hence the ordinate of the middle point of the intercept is

$$-a^2y'/(x'^2-a^2).$$

Again, the equations of the lines joining (x', y') to $(a, 0)$, $(-a, 0)$ are respectively

$$\frac{x-x'}{x'-a}=\frac{y-y'}{y'} \text{ and } \frac{x-x'}{x'+a}=\frac{y-y'}{y'}.$$

Hence the middle point of the intercept made by these lines on $x=0$ is

$$\frac{1}{2}\left\{y'-\frac{x'y'}{x'-a}+y'-\frac{x'y'}{x'+a}\right\}=-\frac{a^2y'}{x'^2-a^2}.$$

Thus the intercepts have the same middle point, whence the proposition follows at once.

28. Take the tangent and normal at P for axes, and let the equation of the conic be

$$ax^2+2hxy+by^2+2fy=0.$$

Then the centre of the conic is easily found to be

$$\left(\frac{hf}{ab-h^2}, \frac{-af}{ab-h^2}\right),$$

and the other extremity of the diameter through P is therefore

$$\left(\frac{2hf}{ab-h^2}, \frac{-2af}{ab-h^2}\right).$$

Let (x', y') be one angular point of the triangle, then the equation of the tangents through (x', y') is

$$(ax^2+2hxy+by^2+2fy)(ax'^2+2hx'y'+by'^2+2fy')-\{ax'x+h(xy'+x'y)+byy'$$
$$+f(y+y')\}^2=0.$$

And, if the extremities of the base be equidistant from the centre of the conic, the abscissa of the middle point of the base is the same as the abscissa of the centre of the conic.

Hence
$$\frac{hf}{ab-h^2}=\frac{(ax'+hy')f}{(ab-h^2)y'+2af},$$

whence
$$x'=2hf/(ab-h^2).$$

Hence the locus of (x', y') is perpendicular to the tangent at P and passes through the other end of the diameter through P.

29. The two tangents to the parabola being at right angles the directrix will pass through their point of intersection O. Also, if S be the focus, the two tangents from O will bisect the angles between the directrix and OS. [In the figure to Art. 98, the tangent RP bisects the angle SRM, and the other tangent from R will bisect the angle SRO.]

Let SK be the perpendicular from S on the directrix; then $SK=2a$.

And, if XOS be θ, KOS will be 2θ, or $\pi-2\theta$. Therefore $OS\sin 2\theta=2a$.

Let P be any fixed point on the axis, and let $SP=c$; then, if x, y are the co-ordinates of P, we have

$$x=OS\cos\theta+c\cos\left(\frac{\pi}{2}-\theta\right)=\frac{a}{\sin\theta}+c\sin\theta,$$

and $$y=OS\sin\theta+c\sin\left(\frac{\pi}{2}-\theta\right)=\frac{a}{\cos\theta}+c\cos\theta.$$

The elimination of θ gives the required result, namely

$$x^2y^2=(a+c)^2(x^2+y^2-c^2-4ac)+2ac^2(a+c)$$
$$+\frac{a^2c^4}{x^2+y^2-4ac-c^2}.$$

The equations of the loci of the foci and of the vertices are obtained by putting $c=0$ and $c=-a$ respectively in the above: the equations are

$$x^2y^2=a^2(x^2+y^2) \text{ and } x^2y^2(x^2+y^2+3a^2)=a^6.$$

30. If the tangents make angles θ and $\theta+\alpha$ respectively with the minor axis of the ellipse, we have

$$p_1^2=a^2\cos^2\theta+b^2\sin^2\theta,$$

and $$p_2^2=a^2\cos^2(\theta+\alpha)+b^2\sin^2(\theta+\alpha),$$

where p_1, p_2 are the perpendiculars from the centre on the tangents.

Now, if x, y be the co-ordinates of the centre of the conic referred to the tangents as axes, we have $x=p_1\operatorname{cosec}\alpha$ and $y=p_2\operatorname{cosec}\alpha$.

Hence
$$2x^2\sin^2\alpha=a^2+b^2+(a^2-b^2)\cos 2\theta,$$

and $$2y^2\sin^2\alpha=a^2+b^2+(a^2-b^2)\cos 2(\theta+\alpha).$$

$$\therefore\ (x^2+y^2)\sin^2\alpha-(a^2+b^2)=(a^2-b^2)\cos(2\theta+\alpha)\cos\alpha,$$

and $$(x^2-y^2)\sin^2\alpha=(a^2-b^2)\sin(2\theta+\alpha)\sin\alpha.$$

Hence $\{(x^2+y^2)\sin^2\alpha-(a^2+b^2)\}^2+\cos^2\alpha\sin^2\alpha(x^2-y^2)^2=(a^2-b^2)^2\cos^2\alpha$,

which is equivalent to

$$\sin^2\alpha(x^2+y^2-p^2)^2-4\cos^2\alpha(x^2y^2\sin^2\alpha-q^4)=0,$$

where $$p^2=a^2+b^2 \text{ and } q^2=ab.$$

31. If t_1, t_2 be the lengths of the tangents drawn from any point (α, β), and r_1, r_2 be the lengths of the parallel semi-diameters; then, from Art. 186, Cor. III.,

$$t_1^2 \cdot t_2^2/r_1^2 \cdot r_2^2 = (\alpha^2/a^2 + \beta^2/b^2 - 1)^2 \text{.................... (i).}$$

Now $$\frac{1}{r^2} = \frac{\cos^2\theta}{a^2} + \frac{\sin^2\theta}{b^2};$$

whence $$a^2(b^2 - r^2)\tan^2\theta + b^2(a^2 - r^2) = 0 \text{....................... (ii).}$$

But, [Art. 113], the directions of the tangents from (α, β) are given by

$$(\alpha^2 - a^2)\tan^2\theta - 2\alpha\beta\tan\theta + \beta^2 - b^2 = 0 \text{.................(iii).}$$

Hence, eliminating θ from (ii) and (iii), we have

$$\{a^2(b^2 - r^2)(\beta^2 - b^2) - b^2(a^2 - r^2)(\alpha^2 - a^2)\}^2 + 4a^2b^2\alpha^2\beta^2(b^2 - r^2)(a^2 - r^2) = 0;$$

$$\therefore r_1^2 r_2^2 = \frac{a^4b^4\{(\alpha^2 + \beta^2)^2 - 2(a^2 - b^2)(\alpha^2 - \beta^2) + (a^2 - b^2)^2\}}{(b^2\alpha^2 + a^2\beta^2)^2}.$$

From (i), $$t_1t_2 + r_1r_2 = (\alpha^2/a^2 + \beta^2/b^2)\,r_1r_2;$$

$$\therefore t_1t_2 + r_1r_2 = \sqrt{\{(\alpha^2 + \beta^2)^2 - 2(a^2 - b^2)(\alpha^2 - \beta^2) + (a^2 - b^2)^2\}},$$

and it is easily proved that

$$SO^2 \cdot HO^2 = (\alpha^2 + \beta^2)^2 - 2(a^2 - b^2)(\alpha^2 - \beta^2) + (a^2 - b^2)^2.$$

Hence $$OP \cdot OQ + CP' \cdot CQ' = OS \cdot OH.$$

[A simple geometrical proof is given in the Solutions of the Cambridge Problems and Riders for 1878.]

32. Take the given perpendicular lines AC and BD for axes, and let the points P, Q be (a, b) and (a', b') respectively.

Let the equations of APB and CQD be respectively

$$lx + my = 1 \text{ and } l'x + m'y = 1.$$

Then, since the lines go through the points (a, b), (a', b') respectively, we have

$$la + mb = 1 \text{.................................. (i),}$$

$$l'a' + m'b' = 1 \text{..................................(ii).}$$

The equations of AD and BC are respectively

$$lx + m'y = 1 \text{.................................. (iii),}$$

$$l'x + my = 1 \text{................................ (iv).}$$

Also, since APB and CQD are at right angles, we have

$$ll' + mm' = 0 \text{..................................(v).}$$

The locus required is obtained by the elimination of l, m, l' and m' from the equations (i), (ii), (iii), (iv) and (v).

From the first four equations it is easy to shew that

$$l = \{b(a'y + b'x) - y(a'y + bx)\}/(bb'x^2 - aa'y^2),$$

$$m = \{a(a'y + b'x) - x(ay + b'x)\}/(aa'y^2 - bb'x^2),$$

$$l' = \{b'(ay+bx) - y(ay+b'x)\}/(bb'x^2 - aa'y^2),$$
$$m' = \{a'(ay+bx) - x(a'y+bx)\}/(aa'y^2 - bb'x^2).$$

Hence, from (v), the required equation is

$$\{b(a'y+b'x) - y(a'y+bx)\}\{b'(ay+bx) - y(ay+b'x)\}$$
$$+\{a(a'y+b'x) - x(ay+b'x)\}\{a'(ay+bx) - x(a'y+bx)\} = 0.$$

If PQ subtends a right angle at the origin, $aa'+bb'=0$, and the above equation may be written

$$(x^2+y^2)[(ay+b'x)(a'y+bx) + aa'(a+a')x + bb'(b+b')y] = 0.$$

The locus is therefore a point-circle, and the conic

$$(ay+b'x)(a'y+bx) + aa'(a+a')x + bb'(b+b')y = 0,$$

which is a rectangular hyperbola since $aa'+bb'=0$.

33. Let the equation of the ellipse be $ax^2+by^2=1$.

Then the foot of the perpendicular from any point (x', y') on its polar with respect to the ellipse is given by

$$ax'x + by'y = 1 \text{ .... (i)},$$
$$ax'(y-y') - by'(x-x') = 0 \text{ .... (ii)}.$$

But if (x', y') be on the fixed diameter $lx+my=0$, we have

$$lx'+my'=0 \text{ .... (iii)}.$$

From (i) and (iii)

$$\frac{x'}{m} = \frac{y'}{-l} = \frac{1}{max - lby};$$

also (ii) may be written

$$\frac{bx}{x'} - \frac{ay}{y'} + a - b = 0,$$

whence $$\left(\frac{bx}{m} + \frac{ay}{l}\right)(max - lby) + a - b = 0,$$

which represents a rectangular hyperbola.

34. Let the equations of the conics be

$$ax^2+by^2=1 \text{ and } a'x^2+b'y^2=1.$$

The equations of the polars of $P(x', y')$ with respect to the conics are

$$axx'+byy'=1 \text{ .... (i)},$$

and $$a'xx'+b'yy'=1 \text{ .... (ii)}.$$

Also, if (x', y') move on the fixed straight line $Ax+By=1$, we have

$$Ax'+By'=1 \text{ .... (iii)}.$$

Eliminating x' and y' from (i), (ii), (iii) we have the equation of Q, namely

$$\begin{vmatrix} ax & by & 1 \\ a'x & b'y & 1 \\ A & B & 1 \end{vmatrix} = 0,$$

that is $$(ab'-a'b)\,xy+B\,(a'-a)\,x+A\,(b-b')\,y=0,$$

which represents a rectangular hyperbola whose asymptotes are parallel to the axes of the given conics.

35. Let the equations of the given conics be

$$ax^2+2hxy+by^2+2gx+2fy+c=0,$$

and $$a'x^2+2h'xy+b'y^2+2g'x+2f'y+c=0.$$

Then the polars of (x', y') with respect to the given conics are

$$(ax'+hy'+g)\,x+(hx'+by'+f)\,y+gx'+fy'+c=0 \quad\text{..........(i)},$$

and $$(a'x'+h'y'+g')\,x+(h'x'+b'y'+f')\,y+g'x'+f'y'+c'=0 \quad\text{...... (ii)}.$$

If the lines (i) and (ii) are parallel, we have

$$\frac{ax'+hy'+g}{a'x'+h'y'+g'}=\frac{hx'+by'+f}{h'x'+b'y'+f'},$$

and therefore the locus of (x', y') is the conic whose equation is

$$(ax+hy+g)(h'x+b'y+f')-(a'x+h'y+g')(hx+by+f)=0.$$

Again, if the lines (i) and (ii) are at right angles, we have

$$(ax'+hy'+g)(a'x'+h'y'+g')+(hx'+by'+f)(h'x'+b'y'+f')=0;$$

hence, in this case also, the locus of (x', y') is a conic.

36. Draw lines through A and B parallel respectively to the polars of B and A, and let these lines meet in K. Then, if OK cut PQ in R, since AK is parallel to OQ and BK to OP, we have

$$AR : QR=KR : OR=BR : PR.$$

Hence we have to prove that the centre of the conic lies on the line OK.

Let the equation of the conic be

$$ax^2+by^2=1,$$

and let A be (x_1, y_1) and B be (x_2, y_2).

Then the equations of OP, OQ are respectively

$$ax_1x+by_1y=1 \text{ and } ax_2x+by_2y=1.$$

Hence the equation of the line joining O to the centre is

$$ax\,(x_1-x_2)+by\,(y_1-y_2)=0.$$

Again, the equations of AK, BK are respectively

$$ax_2\,(x-x_1)+by_2\,(y-y_1)=0 \text{ and } ax_1\,(x-x_2)+by_1\,(y-y_2)=0.$$

Hence the equation of the line joining K to the centre of the conic is

$$ax\,(x_1-x_2)+by\,(y_1-y_2)=0.$$

The centre of the conic is therefore on the line OK.

37. Since the conics have a common director-circle, they have a common centre, C suppose.

Let P be the common point, P' the other extremity of the diameter PCP'. Then, if S, H be the foci, we have

$$CP^2 + SP \cdot PH = a^2 + b^2$$
$$= \text{square of radius of director-circle.}$$

Hence $SP \cdot HP$ is constant. But $SPHP'$ is a parallelogram, and therefore
$$SP' = HP.$$

Hence S moves so that $SP \cdot SP'$ is constant.

The equation of the locus is therefore

$$\{(x-c)^2 + y^2\}\ \{(x+c)^2 + y^2\} = k^4,$$

where $k^2 = SP \cdot SP'$ and $c = CP$. The curve is called a *lemniscate*.

38. Take the two given tangents for axes, and let (a, b) be the given centre of the conics.

Then, if (x, y) be one of the foci, the other focus will be

$$(2a - x,\ 2b - y).$$

Since the product of the perpendiculars from the foci of a conic on any tangent is constant, we have

$$x(2a - x)\sin^2\omega = y(2b - y)\sin^2\omega,$$

where ω is the angle between the axes.

Hence the required locus is the rectangular hyperbola whose equation is

$$x^2 - y^2 - 2ax + 2by = 0.$$

39. The centre is given by

$$ax + hy = 0,\ hx + by = 1;$$

hence
$$y = \frac{a}{ab - h^2}.$$

The line through the centre parallel to the tangent at the origin cuts the conic in points given by

$$ax^2 + \frac{2ah}{ab - h^2}x + \frac{ba^2}{(ab - h^2)^2} - \frac{2a}{ab - h^2} = 0;$$

$$\therefore \left(\frac{x_1 - x_2}{2}\right)^2 = \frac{h^2}{(ab - h^2)^2} - \frac{ab}{(ab - h^2)^2} + \frac{2}{ab - h^2} = \frac{1}{ab - h^2}.$$

Hence $CD^2 = \dfrac{1}{ab - h^2}$, and therefore the product of the focal distances is $\dfrac{1}{ab - h^2}$.

40. Let S be the given focus, and P the point of contact of the given tangent; also let O be the middle point of SP, and let OK be parallel to the given tangent. Then, if S' be the other focus of the conic, and C be middle point of SS', C will be the centre of the conic.

Then, if DCD' be the diameter parallel to the tangent at P,

$$CD^2 = SP \,.\, PS' = 4SO \,.\, OC.$$

Since PS', PS are equally inclined to the given tangent, and OC is parallel to PS', it follows that OC is a fixed straight line. Also CD is parallel to the given tangent, or to OK.

Hence the equation of the locus of D, or D', when referred to OC and OK as axes, is $y^2 = 4ax$, where $a = SO$. The locus is therefore a parabola of which OK is a tangent and OC a diameter; and, since the focal distance is equal to SO, and OS and OC make equal angles with OK, S must be the focus of the parabola.

41. Take the two tangents for axes, and let the equations of the other lines be $x = a$, $y = b$ respectively.

Let the foci be (a, g) and (f, b); then, if (x, y) be the centre, we have

$$2x = a + f \text{ and } 2y = b + g.$$

But, since the axes are tangents, the product of the distances of the foci from one axis is equal to the product of their distances from the other, and therefore $af = bg$.

Hence $$2ax - a^2 = 2by - b^2,$$

so that the locus of the centres is a straight line.

42. This follows at once from question 55, page 144.

43. Let S, H be the foci of the fixed ellipse, and C be the common centre. Let P be the point of contact of the ellipses, and let the tangent at S to the inner ellipse cut the common tangent at P in T. Join CT, cutting SP in V.

Then CT bisects SP, and is therefore parallel to HP. Hence CT and SP make equal angles with the tangent PT; from which it follows that

$$VT = VP = VS,$$

and therefore the angle STP is a right angle.

Hence CT^2 = sum of squares of the semi-axes of the variable ellipse.

But $$CT = CV + VT = CV + VP = \frac{1}{2}(HP + SP) = \text{constant}.$$

Hence *the sum of the squares of the axes of the variable ellipse is constant.*

Now let O, O' be the two foci; then $CS^2 + OS \,.\, O'S$ = sum of squares of semi-axes; hence $OS \,.\, O'S$, and therefore also $OS \,.\, OH$ is constant. Then see 37.

44. By question 23, page 164, if a rectangular hyperbola cut a circle in four points the centre of mean position of the four points is midway between the centres of the curves.

Let then A, B, C, D, E be five points on the circle whose centre is O; and let G be the centre of mean position of the five points, and a, b, c, d, e the centres of mean position of four of the points excluding A, B, C, D, E respectively. Then Aga is a straight line and $AG=4Ga$; so also $BG=4Gb$, $CG=4Gc$, $DG=4Gd$ and $EG=4Ge$.

Hence a, b, c, d, e are on a circle whose radius is one-fourth of the radius of the given circle.

Let A_1, B_1, C_1, D_1, E_1 be the centres of the five rectangular hyperbolas which pass through four of the five points excluding A, B, C, D, E respectively. Then OaA_1 is a straight line, and $OA_1=2Oa$; so also $OB_1=2Ob$, &c. Hence the five points A_1, B_1, C_1, D_1, E_1 are on a circle whose radius is double the radius of the circle on which a, b, c, d, e lie and therefore half the radius of the original circle.

45. Let the equation of the conic be $ax^2+by^2=1$; then the most general equation of the rectangular hyperbola whose asymptotes are parallel to the axes of the conic is

$$xy+gx+fy+c=0.$$

The abscissæ of the points of intersection are given by the equation

$$ax^2+b\left(\frac{gx+c}{x+f}\right)^2=1,$$

or
$$ax^4+2afx^3+(af^2+bg^2-1)x^2+2(bgc-f)x+bc^2-f^2=0.$$

Hence, if (x_1, y_1), &c. be the four points of intersection, we have

$$x_1+x_2+x_3+x_4=-2f,$$

and similarly
$$y_1+y_2+y_3+y_4=-2g.$$

Hence the centre of mean position of the four points of intersection is

$$\left(-\frac{f}{2}, -\frac{g}{2}\right).$$

The centre of the hyperbola is easily seen to be $(-f, -g)$, and therefore the centre of mean position of the four points of intersection is midway between the centres of the two curves.

46. Let the equations of the sides of the triangle be

$$x=0,\ \ y=0,\ \text{ and }\ lx+my+n=0;$$

and let the three parallel lines be

$$x-\alpha=0,\ \ y-\beta=0\ \text{ and }\ lx+my+n-\gamma=0.$$

Then the curve whose equation is

$$xy(lx+my+n)-\lambda(x-\alpha)(y-\beta)(lx+my+n-\gamma)=0$$

will clearly pass through the six points of intersection, whatever the value of λ may be, for the curve meets $x=0$ where $(y-\beta)(lx+my+n-\gamma)=0$, and so for the other sides. But when $\lambda=1$, the curve is a *conic.*

47. Since $$\frac{2}{PO}=\frac{1}{PG}+\frac{1}{PG'},$$

it follows that $\{PGOG'\}$ is harmonic.

Therefore, as GCG' is a right angle, CO and CP are equally inclined to the axes of the conic.

Hence, if CO cut the curve in Q, Q', QP, $Q'P$ will be at right angles to one another.

But all chords which subtend a right angle at P cut the normal at P in the same point, and therefore pass through O, which is the point of intersection of the normal and one such chord.

48. Take the tangent at O for axis of x and the diameter through O for the axis of y; then the equation of the conic will be of the form

$$ax^2+by^2+2ey=0.$$

Let the equation of PP' be $lx+my=1$.

Then OP, OP' will be given by

$$ax^2+by^2+2ey\,(lx+my)=0.$$

The extremity of the diameter through O is $\left(0, -\frac{2e}{b}\right)$, and the equation of the tangent at that point is $y=-\frac{2e}{b}$.

Hence at the points of intersection of OP, OP' and the tangent at O', we have

$$ax^2+\frac{4e^2}{b}-\frac{4e}{b}\left(lx-\frac{2me}{b}\right)=0.$$

Hence $$O'Q \,.\, O'Q'=x_1x_2=(4e^2b+8me^2)/b^2a,$$

from which it follows that m is constant since $O'Q \,.\, O'Q'$ is constant; and when m is constant the line $lx+my=1$ cuts OO' in a fixed point.

49. Take the tangent and normal at P for axes, and let the equation of the conic be

$$ax^2+2hxy+by^2+2fy=0.$$

Let the equation of LM be $y=\eta$.

Then the equation of PL, PM is

$$ax^2+2hxy+by^2+\frac{2f}{\eta}y^2=0.$$

The equation of the bisectors of the angles LPM is

$$\frac{x^2-y^2}{a-b-\frac{2f}{\eta}}=\frac{xy}{h},$$

or

$$h(x^2-y^2)=xy\left(a-b-\frac{2f}{\eta}\right).$$

Hence where the bisectors meet $y=\eta$ the co-ordinates satisfy the relation

$$h(x^2-y^2)=xy(a-b)-2fx.$$

Thus the locus of R is an hyperbola whose asymptotes are parallel to the lines

$$h(x^2-y^2)=xy(a-b);$$

and these lines bisect the angles between the lines

$$ax^2+2hxy+by^2=0,$$

which are parallel to the asymptotes of the original conic.

Hence, as the axes of a conic bisect the angles between its asymptotes, the asymptotes of the locus of R are parallel to the axes of the original conic.

50. Let the equation of the given conic be $ax^2+by^2=1$, and let $(c, 0)$ be the given point in its transverse axis. Then the equation of any chord through $(c, 0)$ is

$$y-m(x-c)=0,$$

and the equation of any conic which touches the given conic at the ends of this chord is

$$ax^2+by^2-1+\lambda\{y-m(x-c)\}^2=0 \quad\text{.....................(i).}$$

Now the centre of the conic (i) is given by

$$ax-\lambda m\{y-m(x-c)\}=0 \quad\text{.......................(ii),}$$

and

$$by+\lambda\{y-m(x-c)\}=0 \quad\text{.......................(iii).}$$

Also, since (i) passes through $(0, 0)$, we have

$$-1+\lambda m^2c^2=0 \quad\text{...............................(iv).}$$

The required locus is found by eliminating λ and m from the equations (ii), (iii) and (iv).

From (ii) and (iii), $m=-ax/by$;

$$\therefore\ mc^2ax-y+m(x-c)=0;$$

$$\therefore\ c^2a^2x^2+by^2+ax(x-c)=0.$$

The locus of the centres is therefore a centric conic.

51. Let $\frac{\pi}{4}-\alpha$, $\frac{\pi}{4}+\alpha$ be the eccentric angles of Q, Q' respectively, and let the equation of the circle QCQ' be

$$x^2+y^2+2gx+2fy+c=0.$$

Then we have the conditions

$$a^2\cos^2\left(\frac{\pi}{4}-\alpha\right)+b^2\sin^2\left(\frac{\pi}{4}-\alpha\right)+2ga\cos\left(\frac{\pi}{4}-\alpha\right)+2fb\sin\left(\frac{\pi}{4}-\alpha\right)+c=0,$$

$$a^2\cos^2\left(\frac{\pi}{4}+\alpha\right)+b^2\sin^2\left(\frac{\pi}{4}+\alpha\right)+2ga\cos\left(\frac{\pi}{4}+\alpha\right)+2fb\sin\left(\frac{\pi}{4}+\alpha\right)+c=0,$$

and $c=0$.

Hence, by addition and subtraction of the first two relations, we have

$$a^2+b^2+2\sqrt{2}ga\cos\alpha+2\sqrt{2}fb\cos\alpha=0,$$

and $$(a^2-b^2)\sin 2\alpha+2\sqrt{2}ga\sin\alpha-2\sqrt{2}fb\sin\alpha=0$$

or $$(a^2-b^2)\cos\alpha+\sqrt{2}ga-\sqrt{2}fb=0.$$

Eliminating $\cos\alpha$, we have

$$a^4-b^4=4(g^2a^2-f^2b^2).$$

But the co-ordinates of the centre of the circle are $-g$, $-f$, and therefore the required locus is the hyperbola whose equation is

$$4(a^2x^2-b^2y^2)=a^4-b^4.$$

52. Let the equation of the tangent at P be $x\cos\alpha+y\sin\alpha-p=0$, and let the equation of the chord through Q, Q', the other points of intersection, be $x\cos\alpha'+y\sin\alpha'-p'=0$.

Then the equation

$$\frac{x^2}{a^2}+\frac{y^2}{b^2}-1+\lambda(x\cos\alpha+y\sin\alpha-p)(x\cos\alpha'+y\sin\alpha'-p')=0,$$

will represent any conic touching the given conic at P and passing through Q and Q'. The above equation will represent a *circle* provided

$$\sin(\alpha+\alpha')=0 \quad\text{.................. (i)},$$

and

$$\frac{1}{a^2}+\lambda\cos\alpha\cos\alpha'=\frac{1}{b^2}+\lambda\sin\alpha\sin\alpha' \quad\text{.................. (ii)}.$$

Also the circle will pass through (0, 0) provided

$$-1+\lambda pp'=0 \quad\text{.................. (iii)}.$$

And, since $x\cos\alpha+y\sin\alpha-p=0$ touches the ellipse,

$$p^2=a^2\cos^2\alpha+b^2\sin^2\alpha \quad\text{.................. (iv)}.$$

From (i) and (ii) $$\lambda = \pm\left(\frac{1}{a^2} - \frac{1}{b^2}\right);$$
hence, from (iii),

$$\frac{a^4b^4}{(a^2-b^2)^2} = \frac{1}{\lambda^2} = p'^2p^2 = p'^2(a^2\cos^2\alpha + b^2\sin^2\alpha) = p'^2(a^2\cos^2\alpha' + b^2\sin^2\alpha').$$

The coordinates of the foot of the perpendicular on QQ' are $p'\cos\alpha'$ and $p'\sin\alpha'$; hence the equation of the required locus is

$$a^2x^2 + b^2y^2 = a^4b^4/(a^2-b^2)^2.$$

53. The conic

$$\lambda\left(\frac{x^2}{a^2} + \frac{y^2}{b^2} - 1\right) - (2xy - c) = 0 \quad \text{.......................(i)}$$

will go through the points of intersection of the two conics.

If $\lambda = c$, the conic (i) will represent the two straight lines

$$\frac{cx^2}{a^2} - 2xy + \frac{cy^2}{b^2} = 0 \quad \text{..............................(ii).}$$

If the curves touch the lines (ii) must be coincident, and therefore $c^2 = a^2b^2$. Thus $c = \pm ab$ is the required condition.

If $c^2 = a^2b^2$, the lines (ii) will be $\frac{x}{a} \pm \frac{y}{b} = 0$, so that the points of contact are on one or other of the equi-conjugates.

The polars of (α, β) with respect to the two curves are

$$\frac{x\alpha}{a^2} + \frac{y\beta}{b^2} = 1 \quad\text{and}\quad \frac{x\beta}{ab} + \frac{y\alpha}{ab} = \pm 1,$$

and these polars meet on the lines

$$\frac{x\alpha}{a^2} + \frac{y\beta}{b^2} \mp \left(\frac{x\beta}{ab} + \frac{y\alpha}{ab}\right) = 0,$$

that is on $$\frac{x}{a} \mp \frac{y}{b} = 0.$$

54. Let the conic which goes through the five points A, B, C, D, E cut the circle ABE in G: then, AB and CD make equal angles with the axes of the conic [Art. 186], and so also do AB and EG; hence EG is parallel to CD, and therefore G and F are coincident. Hence the conic through the five points A, B, C, D, E will also pass through F.

The *directions* of the axes of the conic are known since AB and CD make equal angles with the axes; hence the axes can be drawn when the *centre* of the conic is found.

Since CD and EF are parallel chords, the line joining the middle points of CD and EF is a diameter. To find a second diameter, draw a circle through D, C and E: if this circle cut the conic in a fourth point H, EH and CD make equal angles with the axes of the conic, and therefore EH is parallel to AB, and therefore H is found by drawing through E a line parallel to AB to cut the circle DCE in H. Then the line through the middle points of the two parallel chords AB and EH will be another diameter of the conic. The centre of the conic is thus determined, and the axes can now be drawn.

55. The six points are always on a *conic* whose equation is

$$(ax'x''+by'y''-1)(ax^2+by^2-1)-(axx'+byy'-1)(axx''+byy''-1)=0 \text{...(i)},$$

the equation of the given conic being $ax^2+by^2-1=0$ and P, P' being (x', y') and (x'', y'') respectively [Ex. 3, Art 187].

The conditions that (i) may be a circle are

$$x'y''+x''y'=0 \text{..................................(ii)},$$

and

$$a(by'y''-1)=b(ax'x''-1),$$

or

$$x'x''-y'y''=\frac{1}{a}-\frac{1}{b} \text{.............................. (iii)}.$$

From (ii) and (iii), by squaring and adding,

$$(x'^2+y'^2)(x''^2+y''^2)=\left(\frac{1}{a}-\frac{1}{b}\right)^2,$$

that is $CP^2 . CP'^2=CS^4$, where C is the centre and S a focus of the conic.

It follows from (ii) that CP and CP' make equal angles with the axes; and, since $\frac{y'}{y''}(x''^2+y''^2)=\frac{1}{a}-\frac{1}{b}$, and $\frac{1}{a}-\frac{1}{b}$ is positive if the major axis of the conic is along the axis of x, it follows that y' and y'' have different signs, and hence the points P, P' are on different sides of the transverse axis.

Since $PC : CS=CS : CP'$ and the angles PCS, SCP' are equal, it follows that the angles PSC and $SP'C$ are equal and that $PS : SP'=PC : P'C$. Now let the ellipse become a parabola; then C will be at infinity and from the above relations we see that in the case of a parabola PSP' is a straight line and $PS=SP'$.

56. Let T be (x', y') and T' be (x'', y''), the equation of the parabola being

$$y^2-4ax=0.$$

Then the equations of PQ, $P'Q'$ will be respectively

$$yy'-2a(x+x')=0 \text{ and } yy''-2a(x+x'')=0.$$

Hence the equation of any conic through P, Q, P', Q' is given by

$$\lambda(y^2-4ax)-\{yy'-2a(x+x')\}\{yy''-2a(x+x'')\}=0.$$

If the conic pass through (x', y') we have

$$\lambda(y'^2-4ax')-(y'^2-4ax')\{y'y''-2a(x'+x'')\}=0;$$

$$\therefore \lambda=yy''-2a(x'+x''),$$

and it is obvious that when λ has this value the conic will also pass through (x'', y'').

Hence the six points P, Q, P', Q', T and T' all lie on a conic whose equation is

$$(y^2 - 4ax)\{y'y'' - 2a(x' + x'')\} = \{yy' - 2a(x + x')\}\{yy'' - 2a(x + x'')\}.$$

In order that this conic may be a rectangular hyperbola it is necessary and sufficient that

$$y'y'' - 2a(x' + x'') = y'y'' + 4a^2,$$

or

$$\frac{1}{2}(x' + x'') = -a,$$

which shews that the middle point of TT' must be on the directrix.

57. Let T be the point of intersection of AA', BB' and CC'. Then

$$TA \cdot TA' = TC \cdot TC' = TB \cdot TB',$$

from which it follows that the point T is on the radical axis of any two of the circles OAA', OBB', OCC'. Also, since the point O is common to all three circles, it is on the radical axis of any two of the circles. Hence OT is the radical axis of the three circles.

58. Let the point be (α, β), the equation of the conic being

$$ax^2 + by^2 = 1.$$

Let (x', y') be the middle point of one of the common chords, then the equation of that chord is

$$(x - x')\,ax' + (y - y')\,by' = 0 \text{ ....(i).}$$

Since the extremities of the chord are equidistant from O, the line through (x', y') perpendicular to (i) must pass through (α, β); we therefore have

$$\frac{\alpha - x'}{ax'} = \frac{\beta - y'}{by'}.$$

Hence (x', y') is always on the rectangular hyperbola whose equation is

$$ax(y - \beta) - by(x - \alpha) = 0.$$

59. Take O for origin, and let the equation of the conic be

$$ax^2 + 2hxy + by^2 + 2gx + 2fy + c = 0.$$

Then every conic of the system is included in the equation

$$ax^2 + 2hxy + by^2 + 2gx + 2fy + c + \lambda(x^2 + y^2 - k^2) = 0 \text{ ....(i),}$$

where λ and k are arbitrary.

Now the centre of (i) satisfies the equations

$$ax + hy + g + \lambda x = 0,$$

and

$$hx + by + f + \lambda y = 0.$$

Hence the centre of (i), for all values of λ and k, is on the conic whose equation is

$$y(ax+hy+g)-x(bx+by+f)=0,$$

and this conic is clearly a rectangular hyperbola.

60. Let S, S' be the foci of the conic, and PSP' be the focal chord. Then, if the tangents at P, P' meet in T and the normals in G, the points T, P, G, P' are on a circle, and we have

$$PTG=PP'G=\frac{\pi}{2}-TP'S$$

$$=P'TS, \text{ since } TS \text{ is perpendicular to } PSP'.$$

But we know that $PTS'=P'TS$, and therefore TGS' is a straight line.

61. The equation of the normal at ϕ is

$$\frac{ax}{\cos\phi}-\frac{by}{\sin\phi}=a^2-b^2.$$

Hence $$CG=\frac{a^2-b^2}{a}\cos\phi \text{ ....(i).}$$

If the normals at the points ϕ_1, ϕ_2, ϕ_3, ϕ_4 meet in any point (α, β), these must be the values of ϕ given by

$$\frac{a\alpha}{\cos\phi}-\frac{b\beta}{\sin\phi}=a^2-b^2=c^2.$$

Hence $$(a\alpha-c^2z)^2(1-z^2)=b^2\beta^2z^2 \text{ ....(ii),}$$

where z is put for $\cos\phi$.

Now if z_1, z_2, z_3, z_4 be the roots of (ii),

$$\Sigma z_1=\frac{2ac^2\alpha}{c^4}=\frac{2a\alpha}{c^2}$$

and $$\Sigma\frac{1}{z_1}=\frac{2c^2a\alpha}{a^2\alpha^2}=\frac{2c^2}{a\alpha}.$$

Hence $$\Sigma\frac{1}{z_1}=\frac{4}{\Sigma z},$$

and therefore also, from (i), $\Sigma\dfrac{1}{CG_1}=\dfrac{4}{\Sigma CG_1}$.

62. Let the equation of the ellipse be $ax^2+by^2=1$, and let O be (α, β).

Then, by Art. 196, the conic

$$(a-b)xy+b\alpha y-a\beta x=0 \text{ ....(i)}$$

goes through the points A, B, C, D and O.

Now the centre of (i) is given by

$$(a-b)\,y - a\beta = 0,$$

$$(a-b)\,x + b\alpha = 0.$$

Hence $\frac{x}{\alpha} + \frac{y}{\beta} = 1$, so that the locus of the centre, for different values of a and b, is a straight line.

63. Let the equation of the ellipse be $ax^2 + by^2 = 1$, and let O be (α, β).

Then if the normal at (ξ, η) pass through (α, β) we must have the condition

$$\frac{\xi - \alpha}{a\xi} = \frac{\eta - \beta}{b\eta},$$

or $$b\alpha\eta - a\beta\xi + (a-b)\,\xi\eta = 0 \quad\text{.................(i).}$$

The equation of the line joining C to (ξ, η) is $\frac{x}{\xi} = \frac{y}{\eta}$, and the equation of the line through (ξ, η) which makes an equal angle with the axis is

$$\frac{x-\xi}{\xi} = \frac{y-\eta}{-\eta},$$

or $$x\eta + y\xi - 2\xi\eta = 0 \quad\text{.................(ii).}$$

Now the line (ii) will pass through the point given by

$$\frac{x}{b\alpha} = \frac{y}{-a\beta} = \frac{-2}{a-b}$$

for *all values* of ξ and η which satisfy (i).

Now (i) is the condition that the perpendicular from (ξ, η) on its polar with respect to $ax^2 + by^2 - 1 = 0$ should pass through (α, β), so that the proposition may be enunciated as follows:

'If the perpendicular from a point P on its polar with respect to an ellipse whose centre is C pass through a fixed point O, then will the line through P which makes with the axis of the ellipse an angle equal to that made by PC pass through another fixed point.'

64. Let the equation of the ellipse be $ax^2 + by^2 - 1 = 0$, and let O be (α, β).

Then if the normal at (ξ, η) [or if the perpendicular from (ξ, η) on its polar with respect to the ellipse] pass through O, we have the condition

$$\frac{\xi - \alpha}{a\xi} = \frac{\eta - \beta}{b\eta},$$

or $$a\beta\xi - b\alpha\eta + (b-a)\,\xi\eta = 0 \quad\text{.................(i).}$$

The equation of the line through (ξ, η) making with the axis of the ellipse an angle equal to that made by the perpendicular from (ξ, η) on its polar with respect to the ellipse is

$$\frac{x-\xi}{a\xi}+\frac{y-\eta}{b\eta}=0,$$

or $$ay\xi+bx\eta-(a+b)\,\xi\eta=0 \quad\text{..........(ii)}.$$

Now the line (ii) will pass through the point given by

$$\frac{y}{\beta}=\frac{x}{-\alpha}=\frac{a+b}{a-b}$$

for *all values* of ξ and η which satisfy (i). This proves the theorem.

65. Let the equation of the ellipse be $x^2/a^2+y^2/b^2=1$, and let $P\ (\xi, \eta)$ be a point on it such that the normal at P passes through a fixed point $O\ (\alpha, \beta)$. Then we have the condition

$$\frac{\xi-\alpha}{\frac{\xi}{a^2}}=\frac{\eta-\beta}{\frac{\eta}{b^2}},$$

or $$b^2\beta\xi-a^2\alpha\eta+(a^2-b^2)\xi\eta=0 \quad\text{..........(i)}.$$

The line through P parallel to CP' is

$$\frac{x-\xi}{\xi}=\frac{y-\eta}{\frac{a}{b}\eta};$$

or $$by\xi-ax\eta+(a-b)\,\xi\eta=0 \quad\text{..........(ii)}.$$

Now the line (ii) will pass through the point given by

$$\frac{y}{b\beta}=\frac{x}{a\alpha}=\frac{1}{a+b}$$

for *all values* of ξ and η which satisfy (i). This proves the proposition.

66. Let the eccentric angles of the four points P, Q, R, S the normals at which meet in O be $\theta_1, \theta_2, \theta_3, \theta_4$ respectively. Let Ap, Aq, Ar, As be chords through the vertex A perpendicular to the four normals, and let $\alpha_1, \alpha_2, \alpha_3, \alpha_4$ be the eccentric angles of p, q, r, s respectively.

Then Ap, Aq, Ar, As are parallel to the tangents at P, Q, R, S respectively, and hence

$$\alpha_1=2\theta_1,\ \alpha_2=2\theta_2,\ \alpha_3=2\theta_3 \text{ and } \alpha_4=2\theta_4.$$

But we know that

$$\theta_1+\theta_2+\theta_3+\theta_4=(2n+1)\,\pi. \quad \text{[Art. 198.]}$$

Hence $$\alpha_1+\alpha_2+\alpha_3+\alpha_4=2m\pi,$$

which is the necessary and sufficient condition that the four points p, q, r, s should be on a circle.

67. Let the eccentric angles of the points of contact of the tangents be θ_1, θ_2. Then it is easy to shew that the tangents at θ_1, θ_2 intersect in the point

$$\left\{a\cos\frac{1}{2}(\theta_1+\theta_2)/\cos\frac{1}{2}(\theta_1-\theta_2),\ b\sin\frac{1}{2}(\theta_1+\theta_2)/\cos\frac{1}{2}(\theta_1-\theta_2)\right\},$$

and that the normals intersect in the point

$$\left\{\frac{c^2}{a}\cos\theta_1\cos\theta_2\cos\frac{1}{2}(\theta_1+\theta_2)/\cos\frac{1}{2}(\theta_1-\theta_2),\right.$$
$$\left.-\frac{c^2}{b}\sin\theta_1\sin\theta_2\sin\frac{1}{2}(\theta_1+\theta_2)/\cos\frac{1}{2}(\theta_1-\theta_2)\right\},$$

where $$c^2=a^2-b^2.$$

Since the point of intersection of the tangents is on the ellipse

$$x^2/a^2+y^2/b^2=4,$$

we have $$4\cos^2\frac{1}{2}(\theta_1-\theta_2)=1\ldots\ldots\ldots\ldots(\text{i}).$$

Hence, if (x, y) be the point of intersection of the normals, we have

$$\frac{ax}{c^2}=\pm2\cos\theta_1\cos\theta_2\cos\frac{1}{2}(\theta_1+\theta_2)=\pm\cos\frac{1}{2}(\theta_1+\theta_2)\{\cos(\theta_1+\theta_2)+\cos(\theta_1-\theta_2)\},$$

and

$$-\frac{by}{c^2}=\pm2\sin\theta_1\sin\theta_2\sin\frac{1}{2}(\theta_1+\theta_2)=\pm\sin\frac{1}{2}(\theta_1+\theta_2)\{\cos(\theta_1+\theta_2)-\cos(\theta_1-\theta_2)\};$$

$$\therefore (a^2x^2+b^2y^2)/c^4$$
$$=\cos^2(\theta_1+\theta_2)+\cos^2(\theta_1-\theta_2)+2\cos(\theta_1+\theta_2)\cos(\theta_1-\theta_2)\cos(\theta_1+\theta_2)$$
$$=\frac{1}{4},\ \text{since}\ \cos(\theta_1-\theta_2)=\pm\frac{1}{2},\ \text{from (i)},$$

so that the normals intersect on the ellipse whose equation is

$$4a^2x^2+4b^2y^2=c^4.$$

68. The normals are the perpendiculars from the angular points of the triangle ABC on the opposite sides, and therefore meet in a point. The eccentric angles of two angular points differ by $\frac{2\pi}{3}$; and, by the preceding question, the normals at two points whose eccentric angles differ by $\frac{2\pi}{3}$, meet on the ellipse $$4a^2x^2+4b^2y^2=c^4.$$

69. The tangent to $xy=c$ at the point (x', y') is $xy'+x'y=2c$, and therefore the normal at (x', y') is

$$(x-x')x'-(y-y')y'=0.$$

Hence if the normal at (x, y) pass through (X, Y) we have

$$(X-x)x-(Y-y)y=0,\ \text{where}\ xy=c.$$

Hence x_1, x_2, x_3, x_4 are the four roots of

$$Xx^3 - x^4 - Yxc + c^2 = 0;$$

$$\therefore\ x_1 + x_2 + x_3 + x_4 = X,$$

and similarly $$y_1 + y_2 + y_3 + y_4 = Y.$$

70. The point of intersection of the normals at θ_1, θ_2 is given by

$$ax = c^2 \cos\theta_1 \cos\theta_2 \cos\frac{1}{2}(\theta_1 + \theta_2) \,/\, \cos\frac{1}{2}(\theta_1 - \theta_2)$$

$$-by = c^2 \sin\theta_1 \sin\theta_2 \sin\frac{1}{2}(\theta_1 + \theta_2) \,/\, \cos\frac{1}{2}(\theta_1 - \theta_2).$$

Now for a system of parallel chords

$$\theta_1 + \theta_2 = \text{constant} = 2a \text{ suppose.}$$

Hence $$2ax \sec a / c^2 = \{\cos 2a + \cos(\theta_1 - \theta_2)\} / \cos\frac{1}{2}(\theta_1 - \theta_2)$$

$$2by \operatorname{cosec} a / c^2 = \{\cos 2a - \cos(\theta_1 - \theta_2)\} / \cos\frac{1}{2}(\theta_1 - \theta_2).$$

Hence $$(ax \sec a + by \operatorname{cosec} a)/c^2 \cos 2a = \sec\frac{1}{2}(\theta_1 - \theta_2) \quad \text{............ (i)},$$

and $$(ax \sec a - by \operatorname{cosec} a)/c^2 = \cos(\theta_1 - \theta_2) \sec\frac{1}{2}(\theta_1 - \theta_2) \quad \text{..........(ii)}.$$

Hence, by addition,

$$(ax \cos a - by \sin a)/c^2 = \cos\frac{1}{2}(\theta_1 - \theta_2) \quad \text{..................(iii)}.$$

Eliminating $\theta_1 - \theta_2$ from (i) and (iii) we have the equation of the locus, namely

$$(ax \cos a - by \sin a)(ax \sec a + by \operatorname{cosec} a) = c^4 \cos 2a.$$

71. The equation of the tangent to $4xy = ab$ at (x', y') is

$$2xy' + 2yx' = ab.$$

If this line cut the ellipse at points whose eccentric angles are θ_1, θ_2, the above equation must represent the same straight line as

$$\frac{x}{a}\cos\frac{1}{2}(\theta_1 + \theta_2) + \frac{y}{b}\sin\frac{1}{2}(\theta_1 + \theta_2) = \cos\frac{1}{2}(\theta_1 - \theta_2).$$

Hence $$\cos\frac{1}{2}(\theta_1 + \theta_2)/2ay' = \sin\frac{1}{2}(\theta_1 + \theta_2)/2bx' = \cos\frac{1}{2}(\theta_1 - \theta_2)/ab;$$

$$\therefore\ \sin\frac{1}{2}(\theta_1 + \theta_2)\cos\frac{1}{2}(\theta_1 + \theta_2) = \cos^2\frac{1}{2}(\theta_1 - \theta_2),$$

or $$\sin(\theta_1 + \theta_2) = 1 + \cos(\theta_1 - \theta_2).$$

Now, if (x, y) be the point of intersection of the normals at θ_1, θ_2 we have [see 67]

$$-\frac{ax}{by}=\frac{\cos\theta_1\cos\theta_2}{\sin\theta_1\sin\theta_2}\cot\frac{1}{2}(\theta_1+\theta_2)=\frac{\cos(\theta_1+\theta_2)+\cos(\theta_1-\theta_2)}{\cos(\theta_1+\theta_2)-\cos(\theta_1-\theta_2)}\cot\frac{1}{2}(\theta_1+\theta_2)$$

$$=\frac{\cos(\theta_1+\theta_2)+\sin(\theta_1+\theta_2)-1}{\cos(\theta_1+\theta_2)-\sin(\theta_1+\theta_2)+1}\cot\frac{1}{2}(\theta_1+\theta_2)=1.$$

Hence the normals meet on the diameter whose equation is

$$ax+by=0.$$

72. The normal at (x, y) to the conic

$$x^2/a^2+y^2/b^2=1.$$

will go through (ξ, η) provided

$$\frac{a^2\xi}{x}-\frac{b^2\eta}{y}=a^2-b^2.$$

Hence the abscissae of the four points are the roots of

$$\{a^2\xi-(a^2-b^2)x\}^2(a^2-x^2)=a^2b^2\eta^2x^2 \quad\text{.....................(i).}$$

If the four points are (x_1, y_1), &c. we have from (i)

$$\Sigma x_1=\frac{2a^2\xi}{a^2-b^2},\ \Sigma x_1x_2=\frac{a^2}{(a^2-b^2)^2}\{a^2\xi^2+b^2\eta^2-(a^2-b^2)^2\}.$$

Hence $$\Sigma x_1^2=\frac{2a^2}{(a^2-b^2)^2}\{a^2\xi^2-b^2\eta^2+(a^2-b^2)^2\}.$$

Similarly $$\Sigma y_1^2=\frac{2b^2}{(a^2-b^2)^2}\{b^2\eta^2-a^2\xi^2+(a^2-b^2)^2\}.$$

Now $$\Sigma\frac{1}{p_1^2}=\frac{\Sigma x_1^2}{a^4}+\frac{\Sigma y_1^2}{b^4}$$

$$=\frac{2}{a^2-b^2}\left(\frac{\eta^2}{a^2}-\frac{\xi^2}{b^2}\right)+\frac{2}{a^2}+\frac{2}{b^2}.$$

Hence, if $\Sigma\frac{1}{p_1^2}=\frac{1}{c^2}$, the locus of (ξ, η) is the hyperbola

$$\frac{y^2}{a^2}-\frac{x^2}{b^2}=\frac{a^2-b^2}{2c^2}-\frac{a^4-b^4}{a^2b^2}.$$

73. If the normals at the four points (x_1, y_1), &c. meet in the point (ξ, η); and if r_1, r_2, r_3, r_4 be the lengths of the four normals; then

$$\Sigma r_1^2=\Sigma(\xi-x_1)^2+\Sigma(\eta-y_1)^2=4(\xi^2+\eta^2)-2\xi\Sigma x_1-2\eta\Sigma y_1+\Sigma x_1^2+\Sigma y_1^2.$$

But, from the previous question,

$$\Sigma x_1 = \frac{2a^2\xi}{a^2-b^2},\ \Sigma x_1^2 = \frac{2a^2}{(a^2-b^2)^2}\{a^2\xi^2 - b^2\eta^2 + (a^2-b^2)^2\},$$

$$\Sigma y_1 = \frac{2b^2\eta}{b^2-a^2} \text{ and } \Sigma y_1^2 = \frac{2b^2}{(a^2-b^2)^2}\{-a^2\xi^2 + b^2\eta^2 + (a^2-b^2)^2\}.$$

Hence $$\Sigma r_1^2 = 4(\xi^2+\eta^2) - \frac{4a^2\xi^2}{a^2-b^2} + \frac{4b^2\eta^2}{a^2-b^2}$$

$$+ \frac{2}{a^2-b^2}\{a^2\xi^2 - b^2\eta^2\} + 2a^2 + 2b^2$$

$$= \frac{2\eta^2(2a^2-b^2) - 2\xi^2(2b^2-a^2)}{a^2-b^2} + 2a^2 + 2b^2.$$

Hence the required locus is a conic.

74. The feet of the normals from (f, g) to the ellipse

$$x^2/a^2 + y^2/b^2 - 1 = 0$$

lie on the conic

$$xy(b^2-a^2) + a^2fy - b^2gx = 0.$$

Hence the equation of every conic through the four points of intersection is included in

$$x^2/a^2 + y^2/b^2 - 1 - \lambda\{xy(b^2-a^2) + a^2fy - b^2gx\} = 0 \text{..............(i)}.$$

Hence, for some value of λ, (i) will represent the same conic as

$$(xx'/a^2 + yy'/b^2 - 1)(xx''/a^2 - yy''/b^2 - 1) = 0,$$

$$\therefore \frac{x'x''}{a^2} = \frac{y'y''}{b^2} = \frac{x'+x''}{\lambda b^2a^2g} = \frac{y'+y''}{-\lambda a^2b^2f} = -1.$$

Hence $$\frac{x'x''}{a^2} = \frac{y'y''}{b^2} = -1 \text{ ....(ii)},$$

and $$f(x'+x'') + g(y'+y'') = 0 \text{ ....(iii)}.$$

From (iii) it follows that the middle point of the line joining (x', y') to (x'', y'') is on $fx + gy = 0$; this proves the theorem, since (x', y') and (x'', y'') may be the extremities of any one of the diagonals of the quadrilateral formed by the tangents at the feet of the normals from (f, g).

75. Let the normals to the ellipse $x^2/a^2 + y^2/b^2 - 1 = 0$ at the four points (x_1, y_1), (x_2, y_2), (x_3, y_3) and (x_4, y_4) co-intersect. Then the four extremities of the diagonals constructed as in the question will be $\left(\frac{a^2}{x_1}, \frac{b^2}{y_1}\right)$, &c. To shew that these four points are on a straight line, it will be sufficient to shew

that any three are on a straight line, the conditions for which are of the form

$$\begin{vmatrix} \frac{a^2}{x_1}, & \frac{b^2}{y_1}, & 1 \\ \frac{a^2}{x_2}, & \frac{b^2}{y_2}, & 1 \\ \frac{a^2}{x_3}, & \frac{b^2}{y_3}, & 1 \end{vmatrix} = 0, \text{ or } \begin{vmatrix} \frac{1}{x_1}, & \frac{1}{y_1}, & 1 \\ \frac{1}{x_2}, & \frac{1}{y_2}, & 1 \\ \frac{1}{x_3}, & \frac{1}{y_3}, & 1 \end{vmatrix} = 0 \text{............(i).}$$

But, since the normals at the points (x_1, y_1), &c. co-intersect, the following equations are simultaneously true, namely

$$\frac{a^2x}{x_1} - \frac{b^2y}{y_1} + b^2 - a^2 = 0, \text{ \&c.}$$

By eliminating a^2x, $-b^2y$ and $b^2 - a^2$ from any three of the last four equations we see that the conditions (i) are satisfied.

76. The equation

$$\lambda(x^2/a^2 + y^2/b^2 - 1) - (a^2 - b^2)xy + a^2fy - b^2gx = 0$$

includes all conics through A, B, C, D, the normals at A, B, C, D meeting at (f, g).

If the conic pass through the point $(ae, 0)$, we have

$$\lambda(e^2 - 1) - b^2gae = 0, \text{ or } \lambda = -ga^3e.$$

Also $x = \frac{a}{e}$ is to cut the conic in co-incident points; and therefore

$$-a^3eg\left(\frac{1}{e^2} + \frac{y^2}{b^2} - 1\right) - (a^2 - b^2)\frac{a}{e}y + a^2fy - b^2g\frac{a}{e} = 0$$

has equal roots, the condition for which is that $8a^4g^2 = a^4(ae - f)^2$.

Hence the locus of (f, g) is the two straight lines $ae - x = \pm 2\sqrt{2}y$.

77. As in question 72, the abscissae of the four points the normals at which meet in (ξ, η) are the roots of the equation

$$\{a^2\xi - (a^2 - b^2)x\}^2(a^2 - x^2) = a^2b^2\eta^2x^2.$$

Now substitute $\frac{a - r}{e}$ for x, and we shall obtain an equation whose roots are the values of the distances of the four points from the focus $(ae, 0)$: this equation will be

$$\{a^2e\xi - (a^2 - b^2)(a - r)\}^2\{a^2e^2 - (a - r)^2\}$$
$$= e^2a^2b^2\eta^2(a - r)^2.$$

Hence $SA \cdot SB \cdot SC \cdot SD$

$$= \text{coefficient of } r^0/\text{coefficient of } r^4$$
$$= \frac{b^2}{e^2}\{\eta^2 + (\xi - ae)^2\} = \frac{b^2}{e^2} \cdot SO^2.$$

78. Let the equation to the hyperbola be $xy=c^2$. Then the condition that the normal at (x, y) should pass through the point (ξ, η) is

$$(\xi-x)x-(\eta-y)y=0.$$

Hence the abscissae of the four points are given by the equation

$$(\xi-x)x^3-(\eta x-c^2)c^2=0.$$

Hence $\qquad \Sigma x=\xi$ and $\Sigma x^2=\xi^2$.

Similarly $\qquad \Sigma y=\eta$ and $\Sigma y^2=\eta^2$.

Now the sum of the squares of the lengths of the normals

$$=\Sigma\{(\xi-x)^2+(\eta-y)^2\}$$
$$=4(\xi^2+\eta^2)-2\xi\Sigma x-2\eta\Sigma y+\Sigma x^2+\Sigma y^2$$
$$=4(\xi^2+\eta^2)-2\xi^2-2\eta^2+\xi^2+\eta^2$$
$$=3(\xi^2+\eta^2).$$

79. The point of intersection of the normals at the points whose eccentric angles are θ_1, θ_2 is given by

$$ax=c^2\cos\theta_1\cos\theta_2\cos\frac{1}{2}(\theta_1+\theta_2)/\cos\frac{1}{2}(\theta_1-\theta_2),$$

and $$-by=c^2\sin\theta_1\sin\theta_2\sin\frac{1}{2}(\theta_1+\theta_2)/\cos\frac{1}{2}(\theta_1-\theta_2).$$

But, if the line joining θ_1, θ_2 pass through the point $\left(a\sqrt{\frac{a-b}{a+b}}, 0\right)$ we have

$$\sqrt{\frac{a-b}{a+b}}\cos\frac{1}{2}(\theta_1+\theta_2)=\cos\frac{1}{2}(\theta_1-\theta_2);$$

$$\therefore \frac{a-b}{a+b}\{1+\cos(\theta_1+\theta_2)\}=1+\cos(\theta_1-\theta_2)\ldots\ldots\ldots\ldots\ldots\text{(i).}$$

Now $$\frac{2ax}{c^2}\sqrt{\frac{a-b}{a+b}}=\cos(\theta_1+\theta_2)+\cos(\theta_1-\theta_2);$$

$\therefore$, from (i), $$\frac{x}{c}+\frac{b}{a}=\cos(\theta_1+\theta_2)\ldots\ldots\ldots\ldots\ldots\ldots\text{(ii).}$$

Also $$\frac{2by}{c^2}\sqrt{\frac{a-b}{a+b}}=\{\cos(\theta_1+\theta_2)-\cos(\theta_1-\theta_2)\}\tan\frac{1}{2}(\theta_1+\theta_2);$$

$\therefore$, from (i), $$\frac{y}{c}=\sin(\theta_1+\theta_2)\ldots\ldots\ldots\ldots\ldots\ldots\text{(iii).}$$

Hence the locus of (x, y) is the circle whose equation is

$$\left(x+\frac{bc}{a}\right)^2+y^2=c^2.$$

80. If the normal at (x, y) to $x^2/a^2+y^2/b^2-1=0$ pass through the point (ξ, η), we have

$$\frac{\xi-x}{\frac{x}{a^2}}=\frac{\eta-y}{\frac{y}{b^2}}=\frac{r}{\sqrt{\left(\frac{x^2}{a^4}+\frac{y^2}{b^4}\right)}},$$

where r is the length of the normal from (ξ, η) to the curve.

Hence
$$r_1^2r_2^2r_3^2r_4^2 \,.\, b^8x_1^2x_2^2x_3^2x_4^2$$
$$=(\xi-x_1)^2(\xi-x_2)^2(\xi-x_3)^2(\xi-x_4)^2\{a^4-(a^2-b^2)x_1^2\}$$
$$\{a^4-(a^2-b^2)x_2^2\}\{a^4-(a^2-b^2)x_3^2\}\{a^4-(a^2-b^2)x_4^2\},$$

where x_1, x_2, x_3, x_4 are the roots of

$$\{a^2\xi-(a^2-b^2)x\}^2(a^2-x^2)-a^2b^2\eta^2x^2=0 \quad\text{(i)}.$$

From (i),
$$x_1^2x_2^2x_3^2x_4^2=a^{12}\xi^4/(a^2-b^2)^4 \quad\text{(ii)}.$$

Put $x=\xi-\lambda$ in (i), then

$$\{b^2\xi+(a^2-b^2)\lambda\}^2\{a^2-(\xi-\lambda)^2\}-a^2b^2\eta^2(\xi-\lambda)^2=0;$$
$$\therefore\ (\xi-x_1)^2(\xi-x_2)^2(\xi-x_3)^2(\xi-x_4)^2=\lambda_1^2\lambda_2^2\lambda_3^2\lambda_4^2$$
$$=b^4\xi^4(a^2b^2-b^2\xi^2-a^2\eta^2)^2/(a^2-b^2)^4 \quad\text{(iii)}.$$

Now
$$a^4-(a^2-b^2)x_1^2=a^2(a-ex_1)(a+ex_1);$$

and putting $a-ex_1=\mu$ in (i), we have

$$a^2(\xi-ae+\mu e)^2\{a^2e^2-(a-\mu)^2\}-b^2\eta^2(a-\mu)^2=0.$$

Hence
$$\Pi(a-ex_1)=\text{coefficient of } \mu^0 \div \text{coefficient of } \mu^4$$
$$=-\{(\xi-ae)^2+\eta^2\}\,b^2/e^2.$$

Similarly
$$\Pi(a+ex_1)=-\{(\xi+ae)^2+\eta^2\}\,b^2/e^2.$$

Hence

$$r_1^2r_2^2r_3^2r_4^2=(a^2b^2-b^2\xi^2-a^2\eta^2)^2\{(\xi+ae)^2+\eta^2\}\{(\xi-ae)^2+\eta^2\}/(a^2-b^2)^2.$$

Now, from question 31, the product of the squares of the tangents

$$=(a^2b^2-b^2\xi^2-a^2\eta^2)^2\{(\xi+ae)^2+\eta^2\}\{(\xi-ae)^2+\eta^2\}/(b^2\xi^2+a^2\eta^2)^2.$$

Also the product of the squares of the perpendiculars from (ξ, η) on the asymptotes

$$=(b^2\xi^2+a^2\eta^2)^2/(a^2-b^2)^2.$$

Hence the continued product of the four normals is equal to the continued product of the two tangents and of the two perpendiculars on the asymptotes.

81. Since the lines $x+\lambda y=0$ and $x+\mu y=0$ are in the direction of conjugate diameters the equation of the curve will be

$$A(x+\lambda y)^2+B(x+\mu y)^2=1 \quad\text{(i)}.$$

The condition that either of the lines $x+\lambda y=\pm p$ should touch (i) is easily found to be $Ap^2=1$; and the condition that either of the lines $x+\mu y=\pm q$ should touch is $Bq^2=1$.

Hence the required equation is

$$(x+\lambda y)^2/p^2+(x+\mu y)^2/q^2=1.$$

82. Let T be the point (x', y'); then the equation of PQ is

$$xx'/a^2+yy'/b^2-1=0.$$

All conics through P and Q are included in

$$x^2/a^2+y^2/b^2-1-(xx'/a^2+yy'/b^2-1)(lx+my+n)=0 \text{ ....(i).}$$

Hence (i) is the circle TPQ provided

$$1=lx'+my'+n \text{ ....(ii),}$$

$$\frac{1}{a^2}-\frac{lx'}{a^2}=\frac{1}{b^2}-\frac{my'}{b^2} \text{ ....(iii)}$$

$$0=\frac{ly'}{b^2}+\frac{mx'}{a^2} \text{ ....(iv).}$$

From (ii), (iii), (iv), we have

$$\frac{la^2}{x'(b^2-a^2)}=\frac{mb^2}{y'(a^2-b^2)}=\frac{n}{x'^2+y'^2} \text{ ....(v);}$$

Hence, if $x'^2+y'^2=c^2$,

we have

$$l^2c^2a^4+m^2c^2b^4=n^2(a^2-b^2)^2 \text{ ....(vi).}$$

Now $lx+my+n=0$

is the equation of PQ, and (vi) shews that

$$lx+my+n=0$$

touches the conic

$$\frac{x^2}{a^4}+\frac{y^2}{b^4}=\frac{c^2}{(a^2-b^2)^2}.$$

83. Let the equation of the chord be

$$y=m(x-ae);$$

then the points A, B are

$$\{a, ma(1-e)\} \text{ and } \{-a, -ma(1+e)\}$$

respectively.

Hence the equation of the circle is

$$(x-a)(x+a)+\{y-ma(1-e)\}\{y+ma(1+e)\}=0,$$

or

$$x^2+y^2+2maey-a^2-m^2b^2=0.$$

The circle cuts the ellipse where

$$y^2(a^2-b^2)-2maeb^2y+m^2b^4=0,$$

that is

$$(aey-mb^2)^2=0.$$

Thus the circle cuts the ellipse in two coincident points.

84. The points A, B are the points of intersection of the director-circle and a tangent to the ellipse.

The equation of the director-circle is

$$x^2+y^2=a^2+b^2.$$

Hence, if $lx+my-1=0$ be the equation of the tangent AB, the equation

$$x^2+y^2-a^2-b^2+\lambda(lx+my-1)=0 \text{ ....(i)}$$

is the equation of any circle through the points A and B.

The circle (i) passes through the focus $(ae, 0)$ provided

$$\lambda=2b^2/(lae-1);$$

and, with this value of λ, (i) may be written

$$\left(x+\frac{lb^2}{lae-1}\right)^2+\left(y+\frac{mb^2}{lae-1}\right)^2=a^2+b^2+\frac{(l^2+m^2)b^4}{(lae-1)^2}+\frac{2b^2}{lae-1}.$$

In order that the radius of the circle may be equal to a, it is necessary and sufficient that

$$\frac{(l^2+m^2)b^2}{(lae-1)^2}+\frac{2}{lae-1}+1=0,$$

or

$$(l^2+m^2)b^2+2lae-2+l^2a^2e^2-2lae+1=0,$$

or

$$l^2a^2+m^2b^2=1 \text{ ....(ii)}.$$

But (ii) is the condition that $lx+my=1$ should touch the ellipse

$$x^2/a^2+y^2/b^2=1.$$

85. Let the equations of the parabola and of the circle be

$$y^2-4ax=0 \text{ and } x^2+(y-\beta)^2=c^2$$

respectively.

The line $y=mx+\dfrac{a}{m}$ touches the parabola for all values of m; the line touches the circle provided

$$\left(\beta-\frac{a}{m}\right)^2\Big/(1+m^2)=c^2,$$

or

$$c^2m^4+c^2m^2-(\beta m-a)^2=0.$$

Since the coefficient of m^3 is zero in the above equation, it follows that the sum of the tangents of the angles the four common tangents make with the axis of the parabola is zero.

86. Let $(a \cos \alpha, a \sin \alpha)$ be the point from which the tangents are drawn; then the equation of the tangents will be

$$(x^2/a^2+y^2/b^2-1)(\cos^2\alpha+a^2\sin^2\alpha/b^2-1)-(x\cos\alpha/a+ya\sin\alpha/b^2-1)^2=0.$$

Hence the equation of any conic through the four points in which the tangents cut the directrices is

$$(x^2/a^2+y^2/b^2-1)(\cos^2\alpha+a^2\sin^2\alpha/b^2-1)-(x\cos\alpha/a+ya\sin\alpha/b^2-1)^2$$
$$+\lambda(e^2x^2-a^2)=0 \text{ .... (i).}$$

We have to shew that, for some value of λ, the left member of (i) will have a factor of the form $y-mx$, and may therefore be written in the form

$$(y-mx)\left\{-\frac{y\sin^2\alpha}{b^2}-x\left(\frac{2\sin\alpha\cos\alpha}{b^2}+\frac{m\sin^2\alpha}{b^2}\right)+2a\sin\alpha/b^2\right\}=0\text{...(ii).}$$

Hence, equating the coefficients of x^2 and x and the constant terms in (i) and (ii), we must have

$$2m\sin\alpha\cos\alpha+m^2\sin^2\alpha=\sin^2\alpha-1+e^2+\lambda b^2e^2 \text{ .... (iii),}$$
$$2am\sin\alpha/b^2=-2\cos\alpha/a \text{ .... (iv),}$$
and
$$0=\cos^2\alpha/a^2+\sin^2\alpha/b^2+\lambda \text{ .... (v),}$$

simultaneously true for some values of λ and m and for all values of α.

From (iv) $m=-(1-e^2)\cot\alpha$; and, when this value of m is substituted in (iii), we have $\lambda b^2=e^2\cos^2\alpha-1$, which is equivalent to (v). Thus the values of λ and m which satisfy two of the relations (iii) (iv), and (v) will also satisfy the third relation, and therefore the equation (i) can be written in the form (ii).

Hence two of the four points of intersection of the tangents and directrices are on a straight line through the centre, and the equation of the other two, is, from (ii),

$$x(2\sin\alpha\cos\alpha+m\sin^2\alpha)+y\sin^2\alpha-2a\sin\alpha=0,$$
where
$$m=-(1-e^2)\cot\alpha;$$

and this line cuts the major axis in the point

$$\{2a\sec\alpha/(1+e^2),\ 0\}.$$

87. Let T be the point of intersection of PC and $P'C'$, and let PCQ, $P'C'Q'$ be diameters of the two conics; then, since PP' is parallel to CC', it follows that QQ' must also be parallel to CC'.

Hence
$$\frac{TQ\,.\,TP}{TC^2}=\frac{TQ'\,.\,TP'}{TC'^2}.$$

But, from Art. 186, Cor. III., $TQ\,.\,TP : TC^2=u : u_0$,
and
$$TQ'\,.\,TP' : TC'^2=v : v_0.$$

Hence the locus of T is given by $uv_0=vu_0$. The locus of T clearly passes through the points of intersection of the given conics [as is obvious *à priori*]. When $u=0$ and $v=0$ are similar and similarly situated, the conic $uv_0-u_0v=0$ will be similar and similarly situated to either.

88. Let the equation of the conic be $ax^2+by^2-1=0$. If a circle have double contact with a conic the chord of contact must be parallel to an axis of the conic, for $ax^2+by^2-1+\lambda(lx+my+n)^2=0$ can only represent a circle when $lm=0$. Also a circle will not pass through the four points of intersection of the conic $ax^2+by^2-1=0$ and the lines $(x-\alpha)(y-\beta)=0$. Hence the two chords of contact of the circles in question must be parallel. The equations of the three circles will therefore be of the forms

$$ax^2+by^2-1+\lambda(x-\alpha)^2=0 \quad\text{.........................(i)},$$

$$ax^2+by^2-1+\mu(x-\beta)^2=0 \quad\text{.........................(ii)},$$

and

$$ax^2+by^2-1+\nu(x-\alpha)(x-\beta)=0 \quad\text{.....................(iii)}.$$

But, in order that (i), (ii) and (iii) should represent circles, we must have

$$\lambda=\mu=\nu=b-a.$$

The squares of the tangents to the different circles from any point (x, y) on the conic are respectively

$$(b-a)(x-\alpha)^2,\ (b-a)(x-\beta)^2 \ \text{ and } \ (b-a)(x-\alpha)(x-\beta).$$

Hence
$$t^2t'^2-T^2=0.$$

89. If a circle have double contact with a conic, the chord of contact must be parallel to one or other of the axes of the conic, and the centre of the circle must be on an axis of the conic. Hence if a conic have double contact with each of two circles the chords of contact must either be parallel or perpendicular.

First let the chords of contact be parallel; then the centres of the two circles must be on the same axis and the chords of contact must be perpendicular to this axis.

The equation of the conic must therefore be of either of the forms

$$(x-a)^2+y^2-c^2+\lambda(x-\alpha)^2=0,$$

or

$$(x-b)^2+y^2-d^2+\mu(x-\beta)^2=0.$$

From this it follows that

$$\frac{1+\lambda}{1+\mu}=\frac{1}{1}=\frac{c^2-a^2-\lambda\alpha^2}{d^2-b^2-\mu\beta^2}=\frac{a+\lambda\alpha}{b+\mu\beta}.$$

Hence $\lambda=\mu$, $a-b=\lambda(\beta-\alpha)$ and $b^2+c^2-a^2-d^2=\lambda(\alpha^2-\beta^2)$;

$$\therefore\ \lambda\alpha=\{b^2+c^2-a^2-d^2+(b-a)^2\}/2(b-a).$$

Hence the general equation of the conic is

$$4(b-a)^2\lambda\{\lambda x^2+(x-a)^2+y^2-c^2\}-4(b-a)\{2b^2+c^2-d^2-2ab\}\lambda x$$
$$+\{2b^2+c^2-d^2-2ab\}^2=0.$$

Next suppose that the two chords of contact are at right angles, and that their equations are

$$x\cos\alpha+y\sin\alpha-p=0, \text{ and } x\sin\alpha-y\cos\alpha-q=0.$$

Then the equation of the conic will be of either of the forms

$$(x-a)^2+y^2-c^2+\lambda\,(x\cos a+y\sin a-p)^2=0,$$

and
$$(x-b)^2+y^2-d^2+\mu\,(x\sin a-y\cos a-q)^2=0.$$

We therefore have

$$\frac{1+\lambda\cos^2 a}{1+\mu\sin^2 a}=\frac{1+\lambda\sin^2 a}{1+\mu\cos^2 a}=\frac{\lambda}{-\mu}=\frac{a+\lambda p\cos a}{b+\mu q\sin a}=\frac{\lambda p\sin a}{-\mu q\cos a}=\frac{a^2-c^2+\lambda p^2}{b^2-d^2+\mu q^2}.$$

Whence we have $p\sin a=q\cos a$, and then

$$(b^2-d^2)\,\lambda+(a^2-c^2)\,\mu+\lambda\mu p^2\sec^2 a=0,$$

$$\lambda b+\mu a+\lambda\mu p\sec a=0,$$

and
$$\lambda+\mu+\lambda\mu=0.$$

Eliminating λ and μ we see that $p\sec a$ is constant; we have also

$$\lambda=(b-a)/(p\sec a-b).$$

Hence in this case the general equation of the conic having double contact with the two circles is

$$(x-a)^2+y^2-c^2+\frac{b-a}{p\sec a-b}\{\cos a\,(x-p\sec a)+y\sin a\}^2=0,$$

where $p\sec a$ is known.

In the second part of the question the chords of contact are assumed to be *parallel* and the locus of the extremities of the latus rectum which is parallel to the chords of contact is required.

The general equation of the conic is found as above to be

$$(x+a)^2+y^2-c^2+\lambda x^2-2xa+\frac{a^2}{\lambda}=0,$$

or
$$x^2\,(1+\lambda)+y^2=c^2-a^2-\frac{a^2}{\lambda}.$$

Hence, if (x, y) be an extremity of the latus rectum parallel to $x=0$, we have

$$x^2=\frac{c^2-a^2-\dfrac{a^2}{\lambda}}{1+\lambda}-\left(c^2-a^2-\frac{a^2}{\lambda}\right)=a^2-\frac{c^2\lambda}{1+\lambda},$$

and
$$y^2=\left(c^2-a^2-\frac{a^2}{\lambda}\right)(1+\lambda).$$

Eliminating λ we have the required result.

90. The centre of the first conic is given by

$$(l^2m' - l'^2m)\,x + (l - l')\,mm'y = lm' - l'm,$$

and

$$(l - l')\,mm'x + (m - m')\,mm'y = 0\,;$$

whence

$$\frac{x}{m - m'} = \frac{y}{l' - l} = \frac{-1}{lm' - l'm} \quad \text{.........................(i).}$$

The centre of the second conic is also the point given by (i), and the common centre of the two conics is the point of intersection of the two straight lines $lx + my - 1 = 0$ and $l'x + m'y - 1 = 0$.

Now it is clear that all conics through the points of intersection of two given concentric conics are concentric with the given conics; and therefore the lines $lx + my - 1 = 0$ and $l'x + m'y - 1 = 0$ are *diameters* of all the conics of the system.

The equation of any conic through the points of intersection of the given conics is $\{l^2m' - l'^2m - \lambda ll'(l - l')\}x^2 + 2\{(l - l')\,mm' - \lambda\,(m - m')\,ll'\}\,xy$

$$+ \{(m - m')\,mm' - \lambda\,(m^2l' - m'^2l)\}y^2 - 2\,(lm' - l'm)\,(x + \lambda y) = 0 \text{......(ii).}$$

The lines $lx + my - 1 = 0$ and $l'x + m'y - 1 = 0$ will [Art. 183] be *parallel* to conjugate diameters of (ii), provided

$$ll'\{(m - m')\,mm' - \lambda\,(m^2l' - m'^2l)\} + mm'\,\{l^2m' - l'^2m - \lambda ll'\,(l - l')\}$$
$$= (lm' + l'm)\,\{(l - l')\,mm' - \lambda\,(m - m')\,ll'\},$$

and it is easy to see that this condition is satisfied.

Hence the lines $lx + my - 1 = 0$ and $l'x + m'y - 1 = 0$ are conjugate diameters of (ii) for all values of λ.

91. Let $ax^2 + by^2 - 1 = 0$ be the equation of the conic, and (f, g) be the fixed point.

Then, if (ξ, η) be the middle point of any chord through (f, g), the equation of the chord will be $a\xi\,(x - \xi) + b\eta\,(y - \eta) = 0$.

Hence, for some value of c, the circle $(x - \xi)^2 + (y - \eta)^2 - c^2 = 0$ will cut the conic in points such that $a\xi\,(x - \xi) + b\eta\,(y - \eta) = 0$ will go through two of their points of intersection, and the equation of the line through the other two points will therefore be $a\xi x - b\eta y + n = 0$.

Hence for some values of λ, c, and n, the equations

$$ax^2 + by^2 - 1 + \lambda\,\{(x - \xi)^2 + (y - \eta)^2 - c^2\} = 0,$$

and

$$(a\xi x + b\eta y - a\xi^2 - b\eta^2)\,(a\xi x - b\eta y + n) = 0$$

will represent the same pair of straight lines. Hence, comparing the coefficients of x and y, we have

$$a\xi\,(n - a\xi^2 - b\eta^2)/2\lambda\xi = b\eta\,(n + a\xi^2 + b\eta^2)/2\lambda\eta,$$

whence

$$n = (a\xi^2 + b\eta^2)\,(a + b)\,/\,(a - b).$$

But the chord $a\xi x + b\eta y - a\xi^2 - b\eta^2 = 0$ goes through the fixed point (f, g);

$$\therefore \quad a\xi f + b\eta g - (a\xi^2 + b\eta^2) = 0.$$

And this shews that the line

$$a\xi x - b\eta y + \frac{a+b}{a-b}(a\xi^2 + b\eta^2) = 0$$

passes through the fixed point given by

$$\frac{x}{f} = \frac{y}{-g} = -\frac{a+b}{a-b}.$$

92. Let ABC be the triangle, and let P, Q be the given points. Then, if AP, BQ, CR cut BC, CA, AB respectively in the points A', B', C', we know that $$BA' \,.\, CB' \,.\, AC' = BC' \,.\, AB' \,.\, CA'.$$

So also $$BA'' \,.\, CB'' \,.\, AC'' = BC'' \,.\, AB'' \,.\, CA''.$$

Hence $BA' \,.\, BA'' \,.\, CB' \,.\, CB'' \,.\, AC' \,.\, AC'' = BC' \,.\, BC'' \,.\, AB' \,.\, AB'' \,.\, CA' \,.\, CA''$.

Hence, by the converse of Carnot's theorem, the six points A', B', C', A'', B'' and C'' are on a conic.

93. Let AB, BC, CD be three of the sides of the quadrilateral, and let these lines pass respectively through the fixed points P, O, Q which lie on a straight line. Take the point O for origin, and POQ for the axis of x, and let P be $(\alpha, 0)$ and Q be $(\beta, 0)$; then the equation of AB and CD will be

$$\{y - m_1(x-\alpha)\}\{y - m_2(x-\beta)\} = 0.$$

Hence, if $$ax^2 + 2hxy + by^2 + 2gx + 2fy + c = 0$$

be the equation of the conic, the equation of BC and AD will be included in

$$ax^2 + 2hxy + by^2 + 2gx + 2fy + c + \lambda\{y - m_1(x-\alpha)\}\{y - m_2(x-\beta)\} = 0 \ldots\ldots \text{(i)}.$$

The points where (i) cuts the axis of x are given by the equation

$$ax^2 + 2gx + c + \lambda m_1 m_2 (x-\alpha)(x-\beta) = 0.$$

Hence if (i) pass through the origin we have

$$c + \lambda m_1 m_2 \alpha\beta = 0,$$

and the other point of intersection is given by

$$ax + 2g - \frac{c}{\alpha\beta}\{x - (\alpha+\beta)\} = 0,$$

so that $$x = (2g\alpha\beta + c\alpha + c\beta)/(c - a\alpha\beta),$$

and this is independent of λ, m_1, and m_2. Thus *any* conic through A, B, C, D and O will cut OPQ in another fixed point.

[The theorem follows at once from Art. 320, Ex. 2.]

94. Project the conic into a circle having the point of intersection of PQ and $P'Q'$ for centre. [See Art. 318.] Then, if PP' pass through the fixed point S, it is obvious that QQ' will pass through the fixed point S', where S, S' are on a line through the centre of the circle and are equidistant from the centre. Also, since the angles SPQ' and $SP'Q$ are right angles, it follows that PQ' and $P'Q$ touch a conic of which S (and similarly S') is a focus, and of which the given circle is the auxiliary circle. This proves the proposition, since a conic and its auxiliary circle have double contact with one another, and this relation is unaltered by projection.

95. Let the tangents at the points whose eccentric angles are α, β cut the lines $x^2-a^2=0$ in points P, Q such that PQ is parallel to one of the equi-conjugates.

Then, if (x, y) be the point of intersection of the tangents, $\tan\frac{\alpha}{2}$ and $\tan\frac{\beta}{2}$ are given by

$$\frac{x}{a}\cos\theta+\frac{y}{b}\sin\theta=1,$$

or by
$$\frac{x}{a}-1+2\frac{y}{b}\tan\frac{\theta}{2}-\left(\frac{x}{a}+1\right)\tan^2\frac{\theta}{2}=0 \;\ldots\ldots\ldots\ldots\ldots\ldots\text{(i)}.$$

Hence

$$\tan\frac{\alpha}{2}\tan\frac{\beta}{2}=(a-x)/(a+x) \text{ and } \tan\frac{\alpha}{2}+\tan\frac{\beta}{2}=2ay/b(x+a) \;\ldots\ldots\text{(ii)}.$$

Now, $y=b\tan\frac{\alpha}{2}$ at P; and $y=b\cot\frac{\beta}{2}$ at Q. Hence, if PQ be parallel to one of the equi-conjugates, we have

$$\tan\frac{\alpha}{2}-\cot\frac{\beta}{2}=\pm 2, \text{ or } \tan\frac{\alpha}{2}\tan\frac{\beta}{2}-1=\pm 2\tan\frac{\beta}{2}.$$

Hence, from (ii), $\dfrac{x}{a+x}=\pm\tan\dfrac{\beta}{2}$.

Substituting for $\tan\frac{\beta}{2}$ in (i) we have $2xy=\pm ab$. Hence the locus of the point of intersection is one or other of two rectangular hyperbolas.

96. Let the equation of the rectangular hyperbola be $xy=c^2$, and let L, M, N, R be (x_1, y_1), (x_2, y_2), (x_3, y_3) and (x_4, y_4) respectively.

Then the equations of LM and NR are respectively

$$xy_1y_2+c^2y-c^2(y_1+y_2)=0 \text{ and } xy_3y_4+c^2y-c^2(y_3+y_4)=0.$$

Hence, if P be (ξ, η), the equation of PAa is

$$\frac{x-\xi}{y_1y_2}=\frac{y-\eta}{c^2}=\frac{r}{\sqrt{(y_1^2y_2^2+c^4)}}.$$

Hence, Pa is the value of r given by

$$\xi y_3y_4+ry_1y_2y_3y_4/\sqrt{(y_1^2y_2^2+c^4)}+c^2\eta+c^4r/\sqrt{(y_1^2y_2^2+c^4)}-c^2(y_3+y_4)=0;$$

$$\therefore\ Pa = -\frac{\xi y_3 y_4 + c^2\eta - c^2(y_3+y_4)}{c^4+y_1y_2y_3y_4}\cdot\sqrt{(y_1^2y_2^2+c^4)}$$

$$= -\xi\frac{(y_3-\eta)(y_4-\eta)}{c^4+y_1y_2y_3y_4}\cdot\sqrt{(y_1^2y_2^2+c^4)}.$$

But $$PA = \{\xi y_1y_2 + c^2\eta - c^2(y_1+y_2)\}/\sqrt{(y_1^2y_2^2+c^4)}$$
$$= \xi(y_1-\eta)(y_2-\eta)/\sqrt{(y_1^2y_2^2+c^4)}.$$

Hence

$$PA\,.\,Pa = -\xi^2\,.\,(y_1-\eta)(y_2-\eta)(y_3-\eta)(y_4-\eta)/(c^4+y_1y_2y_3y_4),$$

and the symmetry of this result shews that $PA\,.\,Pa = PB\,.\,Pb = \&c.$

97. Let P be (ξ, η), and let the ordinates of A, B, C be y_1, y_2, y_3 respectively; then y_1, y_2, y_3 are [Art. 106] the roots of

$$y^3 + 4a(\xi - 2a) + 8a^2\eta = 0 \quad\text{.........................}\quad \text{(i)}.$$

$B'C'$ is parallel to PA, and is therefore the tangent at the other extremities of the focal chord through A. Hence the ordinate of the point of contact of $B'C'$ is $-\dfrac{4a^2}{y_1}$, and therefore the equation of $B'C'$ is

$$y = -\frac{y_1}{2a}x - \frac{2a^2}{y_1}.$$

Hence, from Ex. 18, page 31, the area of $A'B'C'$

$$= \Sigma\frac{4a^5}{y_2^2y_3^2}(y_3-y_2) = \frac{4a^5}{y_1^2y_2^2y_3^2}(y_2-y_3)(y_3-y_1)(y_1-y_2).$$

But the area of $ABC = \dfrac{1}{8a}(y_2-y_3)(y_3-y_1)(y_1-y_2)$. [Ex. 21, p. 97.]

Hence $$\triangle A'B'C' : \triangle ABC = 32a^6/y_1^2y_2^2y_3^2$$
$$= a^2/2\eta^2, \text{ from (i)},$$
$$= \text{constant if } \eta \text{ is constant}.$$

98. Let O, O' be the middle points of AP, PB respectively, and let Q be the centre of the circle which touches the three circles whose centres are C, O, O'.

Let $AC = CB = 2a$, and let $CP = 2c$; then $AP = 2a + 2c$, $BP = 2a - 2c$, $OC = a - c$ and $O'C = a + c$.

Also $$OQ - OP = O'Q - O'P = 2a - CQ;$$

$$\therefore\ OQ = 3a + c - CQ \text{ and } O'Q = 3a - c - CQ.$$

Hence, if r, θ be the polar co-ordinates of Q, C being the pole, we have

$$(3a + c - r)^2 = (a - c)^2 + r^2 + 2(a - c)r\cos\theta,$$

and $$(3a - c - r)^2 = (a + c)^2 + r^2 - 2(a + c)r\cos\theta.$$

Hence $$c(3a-r)=-ac+ar\cos\theta;$$
and $$\therefore\ c=ar\cos\theta/(4a-r) \qquad \text{(i)}.$$
Also $$(3a-r)^2+c^2=a^2+c^2+r^2-2cr\cos\theta;$$
$$\therefore\ 4a^2-3ar=-cr\cos\theta \qquad \text{(ii)}.$$

Eliminating c from (i) and (ii) we have

$$(4a-3r)(4a-r)+r^2\cos^2\theta=0,$$
or $$16a^2-16ar+3r^2+r^2\cos^2\theta=0;$$
$$\therefore\ (4a-2r)^2-r^2\sin^2\theta=0;$$
$$\therefore\ \frac{2a}{r}=1\pm\frac{1}{2}\sin\theta.$$

Thus the required locus is two ellipses having the diameter AB for common latus rectum.

99. Let C be the centre and T the point.

Take CS and CT for axes; then the equation of the ellipse will be

$$(x+y\cos\omega)^2/a^2+y^2\sin^2\omega/b^2=1.$$

Let SL, $S'L'$ be perpendicular to the polar of T; and let lines through G, L, G', L' parallel to CT cut SCS' in the points g, l, g', l' respectively.

The proposition will be true provided $VG\,.\,VL=VG'\,.\,VL'$, and therefore provided $Cg\,.\,Cl=Cg'\,.\,Cl'$.

Let T be the point $(0, \lambda)$; then the equation of GVG' will be

$$\lambda x\cos\omega/a^2+\lambda y(1-e^2\cos^2\omega)/b^2=1 \qquad \text{(i)}.$$

The equation of ST is

$$\lambda x+aey=\lambda ae \qquad \text{(ii)}.$$

From (i) and (ii) we have

$$Cg=\frac{a^2e}{\lambda}\,\frac{b^2-\lambda^2(1-e^2\cos^2\omega)}{b^2e\cos\omega-\lambda a(1-e^2\cos^2\omega)}.$$

The equation of SL is easily found to be

$$x+ye^2\cos\omega=ae \qquad \text{(iii)}.$$

From (i) and (iii) wo have

$$Cl=\frac{a^2e}{\lambda}\,\frac{b^2e\cos\omega-\lambda a(1-e^2\cos^2\omega)}{b^2e^2\cos^2\omega-a^2+a^2e^2\cos^2\omega}.$$

Hence $$Cg\,.\,Cl=\frac{a^4e^2}{\lambda^2}\,\frac{b^2-\lambda^2(1-e^2\cos^2\omega)}{b^2e^2\cos^2\omega-a^2+a^2e^2\cos^2\omega}.$$

Since the value of $Cg\,.\,Cl$ would be unaltered by changing e into $-e$, it follows that $Cg\,.\,Cl=Cg'\,.\,Cl'$.

100. Let the equations of BC, CA, AB be

$$l_1x + m_1y - 1 = 0,\quad l_2x + m_2y - 1 = 0,$$

and
$$l_3x + m_3y - 1 = 0$$

respectively; and let the equations of $a_1b_1c_1$ and $a_2b_2c_2$ be

$$\lambda_1x + \mu_1y - 1 = 0$$

and
$$\lambda_2x + \mu_2y - 1 = 0$$

respectively.

Then the general equations of conics circumscribing

$$b_1b_2c_1c_2,\ c_1c_2a_1a_2 \text{ and } a_1a_2b_1b_2$$

are respectively

$$(l_2x + m_2y - 1)(l_3x + m_3y - 1) + k_1(\lambda_1x + \mu_1y - 1)(\lambda_2x + \mu_2y - 1) = 0 \ldots\text{(i)},$$

$$(l_3x + m_3y - 1)(l_1x + m_1y - 1) + k_2(\lambda_1x + \mu_1y - 1)(\lambda_2x + \mu_2y - 1) = 0 \ldots\text{(ii)},$$

and
$$(l_1x + m_1y - 1)(l_2x + m_2y - 1) + k_3(\lambda_1x + \mu_1y - 1)(\lambda_2x + \mu_2y - 1) = 0 \ldots\text{(iii)}.$$

Now
$$k_3(l_3x + m_3y - 1)(l_1x + m_1y - 1) - k_2(l_1x + m_1y - 1)(l_2x + m_2y - 1) = 0$$

is a conic through the points of intersection of (ii) and (iii), and this is two straight lines of which BC is one and the other is given by the equation

$$k_3(l_3x + m_3y - 1) = k_2(l_2x + m_2y - 1) \ldots\ldots\text{(iv)}.$$

Similarly the equations of the other common chords of (iii) and (i) and of (i) and (ii) are respectively

$$k_1(l_1x + m_1y - 1) = k_3(l_3x + m_3y - 1) \ldots\ldots\text{(v)},$$

and
$$k_2(l_2x + m_2y - 1) = k_1(l_1x + m_1y - 1) \ldots\ldots\text{(vi)}.$$

It is obvious that each of the lines (iv), (v), (vi) passes through an angular point of ABC, and that they all meet in a point, namely in the point given by

$$k_1(l_1x + m_1y - 1) = k_2(l_2x + m_2y - 1) = k_3(l_3x + m_3y - 1).$$

CHAPTER XI.

Pages 257—263.

1. Take the two straight lines for axes, and let ω be the angle between them.

Let (ξ, η) be the centre and c the radius of a circle which intercepts fixed lengths $2a$, $2b$ respectively from the axes.

Then the equation of the circle is

$$(x-\xi)^2+(y-\eta)^2+2(x-\xi)(y-\eta)\cos\omega-c^2=0.$$

This circle meets $y=0$ where

$$x=\xi+\eta\cos\omega\pm\sqrt{(c^2-\eta^2\sin^2\omega)}.$$

Hence $$a=\sqrt{(c^2-\eta^2\sin^2\omega)};$$

and similarly $$b=\sqrt{(c^2-\xi^2\sin^2\omega)}.$$

Hence $$a^2-b^2=(\xi^2-\eta^2)\sin^2\omega.$$

Hence the locus of the centre of the circle is the rectangular hyperbola whose equation is

$$x^2-y^2=(a^2-b^2)\operatorname{cosec}^2\omega.$$

2. Take O for origin and OPP', OQQ' for axes, and let the equation of the conic be

$$ax^2+2hxy+2by^2+2gx+2fy+c=0.$$

Then the equation of any other conic through P, P', Q, Q' is

$$ax^2+2hxy+by^2+2gx+2fy+c+\lambda xy=0,$$

or in polars

$$r^2\{a\cos^2\theta+(2h+\lambda)\sin\theta\cos\theta+b\sin^2\theta\}+2r(g\cos\theta+f\sin\theta)+c=0.$$

Hence $$\frac{1}{r_1}+\frac{1}{r_2}=-2(g\cos\theta+f\sin\theta)/c,$$

and is independent of λ.

Hence $$\frac{1}{OR}+\frac{1}{OR'}=\frac{1}{OS}+\frac{1}{OS'},$$

where S, S' are the points where ORR' cuts *any* conic through P, P', Q, Q'.

3. Take the tangent and normal at O as axes, and let the equation of the conic on which O lies be

$$ax^2+2hxy+by^2+2fy=0.$$

Let $$a'x^2+2h'xy+b'y^2+2g'x+2f'y+c'=0$$

be the equation of any other conic through the four given points; then the equation of any other conic through the four points is

$$ax^2+2hxy+by^2+2fy+\lambda(a'x^2+2h'xy+b'y^2+2g'x+2f'y+c')=0\ldots\text{(i)}.$$

The abscissae of the points where $y=0$ cuts the conic (i) are given by

$$ax^2+\lambda(a'x^2+2g'x+c')=0.$$

Hence $\dfrac{1}{x_1}+\dfrac{1}{x_2}=-\dfrac{2g'}{c'}$ and is therefore independent of λ. Thus $\dfrac{1}{OP}+\dfrac{1}{OP'}$ is constant.

4. Let the equation of the hyperbola be $xy=a^2$ and the equation of the circle $$(x-\alpha)^2+(y-\beta)^2-c^2=0.$$

The abscissae of the points of intersection are given by

$$x^2(x-\alpha)^2+(a^2-\beta x)^2-c^2x^2=0;$$

$$\therefore\ x_1+x_2+x_3+x_4=2\alpha.$$

Similarly $$y_1+y_2+y_3+y_4=2\beta.$$

Now if the points (x_3, y_3) and (x_4, y_4) are extremities of a diameter of the hyperbola $x_3+x_4=0$ and $y_3+y_4=0$. Hence in this case

$$x_1+x_2=2\alpha \text{ and } y_1+y_2=2\beta,$$

which shews that the centre of the circle is the middle point of the line joining (x_1, y_1) and (x_2, y_2).

5. We know from Art. 297 that the axes of the parabolas are always parallel to conjugate diameters of any conic of the system.

We also know that in a given ellipse the acute angle between two conjugate diameters is least when they are the equi-conjugates; and that in different ellipses the angle between the equi-conjugates is greatest in that which has the least eccentricity. Hence if the directions of a pair of conjugate diameters are known, the conic has the least eccentricity when they are the equi-conjugates.

6. If TQ, TQ' be the tangents, and V be the middle point of QQ'; then will TV be a diameter and QQ' will be parallel to its conjugate. Hence, as in 5, the eccentricity of the conic will be least when one of the equi-conjugates is TV and the other is parallel to QQ'.

7. The equation of any conic touching the axes in the points $(a, 0)$ $(0, b)$ is

$$\left(\frac{x}{a}+\frac{y}{b}-1\right)^2-2\lambda xy=0;$$

and the condition that $\frac{x}{l}+\frac{y}{m}-1=0$ should touch the conic is

$$\lambda=2\left(\frac{1}{a}-\frac{1}{l}\right)\left(\frac{1}{b}-\frac{1}{m}\right). \quad \text{[Art. 216.]}$$

Now, if (x, y) be the middle point of the intercept made on

$$\frac{x}{l}+\frac{y}{m}-1=0$$

by the axes, we have $2x=l$ and $2y=m$.

Hence the equation of the required locus is

$$\lambda=2\left(\frac{1}{a}-\frac{1}{2x}\right)\left(\frac{1}{b}-\frac{1}{2y}\right).$$

The locus is therefore a rectangular hyperbola unless $\lambda=\frac{2}{ab}$; and when $\lambda=\frac{2}{ab}$ the locus reduces to a straight line, and the original conic is a parabola for this value of λ.

8. Let the equation of the conic be

$$\left(\frac{x}{a}+\frac{y}{b}-1\right)^2-2\lambda xy=0.$$

Let (α, β) be the centre of the circle; then its equation will be

$$x^2+y^2+2xy\cos\omega-2x(\alpha+\beta\cos\omega)-2y(\beta+\alpha\cos\omega)=0.$$

Hence the equation of PQ is

$$\frac{x}{2(\alpha+\beta\cos\omega)}+\frac{y}{2(\beta+\alpha\cos\omega)}=1.$$

In order that PQ should touch the conic we must have [Art. 216]

$$\lambda=2\left\{\frac{1}{a}-\frac{1}{2(\alpha+\beta\cos\omega)}\right\}\left\{\frac{1}{b}-\frac{1}{2(\beta+\alpha\cos\omega)}\right\}.$$

Hence the locus of (x, y) is the hyperbola whose equation is

$$2(x+y\cos\omega)(y+x\cos\omega)(2-\lambda ab)-2a(y+x\cos\omega)-2b(x+y\cos\omega)+ab=0.$$

9. Let $OA=a$ and $OB=b$; then, if the area of the triangle OAB be constant, $ab=\text{constant}=c^2$ suppose.

The equation of any conic touching the axes at A and B is

$$\left(\frac{x}{a}+\frac{y}{b}-1\right)^2-2\lambda xy=0;$$

and if the conic pass through D, (a, b), we have $1-2\lambda ab=0$.

Hence the equation of the conic is

$$\frac{x^2}{a^2}+\frac{y^2}{b^2}+\frac{xy}{ab}-\frac{2x}{a}-\frac{2y}{b}+1=0.$$

The centre of the conic is therefore given by

$$\frac{x}{a}=\frac{y}{b}=\frac{2}{3}.$$

Hence the locus of the centre is the hyperbola whose equation is

$$xy=\frac{4}{9}c^2.$$

10. The general equation of the conics which touch the axes at the fixed points $(a, 0)$, $(0, b)$ is

$$\left(\frac{x}{a}+\frac{y}{b}-1\right)^2-2\lambda xy=0.$$

The equation of the tangent at (x', y') to this conic is

$$\left(\frac{x}{a}+\frac{y}{b}-1\right)\left(\frac{x'}{a}+\frac{y'}{b}-1\right)-\lambda(xy'+x'y)=0.$$

Hence, if the tangent pass through the fixed point (α, β) the point of contact satisfies the relations

$$\left(\frac{\alpha}{a}+\frac{\beta}{b}-1\right)\left(\frac{x'}{a}+\frac{y'}{b}-1\right)-\lambda(\alpha y'+\beta x')=0$$

and $$\left(\frac{x'}{a}+\frac{y'}{b}-1\right)^2-2\lambda x'y'=0.$$

Hence for all values of λ the point of contact is on the curve whose equation is

$$2xy\left(\frac{\alpha}{a}+\frac{\beta}{b}-1\right)\left(\frac{x}{a}+\frac{y}{b}-1\right)-(\alpha y+\beta x)\left(\frac{x}{a}+\frac{y}{b}-1\right)^2=0.$$

The required locus is therefore the conic whose equation is

$$2\left(\frac{\alpha}{a}+\frac{\beta}{b}-1\right)xy-(\alpha y+\beta x)\left(\frac{x}{a}+\frac{y}{b}-1\right)=0.$$

One of the conics of the system is the pair of coincident straight lines

$$\left(\frac{x}{a}+\frac{y}{b}-1\right)^2=0,$$

and any line through (α, β) cuts this conic in coincident points: this explains the presence of $\frac{x}{a}+\frac{y}{b}-1=0$ as part of the locus of the points of contact.

11. Let the equation of the conic be

$$ax^2+2hxy+by^2+2gx+2fy+c=0;$$

and let the equations of the straight lines be $l_1x+m_1y-1=0$, &c.

Then we have four relations of the form

$$Al_1^2+2Hl_1m_1+Bm_1^2+2Gl_1+2Fm_1+C=0 \quad\text{.................(i)},$$

where A, B, C, F, G, H are the co-factors of a, b, c, f, g, h respectively in the determinant

$$\begin{vmatrix} a & h & g \\ h & b & f \\ g & f & c \end{vmatrix}=\Delta.$$

The equation of the polar of (ξ, η) is

$$(a\xi+h\eta+g)\,x+(h\xi+b\eta+f)\,y+g\xi+f\eta+c=0.$$

Hence if the polar of (ξ, η) is the fixed straight line $\lambda x+\mu y+1=0$, we have

$$a\xi+h\eta+g+k\lambda=0,$$
$$h\xi+b\eta+f+k\mu=0,$$

and

$$g\xi+f\eta+c+k=0.$$

Hence

$$\xi\,(\lambda G+\mu F+C)=\lambda A+\mu H+G \quad\text{.......................(ii)}$$

and

$$\eta\,(\lambda G+\mu F+C)=\lambda H+\mu B+F \quad\text{...................... (iii)}.$$

Eliminating A, B, C, F, G, H from the equations (i), (ii) and (iii), we have

$$\begin{vmatrix} l_1^2 & 2l_1m_1 & m_1^2 & 2l_1 & 2m_1 & 1 \\ l_2^2 & 2l_2m_2 & m_2^2 & 2l_2 & 2m_2 & 1 \\ l_3^2 & 2l_3m_3 & m_3^2 & 2l_3 & 2m_3 & 1 \\ l_4^2 & 2l_4m_4 & m_4^2 & 2l_4 & 2m_4 & 1 \\ \lambda & \mu & 0 & 1-\lambda\xi & -\mu\xi & -\xi \\ 0 & \lambda & \mu & -\lambda\eta & 1-\mu\eta & -\eta \end{vmatrix}=0.$$

Multiply the last column by λ and subtract from the last but two, then multiply the last column by μ and subtract from the last but one, and we have the equivalent determinant

$$\begin{vmatrix} l_1^2 & 2l_1m_1 & m_1^2 & 2l_1-\lambda & 2m_1-\mu & 1 \\ l_2^2 & 2l_2m_2 & m_2^2 & 2l_2-\lambda & 2m_2-\mu & 1 \\ l_3^2 & 2l_3m_3 & m_3^2 & 2l_3-\lambda & 2m_3-\mu & 1 \\ l_4^2 & 2l_4m_4 & m_4^2 & 2l_4-\lambda & 2m_4-\mu & 1 \\ \lambda & \mu & 0 & 1 & 0 & -\xi \\ 0 & \lambda & \mu & 0 & 1 & -\eta \end{vmatrix}=0,$$

from which it is obvious that the locus of (ξ, η) is a straight line.

[The theorem can be easily proved by using Trilinear Co-ordinates, as in Art. 281.]

12. Let the equation of the hyperbola be $xy=c^2$. Then the equation of any conic touching the asymptotes of the hyperbola is of the form

$$(lx+my-1)^2=\lambda xy.$$

Hence the conic whose equation is

$$(lx+my-1)^2=\lambda c^2$$

passes through the points of intersection; and this conic is the two straight lines

$$lx+my-1\pm c\sqrt{\lambda}=0,$$

which are clearly parallel to, and equidistant from, the chord of contact whose equation is

$$lx+my-1=0.$$

13. The equation of any one of the conics may be taken to be

$$\frac{x^2}{a^2}+\frac{y^2}{b^2}=1,$$

with the condition $a+b=$ constant $=c$ suppose.

The equation of the polar of (x', y') with respect to the above conic is

$$xx'/a^2+yy'/b^2=1.$$

Hence if the polar is the fixed straight line $Ax+By=1$, we have

$$x'/a^2=A \text{ and } y'/b^2=B.$$

Hence $$\sqrt{\frac{x'}{A}}+\sqrt{\frac{y'}{B}}=a+b=c.$$

Hence the required locus is the parabola whose equation is

$$\sqrt{\frac{x}{A}}+\sqrt{\frac{y}{B}}=c.$$

14. Take two of the tangents for axes and let the equation of the other given tangent be

$$\frac{x}{h}+\frac{y}{k}-1=0.$$

Let the equation of the parabola be

$$\sqrt{\frac{x}{a}}+\sqrt{\frac{y}{b}}=1;$$

then the equation of the line through the points where the parabola touches the axes is

$$\frac{x}{a}+\frac{y}{b}-1=0\text{..................................(i).}$$

Since the parabola touches the line

$$\frac{x}{h}+\frac{y}{k}-1=0,$$

the equation

$$\frac{x}{a}-2\sqrt{\frac{x}{a}}+1=\frac{k}{b}\left(1-\frac{x}{h}\right)$$

must have equal roots, and therefore

$$\frac{h}{a}+\frac{k}{b}-1=0 \text{ .... (ii).}$$

Now (ii) shews that the chord of contact (i) passes through the *fixed point* (h, k); this proves the theorem.

The theorem may be thus enunciated:—TA, TB are tangents to a parabola which are met by any other tangent in the points P, Q respectively, and the parallelogram $TPOQ$ is completed; then the point O will be on AB.

15. Take the two given lines for axes, and let the equation of the parabola be

$$\sqrt{\frac{x}{h}}+\sqrt{\frac{y}{k}}=1;$$

and let the chord of contact pass through the given point (a, b).

Now the equation of the chord of contact is

$$\frac{x}{h}+\frac{y}{k}-1=0;$$

therefore

$$\frac{a}{h}+\frac{b}{k}-1=0 \text{ .... (i).}$$

The focus is given by the equations

$$x^2+y^2+2xy\cos\omega=hx=ky \text{ .... (ii).}$$

Eliminating h and k from (i) and (ii), we have the equation of the locus of the focus, namely

$$ax+by=x^2+y^2+2xy\cos\omega.$$

The locus of the focus is therefore a circle.

16. The axis of the parabola must be parallel to the line $ax-by=0$.

Hence if the axis pass through the fixed point (α, β) its equation must be

$$a(x-\alpha)-b(y-\beta)=0.$$

Now the focus is given by the equations

$$x^2+y^2+2xy\cos\omega=\frac{x}{a}=\frac{y}{b}.$$

Hence as the focus is on the axis it must be the point of intersection of

$$a(x-\alpha)-b(y-\beta)=0$$

and

$$\frac{x}{a}=\frac{y}{b};$$

and therefore the equation of the locus of the focus for different values of a and b is

$$x(x-\alpha)-y(y-\beta)=0,$$

which represents a rectangular hyperbola.

17. Take O for origin, and let the equation of the conic be

$$ax^2+2hxy+by^2+2gx+2fy+c=0.$$

Then the equation of any two lines through O is

$$Ax^2+2Hxy+By^2=0.$$

The conditions that the conic represented by

$$ax^2+2hxy+by^2+2gx+2fy+c+\lambda(Ax^2+2Hxy+By^2)=0$$

should be a circle are

$$a-b+\lambda(A-B)=0 \text{ .... (i)},$$

and

$$h+\lambda H \quad =0 \text{ .... (ii)}.$$

The centre of the conic is given by

$$ax+hy+g+\lambda(Ax+Hy)=0,$$
$$hx+by+f+\lambda(Hx+By)=0.$$

Hence,

$$y(ax+hy+g)-x(hx+by+f)=\lambda H(x^2-y^2)+\lambda(B-A)xy$$
$$=h(y^2-x^2)+(a-b)xy, \text{ from (i) and (ii)}.$$

Therefore $$gy-fx=0.$$

Hence the locus of the centres of all possible circles is the straight line $gy-fx=0$, and this line passes through O and is perpendicular to the polar of O with respect to the given conic.

18. Let QL cut the conic in S, and let RS and QP meet in M.

Then [Art. 213] the polar of M goes through L; also, since T is the pole of PQ, and M is on PQ, the polar of M goes through T. Hence TLK is the polar of M.

But, from Art. 213, the polar of M goes through the point of intersection of QR and PS, and therefore PSK is a straight line.

19. Take the fixed line on which P moves for the axis of x, and AB for the axis of y.

Let AR and BQ meet in V, and AD and QR meet in W.

Then VW is the polar of P with respect to the conic, and VF is the polar of W.

Let $$ax^2+2hxy+by^2+2gx+2fy+c=0$$

be the equation of the conic, and let P be $(\alpha, 0)$.

Then the equation of WV is

$$(ax+hy+g)\alpha+gx+fy+c=0 \text{ .... (i)}.$$

Hence W is $$\left(0, \; -\frac{g\alpha+c}{h\alpha+f}\right),$$

and therefore the equation of VP is

$$-(hx+by+f)\frac{ga+c}{ha+f}+gx+fy+c=0 \quad\text{.................}\quad \text{(ii)}.$$

Eliminating a between (i) and (ii) we have the equation of the locus of V, namely

$$(gx+fy+c)\{(hf-bg)y+hc-gf\}-(ax+hy+g)\{(fg-ch)x+(f^2-bc)y\}=0.$$

Hence the locus of V is a conic.

20. The conic

$$\frac{x^2}{a^2+\lambda}+\frac{y^2}{b^2+\lambda}=1$$

will pass through the point $(a\cos\alpha,\ b\sin\alpha)$ provided

$$\frac{a^2\cos^2\alpha}{a^2+\lambda}+\frac{b^2\sin^2\alpha}{b^2+\lambda}=1,$$

whence $\lambda=0$ or

$$\lambda=-(a^2\sin^2\alpha+b^2\cos^2\alpha).$$

Hence the equation of the *hyperbola* is

$$\frac{x^2}{(a^2-b^2)\cos^2\alpha}+\frac{y^2}{(b^2-a^2)\sin^2\alpha}=1,$$

or

$$\frac{x^2}{\cos^2\alpha}-\frac{y^2}{\sin^2\alpha}=a^2-b^2.$$

21. The tangent at $(x',\ y')$ to the conic

$$\frac{x^2}{a^2+\lambda}+\frac{y^2}{b^2+\lambda}=1$$

is

$$\frac{xx'}{a^2+\lambda}+\frac{yy'}{b^2+\lambda}=1.$$

If this tangent cut the major axis in the fixed point $(c,\ 0)$, we have

$$cx'=a^2+\lambda.$$

But

$$\frac{x'^2}{a^2+\lambda}+\frac{y'^2}{b^2+\lambda}=1,$$

and eliminating λ we have

$$\frac{x'^2}{cx'}+\frac{y'^2}{b^2-a^2+cx'}=1.$$

Hence the locus of $(x',\ y')$ is the circle whose equation is

$$x(b^2-a^2+cx)+cy^2=c(b^2-a^2+cx).$$

22. If the eccentric angles of P, Q be α, β respectively; then, as in 20, we have

$$\lambda=-(a^2\sin^2\alpha+b^2\cos^2\alpha) \text{ and } \mu=-(a^2\sin^2\beta+b^2\cos^2\beta).$$

If P, Q be extremities of conjugate diameters $\alpha=\beta\pm\frac{\pi}{2}$, and therefore

$$\lambda+\mu=-(a^2+b^2).$$

If the tangents at P and Q be at right angles

$$\frac{\cos\alpha\cos\beta}{a^2}+\frac{\sin\alpha\sin\beta}{b^2}=0.$$

Hence $$-\left(\frac{1}{\lambda}+\frac{1}{\mu}\right)=\frac{1+\tan^2\alpha}{a^2\tan^2\alpha+b^2}+\frac{1+\tan^2\beta}{a^2\tan^2\beta+b^2}$$

$$=\frac{a^2+b^2}{a^2b^2}\text{, since }\tan^2\alpha\tan^2\beta=\frac{b^4}{a^4}.$$

23. Let the equation of any one of the ellipses be

$$\frac{x^2}{a^2+\lambda}+\frac{y^2}{b^2+\lambda}=1.$$

Then, if (x, y) be the extremities of one of the equi-conjugate diameters, we have

$$2(x^2+y^2)=a^2+\lambda+b^2+\lambda.$$

Eliminating λ we have

$$\frac{2x^2}{2x^2+2y^2+a^2-b^2}+\frac{2y^2}{2x^2+2y^2-a^2+b^2}=1;$$

$$\therefore\ 2x^2-2y^2=a^2-b^2,$$

or $$\frac{x^2}{a^2-\frac{a^2+b^2}{2}}+\frac{y^2}{b^2-\frac{a^2+b^2}{2}}=1.$$

Thus the required locus is a confocal rectangular hyperbola.

24. The directions of the tangents to $x^2/a^2+y^2/b^2-1=0$ through the points (x, y) are given by $y=mx+\sqrt{(a^2m^2+b^2)}$. Hence, if θ be the angle between the tangents, we have

$$\tan^2\theta=\frac{4(b^2x^2+a^2y^2-a^2b^2)}{(x^2+y^2-a^2-b^2)^2}.$$

The parameters of the two confocals through (x, y) are given by

$$\frac{x^2}{a^2+\lambda}+\frac{y^2}{b^2+\lambda}=1,$$

or $$\lambda^2-\lambda(x^2+y^2-a^2-b^2)+a^2b^2-b^2x^2-a^2y^2=0.$$

Hence, if λ_1, λ_2 be the two parameters, we have

$$\tan^2\theta=\frac{-4\lambda_1\lambda_2}{(\lambda_1+\lambda_2)^2}\quad\text{(i)},$$

which is the required relation between θ, λ_1 and λ_2.

From (i) we have

$$\frac{\tan^2\frac{\theta}{2}}{\left(1-\tan^2\frac{\theta}{2}\right)^2}=\frac{-\lambda_1\lambda_2}{(\lambda_1+\lambda_2)^2},$$

and this equation is clearly satisfied by $\tan^2\frac{\theta}{2}=-\frac{\lambda_1}{\lambda_2}$ or by

$$\tan^2\frac{\theta}{2}=-\frac{\lambda_2}{\lambda_1}\text{..................................(ii)}.$$

Since the normals to the confocals through any point bisect the angles between the tangents drawn from that point to the original conic, the equation of the tangents when referred to the two normals as axes will be

$$y=\pm x\tan\frac{\theta}{2};\text{ or, from (ii),}$$

$$\frac{x^2}{\lambda_1}+\frac{y^2}{\lambda_2}=0.$$

25. Let d_1, d_2 be the lengths of the semi-diameters parallel to OPP' and OQQ' respectively.

Then [Art. 186] $$\frac{OP\,.\,OP'}{OQ\,.\,OQ'}=\frac{d_1^2}{d_2^2}.$$

Also [Art. 229] $$\frac{PP'}{QQ'}=\frac{d_1^2}{d_2^2};$$

$$\therefore\ OP\,.\,OP'\,.\,QQ'=OQ\,.\,OQ'\,.\,PP'.$$

Or thus

The line through O, (α, β) which makes an angle θ with the axis of x cuts $x^2/a^2+y^2/b^2-1=0$ the points whose distances from O are the roots of

$$r^2\left(\frac{\cos^2\theta}{a^2}+\frac{\sin^2\theta}{b^2}\right)+2r\left(\frac{\alpha\cos\theta}{a^2}+\frac{\beta\sin\theta}{b^2}\right)+\frac{\alpha^2}{a^2}+\frac{\beta^2}{b^2}-1=0.$$

Hence

$$\frac{OP\,.\,OP'}{PP'}=\frac{\frac{\alpha^2}{a^2}+\frac{\beta^2}{b^2}-1}{\sqrt{\left\{\left(\frac{\alpha\cos\theta}{a^2}+\frac{\beta\sin\theta}{b^2}\right)^2-\left(\frac{\alpha^2}{a^2}+\frac{\beta^2}{b^2}-1\right)\left(\frac{\cos^2\theta}{a^2}+\frac{\sin^2\theta}{b^2}\right)\right\}}}$$

$$=ab\left(\frac{\alpha^2}{a^2}+\frac{\beta^2}{b^2}-1\right)\Big/\sqrt{\{\cos^2\theta\,(b^2-\beta^2)-2\alpha\beta\sin\theta\cos\theta+\sin^2\theta\,(a^2-\alpha^2)\}}.$$

Now if OPP' touch the conic $x^2/(a^2+\lambda)+y^2/(b^2+\lambda)=1$ we have

$$\beta-\alpha\tan\theta=\sqrt{\{(a^2+\lambda)\tan^2\theta+b^2+\lambda\}},$$

$$\therefore\ \sin^2\theta\,(\alpha^2-a^2)-2\alpha\beta\sin\theta\cos\theta+\cos^2\theta\,(\beta^2-b^2)=\lambda.$$

Hence $$\frac{OP \,.\, OP'}{PP'} = \frac{ab}{\sqrt{(-\lambda)}} \cdot \left(\frac{\alpha^2}{a^2} + \frac{\beta^2}{b^2} - 1\right).$$

Similarly $$\frac{OQ \,.\, OQ'}{QQ'} = \frac{ab}{\sqrt{(-\lambda)}} \cdot \left(\frac{\alpha^2}{a^2} + \frac{\beta^2}{b^2} - 1\right).$$

Hence $$OP \,.\, OP' \,.\, QQ' = OQ \,.\, OQ' \,.\, PP'.$$

26. The tangent at (x', y') to the conic $x^2/(a^2+\lambda) + y^2/(b^2+\lambda) = 1$ will pass through the fixed point (α, β) if

$$\frac{x'\alpha}{a^2+\lambda} + \frac{y'\beta}{b^2+\lambda} = 1,$$

and $$\frac{x'^2}{a^2+\lambda} + \frac{y'^2}{b^2+\lambda} = 1.$$

To find the locus of (x', y') we have to eliminate λ between the last two equations. We have

$$\frac{\dfrac{1}{a^2+\lambda}}{\beta y' - y'^2} = \frac{\dfrac{1}{b^2+\lambda}}{x'^2 - \alpha x'} = \frac{-1}{x'y'(\alpha y' - \beta x')};$$

$$\therefore \quad \frac{\dfrac{1}{\beta y' - y'^2} - \dfrac{1}{x'^2 - \alpha x'}}{a^2 - b^2} = \frac{-1}{x'y'(\alpha y' - \beta x')}.$$

Hence the locus of (x', y') is the cubic curve whose equation is

$$(x^2 + y^2 - \alpha x - \beta y)(\alpha y - \beta x) - (a^2 - b^2)(x - \alpha)(y - \beta) = 0.$$

The cubic clearly passes through the point (α, β), and also through the foci.

27. The tangent to the conic $x^2/(a^2+\lambda) + y^2/(b^2+\lambda) = 1$ at the point (x, y') will be parallel to the fixed line $y = x \tan \alpha$, if

$$\tan \alpha = -x'(b^2+\lambda)/y'(a^2+\lambda).$$

Hence $$\lambda = -\frac{a^2 y' \sin \alpha + b^2 x' \cos \alpha}{y' \sin \alpha + x' \cos \alpha};$$

$$\therefore \; a^2 + \lambda = \frac{(a^2 - b^2)\, x' \cos \alpha}{y' \sin \alpha + x' \cos \alpha} \text{ and } b^2 + \lambda = \frac{(b^2 - a^2)\, y' \sin \alpha}{y' \sin \alpha + x' \cos \alpha}.$$

Substituting for $a^2 + \lambda$ and $b^2 + \lambda$ in the equation

$$x'^2/(a^2+\lambda) + y'^2/(b^2+\lambda) = 1,$$

we have $$(x' \sec \alpha - y' \operatorname{cosec} \alpha)(x' \cos \alpha + y' \sin \alpha) = a^2 - b^2.$$

Hence the locus of (x', y') is the rectangular hyperbola whose equation is

$$x^2 - y^2 + xy(\tan \alpha - \cot \alpha) = a^2 - b^2 \quad \text{.................... (i).}$$

Now the axes of the hyperbola are given by

$$\frac{x^2-y^2}{2}=\frac{2xy}{\tan\alpha-\cot\alpha} \qquad \text{(ii)}.$$

Eliminating (α) from (i) and (ii) we have the equation of the locus of the vertices of the rectangular hyperbolas for different values of α, namely

$$x^2-y^2+\frac{4x^2y^2}{x^2-y^2}=a^2-b^2,$$

i.e.
$$(x^2+y^2)^2=(a^2-b^2)(x^2-y^2),$$

or, in polars,
$$r^2=(a^2-b^2)\cos 2\theta.$$

28. Let ABC be the triangle, and let the sides BC, CA, AB touch the confocal ellipse in A', B', C' respectively.

Let EAF, FBD and DCE be the tangents at A, B, C.

Then the locus of the pole of BC with respect to all the conics of the given confocal system is the line through A' perpendicular to BC [Art. 227]. Hence $A'D$ is perpendicular to BC.

But, since BC and BA make equal angles with FBD [Art. 228], and BC and CA make equal angles with DCE, it follows that D is the centre of one of the escribed circles of the triangle ABC. Hence A', B', C' are the points where the escribed circles touch the sides of ABC.

29. If two conics have double contact the line through the intersection of the tangents at the points of contact and through the middle point of the chord of contact will pass through the centres of both curves. Hence if a conic have double contact with two concentric conics it must be concentric with the other two; and hence both the chords of contact must be diameters, for the conics

$$ax^2+by^2-1=0 \text{ and } ax^2+by^2-1+\lambda(lx+my+n)^2=0$$

are not concentric unless $n=0$.

Let now a conic have double contact with two confocal conics, and let PCP' and QCQ' be the two chords of contact. Let the tangents at P and Q meet in T; then CT will bisect PQ, since TP and TQ are tangents to the same conic; and since TP and TQ are tangents to two confocals, and CT bisects PQ, it follows from Art. 226, Ex. 4, that the tangents TP and TQ are at right angles. Hence the tangents at P, P' and Q, Q' form a rectangle.

30. Let T be (α, β); then the equation of the polar of T with respect to the conic

$$x^2/(a^2+\lambda)+y^2/(b^2+\lambda)=1$$

is
$$x\alpha/(a^2+\lambda)+y\beta/(b^2+\lambda)=1.$$

Hence, if (ξ, η) be the point of intersection of the normals at the extremities of the chord, the equation of the line joining the other two points the normals at which pass through (ξ, η) will be

$$\frac{x}{\alpha}+\frac{y}{\beta}+1=0. \quad \text{[Art. 197].}$$

Also, for same value of μ, the equations

$$\frac{x^2}{a^2+\lambda}+\frac{y^2}{b^2+\lambda}-1-\mu\left(\frac{x\alpha}{a^2+\lambda}+\frac{y\beta}{b^2+\lambda}-1\right)\left(\frac{x}{\alpha}+\frac{y}{\beta}+1\right)=0$$

and
$$xy\,(a^2-b^2)+(b^2+\lambda)\,\eta x-(a^2+\lambda)\,\xi y=0$$

will represent the same conic.

Hence $\mu=1$; and comparing the coefficients of xy, x and y in the two equations, we have

$$\frac{\dfrac{\alpha}{(a^2+\lambda)\beta}+\dfrac{\beta}{\alpha\,(b^2+\lambda)}}{a^2-b^2}=\frac{\dfrac{\alpha}{a^2+\lambda}-\dfrac{1}{\alpha}}{(b^2+\lambda)\eta}=\frac{\dfrac{\beta}{b^2+\lambda}-\dfrac{1}{\beta}}{-(a^2+\lambda)\,\xi};$$

$$\therefore\ \frac{\alpha^2(b^2+\lambda)+\beta^2(a^2+\lambda)}{a^2-b^2}=\frac{\beta\,(\alpha^2-a^2-\lambda)}{\eta}=\frac{\alpha\,(\beta^2-b^2-\lambda)}{-\xi}.$$

Hence each fraction

$$=\frac{\alpha^2b^2+\beta^2a^2+\beta^2(\alpha^2-a^2)+\alpha^2(\beta^2-b^2)}{a^2-b^2+\beta\eta-\alpha\xi}=\frac{\alpha\beta\,(\alpha^2-\beta^2-a^2+b^2)}{\alpha\eta+\beta\xi};$$

$$\therefore\ 2\alpha\beta\,(\alpha\eta+\beta\xi)=(\alpha^2-\beta^2-a^2+b^2)\,(a^2-b^2+\beta\eta-\alpha\xi).$$

Hence the locus required is the straight line whose equation is

$$(\alpha x+\beta y)\,(\alpha^2+\beta^2+a^2-b^2)=(a^2-b^2)\,(\alpha^2-\beta^2-a^2+b^2).$$

31. Let A, B, C be respectively the points of contact of the three tangents $B'AC'$, $C'BA'$, $A'CB'$ to the ellipse $x^2/a^2+y^2/b^2-1=0$; and let α, β, γ be the eccentric angles of A, B, C respectively.

Then, at the point A',

$$x=a\cos\tfrac{1}{2}(\beta+\gamma)/\cos\tfrac{1}{2}(\beta-\gamma)\ \text{ and }\ y=b\sin\tfrac{1}{2}(\beta+\gamma)/\cos\tfrac{1}{2}(\beta-\gamma);$$

and similarly for B' and C'.

Hence, if B' and C' are on the conic

$$x^2/(a^2+\lambda)+y^2/(b^2+\lambda)=1,$$

we have

$$\cos^2\tfrac{1}{2}(\gamma-\alpha)=\frac{a^2}{a^2+\lambda}\cos^2\tfrac{1}{2}(\gamma+\alpha)+\frac{b^2}{b^2+\lambda}\sin^2\tfrac{1}{2}(\gamma+\alpha),$$

and
$$\cos^2\tfrac{1}{2}(\alpha-\beta)=\frac{a^2}{a^2+\lambda}\cos^2\tfrac{1}{2}(\alpha+\beta)+\frac{b^2}{b^2+\lambda}\sin^2\tfrac{1}{2}(\alpha+\beta).$$

Hence $\lambda^2-a^2b^2=(a^2-b^2)\,\lambda\cos(\gamma+\alpha)-(a^2+\lambda)\,(b^2+\lambda)\cos(\gamma-\alpha)$,

and $\lambda^2-a^2b^2=(a^2-b^2)\,\lambda\cos(\alpha+\beta)-(a^2+\lambda)\,(b^2+\lambda)\cos(\alpha-\beta)$;

whence

$$(a^2b^2+2b^2\lambda+\lambda^2)\cos\gamma\cos\alpha+(a^2b^2+2a^2\lambda+\lambda^2)\sin\gamma\sin\alpha+\lambda^2-a^2b^2=0,$$

and $(a^2b^2+2b^2\lambda+\lambda^2)\cos\alpha\cos\beta+(a^2b^2+2a^2\lambda+\lambda^2)\sin\alpha\sin\beta+\lambda^2-a^2b^2=0$;

$$\therefore\ \frac{(a^2b^2+2b^2\lambda+\lambda^2)\cos\alpha}{\cos\frac{1}{2}(\beta+\gamma)}=\frac{(a^2b^2+2a^2\lambda+\lambda^2)\sin\alpha}{\sin\frac{1}{2}(\beta+\gamma)}=-\frac{\lambda^2-a^2b^2}{\cos\frac{1}{2}(\beta-\gamma)}.$$

Hence, at the point A', we have

$$x=\frac{a(a^2b^2+2b^2\lambda+\lambda^2)}{a^2b^2-\lambda^2}\cos\alpha,\ \text{ and }\ y=\frac{b(a^2b^2+2a^2\lambda+\lambda^2)}{a^2b^2-\lambda^2}\sin\alpha.$$

Hence A' is on the conic

$$\frac{x^2}{a^2(a^2b^2+2b^2\lambda+\lambda^2)^2}+\frac{y^2}{b^2(a^2b^2+2a^2\lambda+\lambda^2)^2}=\frac{1}{(a^2b^2-\lambda^2)^2},$$

and the difference of the squares of the semi-axes of this conic is

$$\{a^2(a^2b^2+2b^2\lambda+\lambda^2)^2-b^2(a^2b^2+2a^2\lambda+\lambda^2)^2\}\div(a^2b^2-\lambda^2)^2,$$

which is easily seen to be equal to a^2-b^2.

Hence A' moves on a confocal conic.

32. Let the equations of the ellipse and hyperbola be respectively

$$\frac{x^2}{a^2}+\frac{y^2}{b^2}=1\quad\text{and}\quad\frac{x^2}{a^2-\lambda}+\frac{y^2}{b^2-\lambda}=1.$$

Then the lines $\frac{x^2}{a^2}-\frac{y^2}{b^2}=0$ are coincident with

$$\frac{x^2}{a^2-\lambda}+\frac{y^2}{b^2-\lambda}=0;$$

and therefore $$\lambda(a^2+b^2)=2a^2b^2 \quad\text{.....................(i).}$$

Now the equation of any conic through the extremities of the axes of the ellipse is

$$\frac{x^2}{a^2}+\frac{y^2}{b^2}-1+2\mu xy=0 \quad\text{.....................(ii).}$$

Let (x', y') be any point common to the hyperbola and the conic (ii); then, the tangents at (x', y') will be at right angles, if

$$\frac{x'}{a^2-\lambda}\left(\frac{x'}{a^2}+\mu y'\right)+\frac{y'}{b^2-\lambda}\left(\frac{y'}{b^2}+\mu x'\right)=0,$$

or $$\frac{x'^2}{a^2(a^2-\lambda)}+\frac{y'^2}{b^2(b^2-\lambda)}+\mu x'y'\left(\frac{1}{a^2-\lambda}+\frac{1}{b^2-\lambda}\right)=0 \quad\text{.........(iii).}$$

But at the points of intersection of the hyperbola and (ii) we have

$$\frac{\lambda x'^2}{a^2(a^2-\lambda)}+\frac{\lambda y'^2}{b^2(b^2-\lambda)}-2\mu x'y'=0 \quad\text{.....................(iv).}$$

Now (iv) shews that the relation (iii) is true provided λ be such that

$$\frac{1}{a^2-\lambda}+\frac{1}{b^2-\lambda}+\frac{2}{\lambda}=0,$$

which leads to

$$\lambda(a^2+b^2)-2a^2b^2=0,$$

and this is known to be true.

33. The parameters of the confocals which pass through the point (ξ, η) are the roots of

$$\xi^2/(a^2+\lambda)+\eta^2/(b^2+\lambda)=1,$$

or $$\lambda^2+(a^2+b^2-\xi^2-\eta^2)\lambda+a^2b^2-b^2\xi^2-a^2\eta^2=0.$$

Hence $$\lambda_1\lambda_2=a^2b^2-b^2\xi^2-a^2\eta^2,$$

and
$$\begin{aligned}(\lambda_1-\lambda_2)^2&=(a^2+b^2-\xi^2-\eta^2)^2-4(a^2b^2-b^2\xi^2-a^2\eta^2)\\&=(a^2-b^2)^2-2(a^2-b^2)(\xi^2-\eta^2)+(\xi^2+\eta^2)^2\\&=\{(\xi+ae)^2+\eta^2\}\{(\xi-ae)^2+\eta^2\}.\end{aligned}$$

Hence

$$\lambda_1^2\lambda_2^2(\lambda_1-\lambda_2)^2/(a^2-b^2)^2=(b^2\xi^2+a^2\eta^2-a^2b^2)^2\{(\xi+ae)^2+\eta^2\}\{(\xi-ae)^2+\eta^2\}/(a^2-b^2)^2.$$

But the continued product of the squares of the four normals from (ξ, η) is equal to the expression on the right. [See solution of question 80, page 228.]

Or thus*:—

Let $$S=\frac{x^2}{a^2}+\frac{y^2}{b^2}-1 \quad\text{and}\quad S'=(x-\xi)^2+(y-\eta)^2-r^2.$$

Then $kS+S'=0$ is the equation of any conic through A, B, C, D the points of intersection of $S=0$ and the circle $S'=0$ which has its centre at the point (ξ, η) and is of radius r. Now if k be determined so that $kS+S'=0$ may represent straight lines, these straight lines will be one of the three pairs AB and CD, AC and BD, AD and BC. The equation determining k is the cubic

$$\left(\frac{k}{a^2}+1\right)\left(\frac{k}{b^2}+1\right)(\xi^2+\eta^2-r^2-k)=\left(\frac{k}{a^2}+1\right)\eta^2+\left(\frac{k}{b^2}+1\right)\xi^2,$$

which is equivalent to

$$\frac{\xi^2}{a^2+k}+\frac{\eta^2}{b^2+k}=\frac{r^2}{k}.$$

Now, if λ_1, λ_2 be the parameters of the two conics confocal to $S=0$ which pass through (ξ, η) we shall have

$$\frac{\xi^2}{a^2+k}+\frac{\eta^2}{b^2+k}-1=-\frac{(k-\lambda_1)(k-\lambda_2)}{(a^2+k)(b^2+k)}.$$

* The following interesting investigation is due to Mr A. R. Forsyth, of Trinity College.

Hence the equation to determine k is

$$-r^2 = \frac{k(k-\lambda_1)(k-\lambda_2)}{(a^2+k)(b^2+k)} \quad \text{.........................} \quad \text{(i)}.$$

Now, if the circle *touch* the ellipse, two of the three pairs of lines become coincident, and therefore the cubic in k must in this case have equal roots.

But when

$$k(k-\lambda_1)(k-\lambda_2)+r^2(a^2+k)(b^2+k)=0$$

has equal roots, the common value satisfies also

$$(k-\lambda_1)(k-\lambda_2)+k(k-\lambda_2)+k(k-\lambda_1)+r^2(a^2+k)+r^2(b^2+k)=0;$$

$\therefore$, eliminating r, we have

$$\frac{1}{k}+\frac{1}{k-\lambda_1}+\frac{1}{k-\lambda_2}-\frac{1}{a^2+k}-\frac{1}{b^2+k}=0 \quad \text{.................} \quad \text{(ii)}.$$

Now (ii) is of the fourth degree: there are therefore four values of k depending on λ_1 and λ_2, that is on the point (ξ, η), but not on r, such that the circle will touch the ellipse; and when the values of k are known, the corresponding values of r are given by (i).

But when the circle touches the ellipse, r is the length of the normal. Hence the length of any normal is given by (i), the quantity k occurring in it being one of the roots of (ii).

Hence, if r_1, r_2, r_3, r_4 be the lengths of the four normals, we have

$$r_1^2r_2^2r_3^2r_4^2 = \frac{\Pi(k) \,.\, \Pi(k-\lambda_1) \,.\, \Pi(k-\lambda_2)}{\Pi(a^2+k) \,.\, \Pi(b^2+k)}.$$

It will now be found that

$$\Pi(k)=\lambda_1\lambda_2a^2b^2, \quad \Pi(k-\lambda_1)=\lambda_1(\lambda_1-\lambda_2)(a^2+\lambda_1)(b^2+\lambda_1),$$

$$\Pi(k-\lambda_2)=\lambda_2(\lambda_2-\lambda_1)(a^2+\lambda_2)(b^2+\lambda_2), \quad \Pi(a^2+k)=a^2(a^2+\lambda_1)(a^2+\lambda_2)(b^2-a^2)$$

and
$$\Pi(b^2+k)=b^2(b^2+\lambda_1)(b^2+\lambda_2)(a^2-b^2).$$

Hence
$$r_1^2r_2^2r_3^2r_4^2 = \frac{\lambda_1^2\lambda_2^2(\lambda_1-\lambda_2)^2}{(a^2-b^2)^2}.$$

34. It is known [Art. 187, Ex. 1] that any rectangular hyperbola through the three points P, Q, R will pass through S, the orthocentre of the triangle PQR. It is also known [Art. 213] that if PQ, RS meet in A, QS, PR in B, and PS, QR in C; then ABC is a self-polar triangle with respect to any conic through P, Q, R and S. This proves the theorem, since A, B, C are the feet of the perpendiculars of the triangle PQR.

35. Take the tangents TP, TQ for axes, and let the equation of the conic be
$$(ax+by-1)^2+2\lambda xy=0.$$

The equation of the bisectors of the angles between the axes is $x^2 - y^2 = 0$; and these bisectors cut $ax + by - 1 = 0$ in the points O, O' where

$$O \text{ is } \left(\frac{1}{a+b}, \frac{1}{a+b}\right) \text{ and } O' \text{ is } \left(\frac{1}{a-b}, \frac{1}{b-a}\right).$$

Hence the equation of ROR' is

$$x - \frac{1}{a+b} + k\left(y - \frac{1}{a+b}\right) = 0,$$

and therefore the equation of TR, TR' is

$$\{ax + by - (x + ky)(a+b)/(1+k)\}^2 + 2\lambda xy = 0,$$

or

$$(x-y)^2 (ak - b)^2 + 2\lambda (1+k)^2 xy = 0,$$

which is of the form

$$(x-y)^2 + \mu xy = 0 \quad\text{.................................(i).}$$

Now if two straight lines be cut by any circle whose centre is at the intersection of the lines, two pairs of parallel lines will go through the points of intersection, and each pair of lines will be parallel to one or other of the bisectors of the angles between the original lines.

Hence the bisectors of the angles between the lines

$$(x-y)^2 + \mu xy = 0$$

are parallel to

$$(x-y)^2 + \mu xy + L(x^2 + y^2 + 2xy \cos\omega) = \lambda c^2,$$

provided L is so chosen that

$$(x-y)^2 + \mu xy + L(x^2 + y^2 + 2xy \cos\omega)$$

is a perfect square; and it is obvious that if

$$(x-y)^2 + \mu xy + L(x^2 + y^2 + 2xy \cos\omega)$$

is a perfect square it must be $(x \pm y)^2$. Hence the lines $x \pm y = 0$ are the bisectors of the angles between TR and TR': thus TO and TO' are the bisectors of RTR'.

Similarly, if $SO'S'$ be any line through O' cutting the conic in S, S'; then TO and TO' will be the bisectors of STS'.

Or thus:

Let TO, TO' be the tangents to the confocals through T, O and O' being on PQ. Then TO, TO' are the bisectors of the angle PTQ, and the pole of TO' with respect to any one of the system of confocals lies on TO [Art. 227]. But the pole of TO' with respect to the original conic must be on PQ, since TO' passes through the pole of PQ. *Hence O is the pole of TO'* with respect to the given conic; and similarly O' is the pole of TO.

Now let any line through O cut the conic in R, R' and TO' in K; then the pencil $T\{ROR'K\}$ is harmonic, and TO is perpendicular to TK; therefore TO bisects the angle RTR'.

36. The axis of the first parabola bisects the interior angle between AB and DC; and the axis of the second parabola bisects the interior angle between AB and CE.

Hence the angle between the axis is half the angle DCE, or one-quarter of the angle DE subtends at the centre of the circle.

37. It is known (or easily proved) that when a circle cuts a parabola in four points, the sum of the distances of the four points from the axis of the parabola is zero: the axis of the parabola therefore passes through the centre of mean position of the four points. Hence the point of intersection of the axes of the two parabolas through A, B, C, D is the centre of mean position of A, B, C, D.

Now, since ABC is a maximum triangle in an ellipse, the eccentric angles of A, B, C may be taken to be α, $\alpha+\frac{2\pi}{3}$, and $\alpha+\frac{4\pi}{3}$; and therefore [Art. 186, Ex. 1] the eccentric angle of D will be $2n\pi-3\alpha$.

Hence, if (x, y) be the point of intersection of the axes of the parabolas, we have

$$4x=a\left\{\cos\alpha+\cos\left(\alpha+\frac{2\pi}{3}\right)+\cos\left(\alpha+\frac{4\pi}{3}\right)+\cos(2n\pi-3\alpha)\right\},$$

or $$4x=a\cos 3\alpha;$$

and similarly $$4y=b\sin 3\alpha.$$

Hence the equation of the required locus is

$$\frac{x^2}{a^2}+\frac{y^2}{b^2}=\frac{1}{16}.$$

38. The axis of a parabola through the four points α, β, γ, δ will make equal angles with the join of α, β and of γ, δ. Hence the axes of the two parabolas are parallel to the lines

$$x\cos\frac{1}{2}(\alpha+\beta)+y\sin\frac{1}{2}(\alpha+\beta)=\pm\left\{x\cos\frac{1}{2}(\gamma+\delta)+y\sin\frac{1}{2}(\gamma+\delta)\right\},$$

i.e. parallel to

$$x\cos S+y\sin S=0 \text{ and } x\sin S-y\cos S=0.$$

Also the sum of the perpendiculars on either axis from α, β, γ, δ will be zero. Hence, if the axes be

$$x\cos S+y\sin S-p=0$$

and $$x\sin S-y\cos S-q=0,$$

we have $$\Sigma(a\cos\alpha\cos S+a\sin\alpha\sin S-p)=0;$$

$$\therefore\ p=\frac{a}{4}\Sigma\cos(S-\alpha).$$

Similarly $$q=\frac{a}{4}\Sigma\sin(S-\alpha).$$

Hence the equations of the axes are

$$x \cos S + y \sin S = \frac{a}{4} \Sigma \cos (S - \alpha),$$

and

$$x \sin S - y \cos S = \frac{a}{4} \Sigma \sin (S - \alpha).$$

Since the sum of the perpendiculars from $\alpha, \beta, \gamma, \delta$ on the axis of either of the two parabolas through these points is zero, it follows that the point of intersection of the axes is the centre of mean position of $\alpha, \beta, \gamma, \delta$. And similarly for any other four of the five points $\alpha, \beta, \gamma, \delta, \epsilon$. Whence it follows that all the five points of intersection of pairs of axes lie on the circle whose equation is

$$(5x_0 - 4x)^2 + (5y_0 - 4y)^2 = a^2,$$

where

$$5x_0 = a (\cos \alpha + \cos \beta + \cos \gamma + \cos \delta + \cos \epsilon)$$

and

$$5y_0 = a (\sin \alpha + \sin \beta + \sin \gamma + \sin \delta + \sin \epsilon).$$

39. The first part of the question is proved in Art. 212. To prove the second part, divide the polygon into quadrilaterals whose sides are A, B, C, X; X, D, E, Y; Y, F, G, Z; &c. Let the equations of the sides A, B, C, &c., be $a = 0$, $b = 0$, $c = 0$, &c. Then for points on the conic we have $ac = \lambda bx$, $xe = \mu dy$, $yg = \nu fz$, &c. Hence $a . c . e . g ...$ varies as $b . d . f . h ...$

40. The normals at two consecutive points on a curve intersect in the centre of curvature. But the normals at the ends of $\frac{x}{a} \cos \alpha + \frac{y}{b} \sin \alpha - 1 = 0$ and of $\frac{x}{a \cos \alpha} + \frac{y}{b \sin \alpha} + 1 = 0$ meet in a point. Hence the equation of QR is of the form

$$\frac{x}{a \cos \alpha} + \frac{y}{b \sin \alpha} + 1 = 0;$$

and therefore, if T be (x', y'), we have $x' = -a \sec \alpha$ and $y' = -b \operatorname{cosec} \alpha$. Eliminating α we see that (x', y') is on the curve whose equation is

$$\frac{a^2}{x^2} + \frac{b^2}{y^2} = 1.$$

41. The ordinates of the points of intersection of

$$(x - \alpha)^2 + (y - \beta)^2 - c^2 = 0 \text{ and } y^2 - 4ax = 0$$

are given by the equation

$$y^4 + 4a (4a - 2\alpha) y^2 - 32a^2\beta y + 16a^2 (\alpha^2 + \beta^2 - c^2) = 0.$$

If $4a - 2\alpha$ be positive, the signs in $f(y)$ are $+, +, -, \pm 1$ and the signs in $f(-y)$ are $+, +, +, \pm$. Hence the total number of changes of sign in $f(y)$ and in $f(-y)$ is two, and therefore by Descartes' Rule of Signs there cannot be more than two real roots.

The equation of any conic through the four points of intersection of $y^2 - 4ax = 0$ and $(x - \alpha)^2 + (y - \beta)^2 - c^2 = 0$ is of the form

$$\lambda (y^2 - 4ax) + (x - \alpha)^2 + (y - \beta)^2 - c^2 = 0 \quad \text{.....................(i).}$$

In order that (i) may represent a pair of straight lines, λ must satisfy the equation

$$\begin{vmatrix} 1, & 0, & -\alpha-2a\lambda \\ 0, & \lambda+1, & -\beta \\ -\alpha-2a\lambda, & -\beta, & \alpha^2+\beta^2-c^2 \end{vmatrix}=0,$$

or $$4a^2\lambda^3+4a(a+\alpha)\lambda^2+\lambda(4a\alpha-\beta^2+c^2)+c^2=0.$$

Hence $$\lambda_1+\lambda_2+\lambda_3=-\frac{(a+\alpha)}{a} \quad \text{..........................(ii).}$$

Now when (i) represents a pair of straight lines the equation of the parallel straight line through the vertex is $\lambda y^2+x^2+y^2=0$, which meet the parabola in points where $x=-4a(1+\lambda)$. Hence the sum of the abscissae is constant when the sum of the values of λ is constant, and this from (ii) is the case when α is constant.

42. Let $x=0$, $y=0$, $lx+my+1=0$ be the equations of the given straight lines.

Let the equation of a conic with respect to which the three straight lines form a self-polar triangle be

$$ax^2+2hxy+by^2+2gx+2fy+1=0.$$

Then since $lx+my+1=0$ is the polar of $(0, 0)$, it represents the same straight line as $gx+fy+1=0$, and therefore $g=l$, $f=m$.

Again the polar of $\left(-\frac{1}{l}, 0\right)$ with respect to the conic is $x=0$; hence $x=0$ is the same as $-\frac{ax}{l}-\frac{hy}{l}+lx-1+my+1=0$, whence $h=lm$; and this is a sufficient condition that $y=0$ is the polar of $\left(0, -\frac{1}{m}\right)$.

Hence the general equation of a conic with respect to which the given triangle is self-polar is

$$ax^2+2lmxy+by^2+2lx+2my+1=0 \quad \text{....................(i),}$$

where a and b are arbitrary.

Now the centre of (i) is given by

$$ax+lmy+l=0, \quad lmx+by+m=0.$$

And, if (i) is a rectangular hyperbola, $a+b-2lm\cos\omega=0$. Eliminating a and b from the last three equations we have the equation of the locus of the centres of the rectangular hyperbolas, namely

$$lm(x^2+y^2+2xy\cos\omega)+ly+mx=0.$$

Hence the locus is the circle circumscribing the triangle.

[The theorem can be easily proved geometrically from the fact that two conjugate diameters of a rectangular hyperbola make equal angles with an asymptote. See also the solution of 22, page 311.]

43. The equations of any three tangents to the circle are

$$x\cos\alpha+y\sin\alpha-c=0, \; x\cos\beta+y\sin\beta-c=0, \text{ and } x\cos\gamma+y\sin\gamma-c=0.$$

The vertices of the triangle formed by these tangents are

$$\left\{c\cos\frac{1}{2}(\beta+\gamma)\sec\frac{1}{2}(\beta-\gamma),\ c\sin\frac{1}{2}(\beta+\gamma)\sec\frac{1}{2}\beta-\gamma\right\},\ \&c.$$

These three points will all be on the ellipse provided values of α, β, γ can be found which will satisfy the three simultaneous equations

$$\frac{1}{a^2}\cos^2\frac{1}{2}(\beta+\gamma)+\frac{1}{b^2}\sin^2\frac{1}{2}(\beta+\gamma)=\frac{1}{c^2}\cos^2\frac{1}{2}(\beta-\gamma)\ \ldots\ldots\ldots\ldots(\text{i}),$$

$$\frac{1}{a^2}\cos^2\frac{1}{2}(\gamma+\alpha)+\frac{1}{b^2}\sin^2\frac{1}{2}(\gamma+\alpha)=\frac{1}{c^2}\cos^2\frac{1}{2}(\gamma-\alpha)\ \ldots\ldots\ldots(\text{ii}),$$

and

$$\frac{1}{a^2}\cos^2\frac{1}{2}(\alpha+\beta)+\frac{1}{b^2}\sin^2\frac{1}{2}(\alpha+\beta)=\frac{1}{c^2}\cos^2\frac{1}{2}(\alpha+\beta)\ \ldots\ldots\ldots(\text{iii}).$$

Now (i) can be written in the form

$$\frac{1}{a^2}+\frac{1}{b^2}-\frac{1}{c^2}+\left(\frac{1}{a^2}-\frac{1}{b^2}-\frac{1}{c^2}\right)\cos\beta\cos\gamma+\left(\frac{1}{b^2}-\frac{1}{a^2}-\frac{1}{c^2}\right)\sin\beta\sin\gamma=0.$$

But, since $\frac{1}{c}=\frac{1}{a}+\frac{1}{b}$,

$$\frac{1}{a^2}+\frac{1}{b^2}-\frac{1}{c^2}=-\frac{2}{ab},\ \frac{1}{a^2}-\frac{1}{b^2}-\frac{1}{c^2}=-\frac{2}{bc},\ \text{and}\ \frac{1}{b^2}-\frac{1}{a^2}-\frac{1}{c^2}=-\frac{2}{ac}.$$

Hence (i) and (ii) are equivalent to

$$a\cos\beta\cos\gamma+b\sin\beta\sin\gamma+c=0,$$

and

$$a\cos\gamma\cos\alpha+b\sin\gamma\sin\alpha+c=0;$$

$$\therefore\ \frac{a\cos\gamma}{\sin\beta-\sin\alpha}=\frac{b\sin\gamma}{\cos\alpha-\cos\beta}=\frac{c}{\sin(\alpha-\beta)};$$

$$\therefore\ \frac{a\cos\gamma}{\cos\frac{1}{2}(\alpha+\beta)}=\frac{b\sin\gamma}{\sin\frac{1}{2}(\alpha+\beta)}=\frac{-c}{\cos\frac{1}{2}(\alpha-\beta)};$$

$$\therefore\ \frac{1}{a^2}\cos^2\frac{1}{2}(\alpha+\beta)+\frac{1}{b^2}\sin^2\frac{1}{2}(\alpha+\beta)=\frac{1}{c^2}\cos^2\frac{1}{2}(\alpha-\beta),$$

which is (iii).

Hence if any two of the three equations (i), (ii), (iii) be true, the third will be true also. This proves that if *any* tangent to the circle cut the ellipse in the points A, B, and the other tangents to the circle through A and B meet in C, then C will also be on the ellipse.

44. Since the osculating circle at P passes through Q, the tangent at P and the line PQ make equal angles with the axes.

Also, since the osculating circle at Q passes through P, the tangent at Q and the line QP make equal angles with the axes.

Hence the tangents at P and Q are parallel, and therefore PQ is a *diameter*. Then the diameter PQ makes the same angle with the axis as its conjugate, and therefore PQ must be one of the *equi*-conjugates.

45. If the parabola be $y^2 - 4ax = 0$, all conics which have contact of the third order are included in the equation

$$\lambda (y^2 - 4ax) - \{yy' - 2a(x + x')\}^2 = 0.$$

This is a rectangular hyperbola if $\lambda - y'^2 - 4a^2 = 0$, and the equation of the hyperbola is

$$(y^2 - 4ax)(y'^2 + 4a^2) - \{yy' - 2a(x + x')\}^2 = 0,$$

or

$$ay^2 - ax^2 + y'xy - x(y'^2 + 4a^2 + 2ax') + x'y'y - ax'^2 = 0.$$

The centre is given by

$$2ay + y'x + x'y' = 0,$$

and

$$-2ax + y'y - y'^2 - 2ax' - 4a^2 = 0.$$

Hence the centre is given by

$$x + x' + 2a = 0 \text{ and } y = y';$$

and therefore the equation of the locus of the centre for different values of x' and y' which satisfy $y'^2 - 4ax' = 0$ is

$$y^2 + 4a(x + 2a) = 0.$$

Hence the locus is an equal parabola similarly placed with respect to the directrix of the original parabola.

46. Let the normals at P and Q meet in O and cut the conic again in P', Q' respectively; and let the tangents at P, Q meet $Q'OQ$, $P'OP$ respectively in T, T'.

Then, since PQ, PQ' make equal angles with POP', and PT, POP' are at right angles, the range $\{TQOQ'\}$ is harmonic.

Hence the polar of T goes through O; the polar of T also goes through P, and therefore $T'POP'$ is the polar of T.

Then, since the polar of T goes through T', the polar of T' will go through T. The polar of T' also goes through Q; and therefore $TQOQ'$ is the polar of T'. Hence $\{T'POP'\}$ is harmonic, and therefore, as QT', QO are at right angles, QP and QP' make equal angles with QOQ'.

47. The co-ordinates of the point at a distance r from (x', y') measured along the normal inwards are given by

$$\frac{x - x'}{\dfrac{x'}{a^2}} = \frac{y - y'}{\dfrac{y'}{b^2}} = \frac{r}{-\sqrt{\left(\dfrac{x'^2}{a^4} + \dfrac{y'^2}{b^4}\right)}} = -pr,$$

where p is the perpendicular from the centre on the tangent at (x', y').

The radius of curvature $= CD^2/p$.

Hence, at the centre of curvature,

$$x=\frac{x'}{a^2}(a^2-CD^2)=\frac{x'}{a^2}(a^2+\lambda)\quad \text{[See 20]},$$

and

$$y=\frac{y'}{b^2}(b^2-CD^2)=\frac{y'}{b^2}(b^2+\lambda).$$

Also the pole of

$$\frac{xx'}{a^2}+\frac{yy'}{b^2}=1$$

with respect to

$$\frac{x^2}{a^2+\lambda}+\frac{y^2}{b^2+\lambda}=1$$

is easily found to be

$$\left\{\frac{x'}{a^2}(a^2+\lambda),\ \frac{y'}{b^2}(b^2+\lambda)\right\}.$$

48. Let the tangents at A, B, C be QAR, RBP and PCQ respectively.

Then, since AB and AC touch a confocal, AB and AC must be equally inclined to QAR. Similarly BC, BA must be equally inclined to RBP, and CB, CA equally inclined to PCQ. Hence ABC is the pedal triangle of PQR.

Hence PA is perpendicular to QAR, and therefore PA touches the confocal hyperbola through A.

Again, as in Example 28, PA' is perpendicular to $BA'C$, and therefore PA' touches the confocal hyperbola through A'.

But, since ABC is the pedal triangle of PQR, the angle CBP is equal to PQR, and therefore BPA' = complement of $PQR=QPA$. And, since the angles BPA' and QPA are equal, and PB, PQ touch an ellipse, PA' and PA will touch the *same* confocal ellipse. Hence the confocal hyperbolas through A and through A' are the *same*.

Or thus:—

We have found in question 31, that if A', B', C' be on the conic

$$x^2/a^2+y^2/b^2=1,$$

and A, B, C on the confocal conic $x^2/(a^2+\lambda)+y^2/(b^2+\lambda)=1$; then, if the co-ordinates of A' be $a\cos\alpha$, $b\sin\alpha$, the co-ordinates of A will be

$$\frac{a\,(a^2b^2+2b^2\lambda+\lambda^2)}{a^2b^2-\lambda^2}\cos\alpha,\ \frac{b\,(a^2b^2+2a^2\lambda+\lambda^2)}{a^2b^2-\lambda^2}\sin\alpha.$$

The confocal hyperbola through A' will be [question 20]

$$\frac{x^2}{\cos^2\alpha}-\frac{y^2}{\sin^2\alpha}=a^2-b^2.$$

The condition that this hyperbola should pass through A is that

$$a^2(a^2b^2+2b^2\lambda+\lambda^2)^2-b^2(a^2b^2+2a^2\lambda+\lambda^2)^2=(a^2-b^2)(a^2b^2-\lambda^2)^2,$$

and it is easily seen that this condition is satisfied.

49. Let the equation of the hyperbolas be

$$x^2-y^2=\alpha^2-\beta^2 \quad\text{.......... (i)},$$

and

$$xy-\beta x-\alpha y=0 \quad\text{.......... (ii)}.$$

Then equation of any circle through the centre of (i) will be

$$x^2+y^2+2gx+2fy=0 \quad\text{.......... (iii)}.$$

The abscissae of the points in which (ii) and (iii) intersect are given by

$$x^2(x-\alpha)^2+2gx(x-\alpha)^2+\beta^2x^2+2f\beta x(x-\alpha)=0,$$

or

$$x^3-2(\alpha-g)x^2+(\alpha^2+\beta^2-4\alpha g+2\beta f)x+2g\alpha^2-2f\alpha\beta=0 \quad\text{...... (iv)},$$

and the ordinates are given by

$$y^3-\text{..............}+2f\beta^2-2g\alpha\beta=0 \quad\text{.......... (v)}.$$

These points being (x_1, y_1), (x_2, y_2) and (x_3, y_3), the conditions that they form a conjugate triad with respect to (i) are easily found to be

$$x_2x_3-y_2y_3=x_3x_1-y_3y_1=x_1x_2-y_1y_2=\alpha^2-\beta^2.$$

Now

$$x_2x_3-y_2y_3=2(f\beta-g\alpha)\left(\frac{\alpha}{x_1}+\frac{\beta}{y_1}\right)$$

$$=2(f\beta-g\alpha), \text{ since } (x_1, y_1) \text{ is on (ii)}.$$

Hence the necessary and sufficient conditions that PQR may be a self-polar triangle with respect to (i) is that

$$2f\beta-2g\alpha=\alpha^2-\beta^2.$$

But this condition is *not satisfied* by *any* circle through (0, 0) but only by circles whose centres are on the fixed line

$$2\alpha x-2\beta y-\alpha^2+\beta^2=0.$$

50. Let the equation of the hyperbola be $2xy=c^2$, and the equation of the first circle

$$x^2+y^2+2gx+2fy=0.$$

Then if (x_1, y_1), &c. be the four points A, B, C, D respectively, the equation of the second circle will be

$$x^2+y^2+2x_4x+2y_4y=0 \quad\text{.......... (i)}.$$

A', the point of intersection of the tangents at (x_2, y_2) and (x_3, y_3), will be found to be

$$\left(\frac{2x_2x_3}{x_2+x_3},\ \frac{c^2}{x_2+x_3}\right),$$

and this point is on (i) if

$$c^4(x_2+x_3+x_4)+4x_2x_3x_4(x_2x_3+x_3x_4+x_4x_2)=0 \quad\text{.......... (ii)}.$$

Now the abscissae of the points of intersection of the hyperbola and the first circle are given by the equation

$$4x^4 + 8gx^3 + 4fc^2x + c^4 = 0.$$

Hence $$4x_1x_2x_3x_4 = c^4,$$

and $$x_1(x_2 + x_3 + x_4) + x_2x_3 + x_3x_4 + x_4x_2 = 0,$$

and these shew that (ii) is true.

CHAPTER XII.

Pages 271—275.

1. Let the equation of the parabola be $y^2 - 4ax = 0$, and let P be (x', y'); then the equation of NM is

$$\frac{x}{x'} + \frac{y}{y'} - 1 = 0, \text{ or } 4ax + yy' - y'^2 = 0.$$

Hence the equation of the envelope is $-16ax = y^2$.

2. Let the equation of the line be $\frac{x}{a} + \frac{y}{a-c} = 1$, where c is the given difference of the intercepts. The equation may be written $a^2 - a(c + x + y) + cx = 0$; hence the equation of the envelope is $4cx = (x + y + c)^2$, which represents a parabola.

3. Take the fixed straight lines for axes, and let the equation of the line be $\frac{x}{a} + \frac{y}{b} = 1$. Then $ab = \text{constant} = c^2$ suppose. Hence we have

$$a^2y - ac^2 + xc^2 = 0,$$

and therefore the envelope is the hyperbola whose equation is $4xy = c^2$.

4. The equation of PD is

$$\frac{x}{a}\cos\left(\phi + \frac{\pi}{4}\right) + \frac{y}{b}\sin\left(\phi + \frac{\pi}{4}\right) = \cos\frac{\pi}{4},$$

or
$$\left(\frac{x}{a} + \frac{y}{b}\right)\cos\phi - \left(\frac{x}{a} - \frac{y}{b}\right)\sin\phi = 1,$$

or
$$\left(\frac{x}{a} + \frac{y}{b} - 1\right) - 2\left(\frac{x}{a} - \frac{y}{b}\right)\tan\frac{\phi}{2} - \left(\frac{x}{a} + \frac{y}{b} + 1\right)\tan^2\frac{\phi}{2} = 0.$$

Hence the equation of the envelope is

$$\left(\frac{x}{a} + \frac{y}{b} - 1\right)\left(\frac{x}{a} + \frac{y}{b} + 1\right) + \left(\frac{x}{a} - \frac{y}{b}\right)^2 = 0;$$

$$\frac{x^2}{a^2} + \frac{y^2}{b^2} = \frac{1}{2}.$$

The co-ordinates of the middle points of NP and MD are $a\cos\phi$, $\frac{b}{2}\sin\phi$ and $-a\sin\phi$, $\frac{b}{2}\cos\phi$. Hence the equation of the line joining them is

$$\frac{x}{a}(\sin\phi-\cos\phi)-\frac{2y}{b}(\cos\phi+\sin\phi)=-1,$$

or
$$\left(\frac{x}{a}+\frac{2y}{b}-1\right)-2\left(\frac{x}{a}-\frac{2y}{b}\right)\tan\frac{\phi}{2}-\left(\frac{x}{a}+\frac{2y}{b}+1\right)\tan^2\frac{\phi}{2}=0.$$

Hence the equation of the envelope is

$$\left(\frac{x}{a}+\frac{2y}{b}+1\right)\left(\frac{x}{a}+\frac{2y}{b}-1\right)+\left(\frac{x}{a}-\frac{2y}{b}\right)^2=0,$$

or
$$\frac{2x^2}{a^2}+\frac{8y^2}{b^2}=1.$$

5. Let the straight lines meet in O, and let $OA=a$, $OA'=a'$, $AB=l$, $A'B'=l'$; also let $AP=kl$, then will $A'P'=kl'$.

Hence the equation of the line PP' is

$$\frac{x}{a+kl}+\frac{y}{a'+kl'}=1,$$

or
$$k^2ll'+k(al'+a'l-xl'-yl)+aa'-xa'-ya=0.$$

Hence the equation of the envelope is

$$4ll'(aa'-xa'-ya)=(xl'+yl-al'-a'l)^2.$$

The envelope is therefore a parabola, and it is easy to verify that the axes are tangents.

6. Take OAP and OBQ for the axes of x and y respectively. Let $OA=a$, and $OB=b$, and let $AP \cdot BQ=c^2$.

Then the equation of PQ will be

$$\frac{x}{a+k}+\frac{y}{b+\frac{c^2}{k}}=1,$$

or
$$k^2(b-y)+k(c^2+ab-bx-ay)+c^2(a-x)=0.$$

Hence the equation of the envelope is

$$4(a-x)(b-y)=(c^2+ab-bx-ay)^2.$$

Thus the envelope is a conic touching the lines $x=a$ and $y=b$.

7. Take the two given straight lines for axes; then the equation of the system of circles is

$$x^2+2xy\cos\omega+y^2-2ax-2ay+a^2=0.$$

Let (a, β) be the given point; then the polar is

$$x\alpha+(x\beta+y\alpha)\cos\omega+y\beta-a(x+y+\alpha+\beta)+a^2=0.$$

Hence the equation of the envelope of the polar, for different values of a, is

$$4\{x(\alpha+\beta\cos\omega)+y(\beta+\alpha\cos\omega)\}=(x+y+\alpha+\beta)^2.$$

Thus the envelope is a parabola.

8. Let ϕ, ϕ' be the eccentric angles of the extremities of the chord, and let $b\sin\phi+b\sin\phi'=2kb$.

Then the equation of the chord is

$$\frac{x}{a}\cos\frac{\phi+\phi'}{2}+\frac{y}{b}\sin\frac{\phi+\phi'}{2}=\cos\frac{\phi-\phi'}{2}.$$

with the condition $$\sin\frac{\phi+\phi'}{2}\cos\frac{\phi-\phi'}{2}=k.$$

Hence $$\frac{x}{a}\sin\frac{\phi+\phi'}{2}\cos\frac{\phi+\phi'}{2}+\frac{y}{b}\sin^2\frac{\phi+\phi'}{2}=k,$$

or $$\left(\frac{y}{b}-k\right)\tan^2\frac{\phi+\phi'}{2}+\frac{x}{a}\tan\frac{\phi+\phi'}{2}-k=0.$$

Hence the envelope is the parabola whose equation is

$$\frac{x^2}{a^2}+4k\left(\frac{y}{b}-k\right)=0.$$

9. Let the equation of the fixed tangent be $y=m_1x+\frac{a}{m_1}$, and let $y=mx+\frac{a}{m}$ be the equation of any other tangent PT. Then the co-ordinates of T are $\frac{a}{mm_1}$ and $\frac{a}{m}+\frac{a}{m_1}$; and therefore the equation of TQ is

$$\left(y-\frac{a}{m}-\frac{a}{m_1}\right)m+x-\frac{a}{mm_1}=0,$$

or $$m^2\left(y-\frac{a}{m_1}\right)+m(x-a)-\frac{a}{m_1}=0.$$

Hence the envelope is the parabola whose equation is

$$4\frac{a}{m_1}\left(y-\frac{a}{m_1}\right)+(x-a)^2=0.$$

10. Take the given straight line for axis of x, and let the conic be given by the general equation.

Let P be $(\alpha, 0)$; then the equation of the polar of P is

$$x(a\alpha+g)+y(h\alpha+f)+g\alpha+c=0.$$

Hence the equation of the line through P parallel to its polar is

$$(x-\alpha)(a\alpha+g)+y(h\alpha+f)=0,$$

or

$$\alpha^2 \cdot a-\alpha(ax+hy-g)-(gx+fy)=0.$$

Hence the envelope for different values of α is the parabola whose equation is

$$(ax+hy-g)^2+4a(gx+fy)=0.$$

11. If the corner C fall on C', the line of the crease will bisect CC' at right angles, in Y suppose.

Hence the locus of Y is a fixed line bisecting the page; and, since the locus of the foot of the perpendicular from C on the line of the crease is a straight line, the line of the crease must envelope a parabola of which C is a focus.

12. Let P be one of the points of intersection of the fixed and moving ellipses; then OP must make equal angles with the major axes of the two ellipses, whence it follows that a common chord which is not a diameter must subtend a right angle at the centre. The envelope is therefore a circle from Art. 138, Ex. 3.

13. Let S be the fixed point, and AB the fixed straight line. Draw SA perpendicular to AB; and through A draw AC such that the angle SAC is equal to the given angle. Let SPQ be any position of the angle, P being on AB, and let PQ meet AC in Y; then, since $SAC=SPQ$, S, A, P, Y are all on a circle, and therefore SYP is a right angle.

Hence PQ moves so that the foot of the perpendicular on it from the fixed point S always lies on the fixed line AC: the envelope is therefore a parabola whose focus is S and of which AC is the tangent at the vertex.

14. Take for axes two conjugate diameters one of which is parallel to the given straight line. Let the equation of the ellipse be

$$ax^2+by^2=1,$$

and the equation of the straight line $x=\alpha$. Then the equation of the chord whose middle point is (α, β) is

$$a\alpha(x-\alpha)+b\beta(y-\beta)=0.$$

Hence the envelope, for different values of β, is the parabola whose equation is

$$by^2+4a\alpha(x-\alpha)=0.$$

15. Suppose the conic to be an ellipse, and let θ, ϕ_1, ϕ_2 be the eccentric angles of O, P, Q respectively.

Then, since OP, OQ are in given directions, $\theta+\phi_1$, and $\theta+\phi_2$ are both constant. Hence $\phi_1 \sim \phi_2 = \text{constant} = 2\alpha$ suppose.

The equation of PQ is

$$\frac{x}{a}\cos\frac{1}{2}(\phi_1+\phi_2)+\frac{y}{b}\sin\frac{1}{2}(\phi_1+\phi_2)=\cos\frac{1}{2}(\phi_1-\phi_2)=\cos a.$$

Hence the line touches the ellipse

$$\frac{x^2}{a^2\cos^2 a}+\frac{y^2}{b^2\cos^2 a}=1$$

at the point whose eccentric angle is $\frac{1}{2}(\phi_1+\phi_2)$.

The above applies to the case of the hyperbola, the eccentric angle of any point (x, y) being the *imaginary* angle given by

$$x=a\cos\theta,\ y=\sqrt{-1}\,.\,b\sin\theta.$$

If the conic is a parabola, and y_1, y_2, y_3 be the ordinates of O, P, Q respectively, then y_1+y_2 and y_1+y_3 are both constant since OP, OQ are drawn in constant directions, and therefore $y_2-y_3=\text{constant}=2c$.

Hence the equation of PQ is

$$y(y_2+y_3)-4ax-y_2y_3=0,$$

or

$$2y(y_2-c)-4ax-y_2(y_2-2c)=0,$$

whence the envelope is the parabola whose equation is

$$y^2-4ax+c^2=0.$$

16. Let the equation of the circle be $x^2+y^2-c^2=0$, and let the equation of PQ be $lx+my-1=0$. Then the equation of the diameters through P and Q is

$$x^2+y^2-c^2(lx+my)^2=0.$$

Hence, if these diameters are conjugate, we have

$$a^2(1-c^2l^2)+b^2(1-c^2m^2)=0,$$

or

$$a^2c^2l^2+b^2c^2m^2=a^2+b^2 \quad \text{(i)}.$$

The envelope of $lx+my=1$ with the condition (i) is

$$\frac{x^2}{a^2c^2}+\frac{y^2}{b^2c^2}=\frac{1}{a^2+b^2}.$$

Thus the envelope is a similar and similarly situated ellipse. The envelope will be the original ellipse provided $c^2=a^2+b^2$, that is provided the circle is the director-circle.

17. Let the equation of the straight line be

$$lx+my+1=0,$$

and let the n fixed points be (x_1, y_1), &c.

Then, by supposition,

$$\Sigma (lx_1 + my_1 + 1)^2 = \text{constant} = nc^2;$$

$$\therefore\ l^2\Sigma x_1^2 + 2lm\Sigma x_1y_1 + m^2\Sigma y_1^2 + 2l\Sigma x_1 + 2m\Sigma y_1 + n(1 - c^2) = 0,$$

which from Art. 239 is the tangential equation of a conic.

18. Take AB, AC for axes of x and y respectively, and let the equation of BC be

$$bx + cy = 1.$$

Let the moving line cut BC, CA, AB respectively in L, M, N, and let the equation of LMN be

$$lx + my = 1.$$

Then the abscissae of N and L will be $\frac{1}{l}$ and $\frac{m-c}{mb-lc}$ respectively. Hence

$$MN : ML = \frac{1}{l} : \frac{m-c}{mb-lc} = \text{constant} = 1 : k.$$

Therefore

$$l(m-c) = k(mb - lc);$$

$$\therefore\ lm - (1-k)\,lc - kbm = 0 \qquad \text{(i).}$$

We have to find the envelope of

$$lx + my - 1 = 0$$

with the condition (i).

The directions of the two lines through (x, y) are given by

$$lm - \{(1-k)\,cl + kbm\}\,(lx + my) = 0.$$

The lines will therefore coincide if

$$4k(1-k)\,bcxy = \{kbx + (1-k)\,cy - 1\}^2.$$

This is equivalent to

$$\sqrt{kbx} + \sqrt{(1-k)\,cy} = 1;$$

the envelope is therefore a parabola touching the sides of the triangle.

19. Take OA, OB for axes, and let the given fixed point be (α, β).

The equation of any circle through O is

$$x^2 + 2xy\cos\omega + y^2 + lx + my = 0.$$

Hence the equation of PQ is

$$\frac{x}{l} + \frac{y}{m} + 1 = 0 \qquad \text{(i).}$$

But, since the circle goes through (α, β), we have

$$\alpha^2 + 2\alpha\beta\cos\omega + \beta^2 + l\alpha + m\beta = 0 \qquad \text{(ii).}$$

The directions of the two lines through (x, y) are given by

$$lm(\alpha^2 + 2\alpha\beta\cos\omega + \beta^2) - (ly + mx)(l\alpha + m\beta) = 0.$$

Hence the equation of the envelope is

$$4\alpha\beta xy = (\alpha x + \beta y - \alpha^2 - 2\alpha\beta\cos\omega - \beta^2)^2,$$

or

$$\sqrt{\alpha x} + \sqrt{\beta y} + \sqrt{(\alpha^2 + 2\alpha\beta\cos\omega + \beta^2)} = 0.$$

Hence the envelope is a parabola which touches the axes.

20. Let the equation of the ellipse be

$$ax^2 + by^2 - 1 = 0;$$

then, from Art. 197, if the equation of RS be

$$lx + my - 1 = 0,$$

the equation of PQ will be

$$\frac{ax}{l} + \frac{by}{m} + 1 = 0.$$

Hence we have to find the envelope of

$$lx + my - 1 = 0$$

with the condition

$$\frac{a\alpha}{l} + \frac{b\beta}{m} + 1 = 0,$$

(α, β) being the fixed point through which PQ passes.

From Art. 219, Ex. 1, or Art. 239, Ex. 2, the equation of the required envelope is

$$\sqrt{a\alpha x} + \sqrt{b\beta y} = 1.$$

Thus RS envelopes a parabola touching the axes.

21. Let the equation of the hyperbola be

$$x^2 - y^2 - a^2 = 0.$$

The equation of the chord whose middle point is (α, β) is

$$(x - \alpha)\,\alpha - (y - \beta)\,\beta = 0.$$

Hence the equation of the line through the middle point of the chord and perpendicular to it is

$$(x - \alpha)\,\beta + (y - \beta)\,\alpha = 0.$$

Now if the extremities of the chord are on a circle whose centre is $(f, 0)$, the line which bisects the chord perpendicularly must pass through $(f, 0)$; we therefore have the condition

$$(f - \alpha)\,\beta - \alpha\beta = 0,$$

whence $\beta = 0$ or $2\alpha = f$. If $\beta = 0$ the line

$$(x - \alpha)\,\alpha - (y - \beta)\,\beta = 0$$

is parallel to an axis; and the envelope of

$$\left(x-\frac{f}{2}\right)\frac{f}{2}-(y-\beta)\beta=0$$

is the parabola whose equation is

$$(2x-f)f=y^2.$$

22. Take the line on which the centres lie for the axis of x, and the conjugate diameter for the axis of y; then the equation of any ellipse of the system is

$$\frac{(x-\alpha)^2}{a^2}+\frac{y^2}{b^2}-1=0,$$

where a and b are known.

The polar of the fixed point (x_1, y_1) is

$$\frac{xx_1}{a^2}+\frac{yy_1}{b^2}-1-\frac{\alpha(x+x_1)}{a^2}+\frac{\alpha^2}{a^2}=0.$$

Hence the required envelope is the parabola whose equation is

$$4\left(\frac{xx_1}{a^2}+\frac{yy_1}{b^2}-1\right)=\frac{(x+x_1)^2}{a^2}.$$

23. Let C, C' be the centres of the two circles of which C is fixed, and let O be the fixed point. Then the locus of C' is a circle whose centre is O. The radical axis bisects CC' at right angles, and therefore the locus of the foot of the perpendicular from C on the radical axis is a circle whose centre is the middle point of CO. Hence the line envelopes a conic of which C is one focus and whose centre is the middle point of CO; hence O is the other focus.

24. Let the equation of the ellipse be $ax^2+by^2=1$.

Let (x_1, y_1) be the pole of the chord joining the extremities of a pair of the diameters; then the equation of the chord is

$$ax_1x+by_1y=1.$$

Hence the equation of the corresponding pair of radii vectores is

$$ax^2+by^2=(ax_1x+by_1y)^2;$$

and, since the sum of the angles these two lines make with the major axis is a right angle, we have $a-a^2x_1^2=b-b^2y_1^2$.

Hence the locus of (x_1, y_1) is the concentric and co-axial hyperbola whose equation is $a^2x^2-b^2y^2=a-b$.

The envelope of the chords is the envelope of $ax_1x+by_1y-1=0$ with the condition $a^2x_1^2-b^2y_1^2-(a-b)=0$.

The envelope is therefore the rectangular hyperbola whose equation is

$$x^2-y^2=\frac{1}{a-b}.$$

25. Let (ka, kb) be any point on the equi-conjugate $bx=ay$; then the equation of the lines joining this point to the extremities of the axis major is easily found to be

$$k^2(bx-ay)^2-a^2(y-kb)^2=0.$$

Hence $$\frac{x^2}{a^2}+\frac{y^2}{b^2}-1-\lambda\{k^2(bx-ay)^2-a^2(y-kb)^2\}=0$$

will, for some value of λ, represent PQ and $y=0$. The coefficient of x^2 must therefore be zero, whence

$$1-\lambda a^2b^2k^2=0.$$

The equation of PQ is then found to be

$$2abk^2x+a^2y-2a^2bk=0;$$

and therefore the envelope, for different values of k, is the rectangular hyperbola whose equation is $2xy=ab$.

26. Let the equation of PNP' be $2x-a=0$. Let a parabola through C, P, P' cut the ellipse at the extremities of the chord whose equation is $lx+my+n=0$. Then the equation of the parabola will be

$$\frac{x^2}{a^2}+\frac{y^2}{b^2}-1-\lambda(2x-a)(lx+my+n)=0 \qquad\qquad\text{(i)}.$$

Since (i) goes through $(0, 0)$, we have $-1+\lambda an=0$; also, since (i) is a parabola, we have $$\left(\frac{1}{a^2}-2l\lambda\right)\frac{1}{b^2}=\lambda^2m^2.$$

Hence, eliminating λ, we have

$$n^2-2aln-b^2m^2=0.$$

The equation of the chord may therefore be written

$$n^2x+2an(my+n)-b^2xm^2=0.$$

Hence the envelope is

$$b^2x(x+2a)+a^2y^2=0,$$

or $$\frac{(x+a)^2}{a^2}+\frac{y^2}{b^2}=1.$$

27. Take the line parallel to and midway between the given parallel lines for axis of y, and the perpendicular line through the fixed point for axis of x; and let the equations of the parallel lines be $x=\pm a$, and let the fixed point be $(c, 0)$.

Then the line $y=m(x-c)$ cuts the parallel lines in the points

$$\{a,\ m(a-c)\} \text{ and } \{-a,\ -m(a+c)\}.$$

Hence the equation of the circle on PQ as diameter is

$$(y+mc-am)(y+mc+am)+x^2-a^2=0;$$

$$\therefore\ x^2+y^2-a^2+2cym+(c^2-a^2)m^2=0.$$

Hence two circles pass through any particular point (x, y); and when (x, y) is a point on the envelope the two circles will coincide. The equation of the envelope is therefore

$$(c^2-a^2)(x^2+y^2-a^2)=c^2y^2,$$

or

$$\frac{x^2}{a^2}+\frac{y^2}{a^2-c^2}=1.$$

28. Take the axes of x parallel to the chords and let the conic be given by the general equation.

The chord $y=\lambda$ cuts the conic in points given by

$$ax^2+2x(h\lambda+g)+b\lambda^2+2f\lambda+c=0.$$

Hence the equation of the circle of which the chord is a diameter is

$$x^2+2x\frac{h\lambda+g}{a}+\frac{b\lambda^2+2f\lambda+c}{a}+(y-\lambda)^2=0.$$

Hence the equation of the envelope of the circles, for different values of λ, is

$$(a+b)(ax^2+ay^2+2gx+c)=(hx-ay+f)^2.$$

The required envelope is therefore a conic.

29. Let y_1, y_2 be the ordinates of the extremities of the chord, then the equation of the chord is

$$y(y_1+y_2)-4ax-y_1y_2=0;$$

also the equation of the circle of which the chord is a diameter is

$$(y-y_1)(y-y_2)+(x-x_1)(x-x_2)=0.$$

Where the circle meets the parabola we have

$$16a^2(y-y_1)(y-y_2)+(y^2-y_1^2)(y^2-y_2^2)=0;$$

$$\therefore\ (y-y_1)(y-y_2)\{16a^2+(y+y_1)(y+y_2)\}=0.$$

Hence, in order that the circle may touch the parabola, the roots of

$$16a^2+(y+y_1)(y+y_2)=0$$

must be equal; and the condition for this is

$$4(16a^2+y_1y_2)=(y_1+y_2)^2,$$

or

$$y_1\sim y_2=8a.$$

Hence the equation of the chord may be written

$$2y(y_2+4a)-4ax-y_2(y_2+8a)=0,$$

and the equation of the envelope is therefore

$$4a(x-2y)=(4a-y)^2,$$

or

$$y^2=4a(x-4a).$$

The required envelope is therefore an equal parabola.

30. Take AP for axis of x, and a perpendicular line through A for axis of y, and let $AP=a$.

Let
$$x \cos \alpha + y \sin \alpha - p = 0$$
be the equation of any possible directrix; then
$$(-p \cos \alpha,\ -p \sin \alpha)$$
is the corresponding focus, and the equation of the corresponding parabola will therefore be
$$(x+p\cos\alpha)^2+(y+p\sin\alpha)^2=(x\cos\alpha+y\sin\alpha-p)^2.$$

But the parabola cuts $y=0$ in the point $(a, 0)$; hence
$$a^2\sin^2\alpha+4ap\cos\alpha=0 \quad\text{.............................(i).}$$

The equation of the directrix may therefore be written
$$4ax\cos^2\alpha+4ay\sin\alpha\cos\alpha+a^2\sin^2\alpha=0.$$

Hence the envelope is the parabola whose equation is $y^2=ax$.

31. Take the two fixed diameters to which the bisectors are parallel for axes, and let the equation of the conic be
$$ax^2+2hxy+by^2=1.$$

The equation of the two tangents from (x_1, y_1) is
$$(ax^2+2hxy+by^2-1)(ax_1^2+2hx_1y_1+by_1^2-1)$$
$$-(ax_1x+hx_1y+hy_1x+by_1y-1)^2=0.$$

The two tangents are therefore parallel to
$$\{(ab-h^2)\,y_1^2-a\}\,x^2+2\,\{(h^2-ab)\,x_1y_1-h\}\,xy+\{(ab-h^2)\,x_1^2-b\}\,y^2=0.$$

By supposition the tangents make equal angles with the axes; hence
$$(h^2-ab)\,x_1y_1-h=0 \quad\text{...............................(i).}$$

The equation of the chord of contact
$$(ax+hy)\,x_1+(hx+by)\,y_1-1=0$$
may be written
$$(h^2-ab)\,(ax+hy)\,x_1^2+h\,(hx+by)-(h^2-ab)\,x_1=0.$$

Hence the equation of the envelope is
$$4h\,(ax+hy)\,(hx+by)=h^2-ab.$$

Thus the envelope is an hyperbola whose asymptotes are $ax+hy=0$ and $hx+by=0$, and these are the conjugates of the axes.

32. Take AB, AC for axes, and let the conic be given by the general equation.

The equation of the polar of any point (x_1, y_1) is
$$x\,(ax_1+hy_1+g)+y\,(hx_1+by_1+f)+gx_1+fy_1+c=0.$$

Hence the equation of the line through A and the middle point of QQ' is

$$x(ax_1+hy_1+g)-y(hx_1+by_1+f)=0.$$

Hence the condition that this line should pass through P is

$$x_1(ax_1+hy_1+g)-y_1(hx_1+by_1+f)=0.$$

Hence the locus of P is the conic whose equation is

$$ax^2-by^2+gx-fy=0.$$

The polar of P is the same line as that represented by the equation

$$lx+my+n=0,$$

provided

$$ax_1+hy_1+g+\lambda l=0,$$

$$hx_1+by_1+f+\lambda m=0,$$

and

$$gx_1+fy_1+c+\lambda n=0.$$

Whence
$$\frac{x_1}{lA+mH+nG}=\frac{y_1}{lH+mB+nF}=\frac{1}{lG+mF+nC}.$$

But
$$x_1(ax_1+hy_1+g)-y_1(hx_1+by_1+f)=0;$$

$$\therefore\ l(lA+mH+nG)-m(lH+mB+nF)=0\ \ldots\ldots\ldots\ldots\ldots(ii),$$

and from Art. 239 the envelope of

$$lx+my+n=0$$

with condition (ii) is a conic.

33. Let the polar equation of the conic referred to the focus as pole be

$$\frac{l}{r}=1+e\cos\theta.$$

The equation of the line joining the points whose vectorial angles are $\alpha+\beta$ and $\alpha-\beta$ is

$$\frac{l}{r}=e\cos\theta+\sec\beta\cos(\theta-\alpha)\ \ldots\ldots\ldots\ldots\ldots\ldots\ldots\ldots(i).$$

The sum of the reciprocals of the focal distances of the extremities of the chord whose equation is (i) is

$$\frac{1}{l}\{2+e\cos(\alpha+\beta)+e\cos(\alpha-\beta)\}.$$

Hence, if the harmonic mean of the focal distances is constant,

$$\cos\alpha\cos\beta=\text{constant}=k\ \text{suppose}.$$

Hence (i) may be written

$$\frac{lk}{r}=ke\cos\theta+\cos\alpha\cos(\theta-\alpha),$$

or $$\frac{2kl}{r} = (1+2ke)\cos\theta + \cos(\theta - 2\alpha)\text{.....................(ii)}.$$

But the line whose equation is (ii) touches the ellipse whose equation is

$$\frac{2kl}{r} = 1 + (1+2ke)\cos\theta$$

at the point whose vectorial angle is 2α.

Thus the envelope is a conic having *one* focus in common with the given conic.

34. Let the equation of any chord which subtends a right angle at the focus of $y^2 - 4ax = 0$ be

$$lx + my - 1 = 0.$$

Referred to parallel axes through the focus the equations of the parabola and the line will be

$$y^2 - 4a(x+a) = 0 \text{ and } lx + my + la - 1 = 0$$

respectively.

Hence the equation of the straight lines joining the origin to the points of intersection is

$$y^2 - 4ax\frac{lx+my}{1-la} - 4a^2\left(\frac{lx+my}{1-la}\right)^2 = 0.$$

Since the lines are at right angles to one another, we have

$$(1-la)^2 - 4al(1-la) - 4a^2(l^2+m^2) = 0,$$

or $$l^2a^2 - 4m^2a^2 - 6al + 1 = 0.$$

Hence the directions of the tangents which pass through (x, y) are given by $$l^2a^2 - 4m^2a^2 - 6al(lx+my) + (lx+my)^2 = 0.$$

Hence the equation of the envelope is

$$(a^2 + x^2 - 6ax)(y^2 - 4a^2) = (xy - 3ay)^2,$$

or $$2y^2 + (x-3a)^2 = 8a^2.$$

35. The simplest proof of this theorem is found by reciprocating with respect to the fixed point. The theorem then becomes :—

'The locus of the point of intersection of two tangents to a parabola which cut at a constant angle is a conic having double contact with the parabola.' [See Art. 107, Ex. 1.]

Or thus:

Take the tangent and the normal at the fixed point O for axes, and let the equation of the conic be

$$ax^2 + 2hxy + by^2 + 2fy = 0.$$

Let $lx+my-1=0$ be the equation of any one of the chords; then the equation of the lines joining its extremities to the origin is

$$ax^2+2hxy+by^2+2fy\,(lx+my)=0.$$

The angle between these lines is

$$\tan^{-1} 2\sqrt{\{(h+fl)^2-a\,(b+2fm)\}}/(a+b+2fm).$$

Hence, if the chord subtend at O the angle α, we have

$$4\,(h+fl)^2-4a\,(b+2fm)-\tan^2\alpha\,(a+b+2fm)^2=0 \ldots\ldots\ldots\ldots\text{(i)}.$$

It therefore follows from Art. 239 that the envelope of the chord is a conic, except when $\alpha=\frac{\pi}{2}$, in which case the chord always passes through the fixed point

$$\left(0,-\frac{2f}{a+b}\right).$$

We will now shew that the envelope has double contact with the given conic, the chord of contact being the polar of $\left(0,\frac{-2f}{a+b}\right)$ whose equation is

$$2hx+(b-a)\,y+2f=0.$$

The line $lx+my-1=0$ will touch the conic

$$ax^2+2hxy+by^2+2fy+\lambda\,\{2hx+(b-a)\,y+2f\}^2=0,$$

provided

$$ax^2+2hxy+by^2+2fy\,(lx+my)+\lambda\,\{2hx+(b-a)\,y+2f\,(lx+my)\}^2=0$$

is a perfect square in x and y, the condition for which is

$$\{a+4\lambda\,(h+fl)^2\}\,\{b+2fm+\lambda\,(b-a+2fm)^2\}=\{h+fl+2\lambda\,(h+fl)\,(b-a+2fm)\}^2,$$

or
$$(h+fl)^2-a\,(b+2fm)-a\lambda\,\{4\,(h+fl)^2+(b-a+2fm)^2\}=0,$$

or
$$(h+fl)^2-a\,(b+2fm)-\frac{a\lambda}{1-4a\lambda}(a+b+2fm)^2=0 \ldots\ldots\ldots\ldots\text{(ii)}.$$

Now (ii) coincides with (i) provided

$$\frac{a\lambda}{1-4a\lambda}=\frac{\tan^2\alpha}{4},\ \text{ or }\ 4a\lambda=\sin^2\alpha.$$

Hence the envelope of $lx+my-1=0$ with condition (i) is the conic

$$4a\,\{ax^2+2hxy+by^2+2fy\}+\sin^2\alpha\,\{2hx+(b-a)\,y+2f\}^2=0.$$

Thus the envelope is a conic having double contact with the given conic.

36. Take the fixed point for origin, and let the equation of the circle be

$$x^2+y^2+2gx+c=0.$$

Let $lx+my-1=0$ be a line joining extremities of a pair of the perpendicular chords; then the lines given by

$$x^2+y^2+2gx(lx+my)+c(lx+my)^2=0$$

must be at right angles, the condition for which is

$$2+2gl+c(l^2+m^2)=0 \qquad\text{(i)}.$$

The envelope of $lx+my-1=0$ with condition (i) is then found to be

$$(2c-g^2)y^2+2cx^2+2cgx+c^2=0 \qquad\text{(ii)}.$$

Writing (ii) in the form

$$(g^2-2c)(x^2+y^2)=(gx+c)^2,$$

we see that the origin is one focus. Also the centre of (ii) is clearly midway between the origin and the centre of the circle; and therefore as the origin is one focus the centre is the other.

37. Let $y^2-4ax=0$ be the equation of the parabola, and let (α, β) be the point S; then the equation of the parabola referred to parallel axes through (α, β) is $(y+\beta)^2=4a(x+\alpha)$.

The polar of $(0, 0)$ is $\beta y-2ax+\beta^2-4a\alpha=0$; therefore the equation of SC is $2ay+\beta x=0$. Hence SC meets the axis $y+\beta=0$ in the point $(2a, -\beta)$.

Let $lx+my+1=0$ be the equation of any chord which subtends a right angle at S; then the lines given by

$$\{y-\beta(lx+my)\}^2+4ax(lx+my)-4a\alpha(lx+my)^2=0$$

are at right angles, and therefore

$$(1-\beta m)^2+\beta^2l^2+4al-4a\alpha(l^2+m^2)=0 \qquad\text{(i)}.$$

From Art. 241 the centre of the conic whose tangential equation is (i) is $(2a, -\beta)$: thus the centre of the conic (i) is C.

Note. *To find the centre of the conic whose tangential equation is given.*

The centre of a conic is always on the line midway between any pair of parallel tangents. Now the two tangents parallel to the axis of y are $l_1x+1=0$ and $l_2x+1=0$, where l_1, l_2 are the roots of $al^2+2gl+c=0$. Hence the centre is on the line $2x+\frac{1}{l_1}+\frac{1}{l_2}=0$, or $2x-\frac{2g}{c}=0$; and similarly on the line $2y-\frac{2f}{c}=0$. Therefore the centre is the point $\left(\frac{g}{c}, \frac{f}{c}\right)$.

38. Let the equation of the conic be $ax^2+by^2=1$, and let O be (α, β).

Transfer the origin to O; then the equation will be

$$ax^2+by^2+2(a\alpha x+b\beta y)+a\alpha^2+b\beta^2-1=0.$$

Let the equation of PQ, one of the chords, be $lx+my+1=0$; then the equation of OP, OQ will be

$$ax^2+by^2-2(a\alpha x+b\beta y)(lx+my)+(a\alpha^2+b\beta^2-1)(lx+my)^2=0.$$

Since OP, OQ are at right angles we have

$$a+b-2a\alpha l-2b\beta m+(a\alpha^2+b\beta^2-1)(l^2+m^2)=0 \text{............ (i).}$$

The directions of the two chords which pass through (x, y) are given by

$$(a+b)(lx+my)^2+2(a\alpha l+b\beta m)(lx+my)+(a\alpha^2+b\beta^2-1)(l^2+m^2)=0.$$

Hence if (x, y) be on the envelope we must have

$$\{(a+b)x^2+2a\alpha x+a\alpha^2+b\beta^2-1\}\{(a+b)y^2+2b\beta y+a\alpha^2+b\beta^2-1\}$$
$$=\{(a+b)xy+a\alpha y+b\beta x\}^2;$$

$$\therefore (a+b)(a\alpha^2+b\beta^2-1)(x^2+y^2)-(a\alpha y-b\beta x)^2$$
$$+2(a\alpha x+b\beta y)(a\alpha^2+b\beta^2-1)+(a\alpha^2+b\beta^2-1)^2=0,$$

or $(a\alpha x+b\beta y+a\alpha^2+b\beta^2-1)^2=(x^2+y^2)\{a^2\alpha^2+b^2\beta^2-(a+b)(a\alpha^2+b\beta^2-1)\}$,

from which it is evident that the origin is a focus and that the equation of the corresponding directrix is

$$a\alpha x+b\beta y+a\alpha^2+b\beta^2-1=0:$$

the directrix is therefore the polar of O with respect to the original conic.

The centre of the conic whose tangential equation is (i) is

$$\left(\frac{-a\alpha}{a+b}, \frac{-b\beta}{a+b}\right) \text{ [see 37];}$$

and therefore the other focus is

$$\left(\frac{-2a\alpha}{a+b}, \frac{-2b\beta}{a+b}\right).$$

Hence the second focus is fixed if the ratio $a : b$ is fixed: the envelopes corresponding to a system of concentric similar and similarly situated conics are therefore confocal.

39. Let $lx+my-1=0$ be the equation of PQ; then the equation of RS will be

$$\frac{x}{(a^2+\lambda)l}+\frac{y}{(b^2+\lambda)m}+1=0;$$

$$\therefore b^2mx+a^2ly+a^2b^2lm+\lambda(mx+ly+a^2lm+b^2lm)+\lambda^2lm=0.$$

Hence the envelope of RS, for different values of λ, is the parabola whose equation is

$$4lm(b^2mx+a^2ly+a^2b^2lm)=(mx+ly+a^2lm+b^2lm)^2,$$

or

$$\left(\frac{x}{l}-\frac{y}{m}+a^2-b^2\right)^2+4\frac{xy}{lm}=0.$$

Hence the envelope touches the axes, the equation of the chord of contact being

$$\frac{x}{l}-\frac{y}{m}+a^2-b^2=0.$$

40. Let a, b be the radii of the circles and $2c$ the distance between their centres. Take the origin midway between the centres of the circles, and let the line joining the centres be the axis of x.

Then the lengths of the intercepts made by the circles on the line whose equation is $lx+my-1=0$ are respectively

$$\sqrt{\left\{a^2-\frac{(lc-1)^2}{l^2+m^2}\right\}},$$

and

$$\sqrt{\left\{b^2-\frac{(lc+1)^2}{l^2+m^2}\right\}}.$$

Hence, if the intercepts be in the ratio $k:1$, we have

$$a^2(l^2+m^2)-(lc-1)^2=k^2b^2(l^2+m^2)-k^2(lc+1)^2,$$

and therefore the line will envelope a conic.

If $k=1$ we have

$$(a^2-b^2)(l^2+m^2)+4cl=0,$$

whence the envelope of $lx+my-1=0$ is at once found to be the parabola whose equation is

$$4c^2y^2=(a^2-b^2)(4cx+a^2-b^2).$$

41. Let the equation of the hyperbola be $x^2-y^2-a^2=0$, and let (α, β) be the fixed point O. Then the equation referred to O will be

$$(x+\alpha)^2-(y+\beta)^2-a^2=0.$$

Then, if $lx+my-1=0$ be the equation of any straight line which subtends a right angle at O, the lines represented by

$$x^2-y^2+2(\alpha x-\beta y)(lx+my)+(\alpha^2-\beta^2-a^2)(lx+my)^2=0$$

will be at right angles, the condition for which is

$$(\alpha^2-\beta^2-a^2)(l^2+m^2)+2\alpha l-2\beta m=0.$$

Hence the directions of the two chords which pass through (x, y) are given by

$$(\alpha^2-\beta^2-a^2)(l^2+m^2)+2(\alpha l-\beta m)(lx+my)=0.$$

Hence if the two chords through (x, y), which subtend a right angle at O, are at right angles, we have

$$\alpha^2-\beta^2-a^2+\alpha x-\beta y=0,$$

so that (x, y) is on the polar of O.

42. Let the equation of PQ be $lx+my-1=0$; the equation of AP, AQ will be

$$y^2-4ax(lx+my)=0.$$

Since AP, AQ make an angle of $\frac{\pi}{4}$ with one another we have

$$(1-4al)^2=4\{4a^2m^2+4al\},$$

or

$$16a^2l^2-16a^2m^2-24al+1=0.$$

Hence the directions of the two chords through (x, y) are given by

$$16a^2l^2 - 16a^2m^2 - 24al(lx+my) + (lx+my)^2 = 0.$$

Hence the equation of the envelope is

$$(x^2 - 24ax + 16a^2)(y^2 - 16a^2) = (xy - 12ay)^2,$$

or

$$\frac{y^2}{16a^2} + \frac{(x-12a)^2}{128a^2} = 1.$$

43. Take the given point for origin, and the given straight line for axis of x, and let the conic be given by the general equation. Then, if

$$lx + my - 1 = 0$$

be the equation of the line joining a pair of the points, the lines represented by

$$ax^2 + 2hxy + by^2 + 2(gx+fy)(lx+my) + c(lx+my)^2 = 0$$

must be equally inclined to the axis of x; and therefore

$$h + gm + fl + clm = 0 \text{..............................} \text{(i)}.$$

We have therefore to find the envelope of $lx + my - 1 = 0$ with the condition (i).

The equation of the two lines through (x, y) is given by

$$clm + (gm + fl)(lx + my) + h(lx+my)^2 = 0 \text{..............} \text{(ii)}.$$

The lines given by (ii) are coincident provided

$$4(hx^2 + fx)(hy^2 + gy) = (2hxy + gx + fy + c)^2,$$

or

$$4(fg - hc)xy = (gx + fy + c)^2.$$

Hence the chords envelops a conic which touches the axes of coordinates: the origin must therefore be on the director-circle of the conic.

44. Take the fixed point O for origin, and let $S_1 = 0$ and $S_2 = 0$ be the equations of any two of a system of conics which pass through four fixed points; then the equation of any other conic of the system is $S_1 + \lambda S_2 = 0$.

Let $lx + my + 1 = 0$ be the equation of any chord of $S_1 + \lambda S_2 = 0$ which subtends a right angle at O; then the straight lines represented by the equation

$$(a_1 + \lambda a_2)x^2 + 2(h_1 + \lambda h_2)xy + (b_1 + \lambda b_2)y^2$$
$$- 2\{(g_1 + \lambda g_2)x + (f_1 + \lambda f_2)y\}(lx + my) + (c_1 + \lambda c_2)(lx+my)^2 = 0$$

must be at right angles to one another. Hence we have

$$a_1 + b_1 - 2lg_1 - 2mf_1 + c_1(l^2 + m^2)$$
$$+ \lambda\{a_2 + b_2 - 2lg_2 - 2mf_2 + c_2(l^2 + m^2)\} = 0 \text{.................} \text{(i)}.$$

Now (i) is the tangential equation of a conic, and therefore the chords of $S_1 + \lambda S_2 = 0$ which subtend a right angle at O envelope a conic. Also the system of conics of which (i) is the tangential equation will clearly all touch the four common tangents of the conics whose tangential equations are respectively

$$a_1+b_1-2lg_1-2mf_1+c_1(l^2+m^2)=0,$$

and

$$a_2+b_2-2lg_2-2mf_2+c_2(l^2+m^2)=0.$$

45. Take AB, AC for axes and let B, C be $(g, 0)$, $(0, f)$ respectively, and let D be (x_1, y_1).

The general equation of a conic through A, B, C is readily found to be

$$ax^2+2hxy+by^2-agx-bfy=0.$$

The conic will also pass through (x_1, y_1) if the three unknown constants a, h and b satisfy the relation

$$a(x_1^2-gx_1)+2hx_1y_1+b(y_1^2-fy_1)=0 \quad\text{..................(i).}$$

The equation of the tangent at B is

$$2agx+2hgy-ag(x+g)-bfy=0.$$

Hence if the equation of PQ be $lx+my+1=0$, we have

$$mag^2+2hg-bf=0 \quad\text{.......................... (ii).}$$

We have similarly

$$lbf^2+2hf-ag=0 \quad\text{.......................... (iii).}$$

From (i), (ii), (iii) we have

$$\begin{vmatrix} x_1^2-gx_1, & x_1y_1, & y_1^2-fy_1 \\ mg^2\ , & g\ , & -f \\ -g\ , & f\ , & lf^2 \end{vmatrix}=0 \quad\text{..................(iv).}$$

Now (iv), being of the second degree in l and m, shews that $lx+my+1=0$ envelopes a conic.

46. Take the fixed point O for origin, and let the equation of the circle be $x^2+y^2+2gx+c=0$. Let $lx+my+1=0$ be the equation of PQ; then to find the distances of P and Q from O we have to eliminate θ between

$$r^2+2gr\cos\theta+c=0 \text{ and } rl\cos\theta+rm\sin\theta+1=0.$$

The result is

$$(lr^2+cl-2g)^2-4g^2m^2r^2+m^2(r^2+c)^2=0.$$

Hence, if $OP \cdot OQ=\text{constant}=a^2$, we have

$$(c^2-a^4)(l^2+m^2)-4cgl+4g^2=0 \quad\text{.......................(i).}$$

Also, if $OP^2+OQ^2=\text{constant}=b^2$, we have

$$(b^2+2c)l^2+(b^2+2c-4g^2)m^2-4gl=0 \quad\text{................ (ii).}$$

The Cartesian equation of the envelope whose tangential equation is (i) is easily found to be

$$(c^2-a^4+4cgx+4g^2x^2)(c^2-a^4+4g^2y^2)-(2cgy+4g^2xy)^2=0,$$

which may be written

$$(c^2 - a^4 + 2cgx)^2 = 4a^4g^2(x^2 + y^2).$$

Hence in the first case the locus is a conic of which the origin is one focus.

The Cartesian equation of the conic whose tangential equation is (ii) is easily found to be

$$(b^2 + 2c - 4g^2)(b^2 + 2c - 4gx) = 4g^2y^2.$$

Hence in the second case the envelope is a parabola.

47. Take the diameter AA' for the axis of x, and let the equation of the circle be $x^2 + y^2 - a^2$, and let A, A' be $(\pm c, 0)$.

Let α, θ_1, θ_2 be the angular co-ordinates of P, Q, R respectively. Then the equations of PQ, PR will be respectively

$$x\cos\frac{1}{2}(\alpha+\theta_1) + y\sin\frac{1}{2}(\alpha+\theta_1) = a\cos\frac{1}{2}(\alpha-\theta_1),$$

and $$x\cos\frac{1}{2}(\alpha+\theta_2) + y\sin\frac{1}{2}(\alpha+\theta_2) = a\cos\frac{1}{2}(\alpha-\theta_2).$$

Since the lines PQ, PR pass through the points $(c, 0)$, $(-c, 0)$ respectively, we have

$$c\cos\frac{1}{2}(\alpha+\theta_1) = a\cos\frac{1}{2}(\alpha-\theta_1) \text{ and } -c\cos\frac{1}{2}(\alpha+\theta_2) = a\cos\frac{1}{2}(\alpha-\theta_2),$$

whence $$\tan\frac{1}{2}\alpha\tan\frac{1}{2}\theta_1 = \frac{c-a}{c+a} \text{ and } \tan\frac{1}{2}\alpha\tan\frac{1}{2}\theta_2 = \frac{c+a}{c-a};$$

$$\therefore \tan\frac{1}{2}\theta_1 \Big/ \tan\frac{1}{2}\theta_2 = (c-a)^2/(c+a)^2 \quad\text{.................. (i).}$$

Now the equation of QR is

$$x\cos\frac{1}{2}(\theta_1+\theta_2) + y\sin\frac{1}{2}(\theta_1+\theta_2) = a\cos\frac{1}{2}(\theta_1-\theta_2),$$

or $$x - a + y\left(\tan\frac{1}{2}\theta_1 + \tan\frac{1}{2}\theta_2\right) - (x+a)\tan\frac{1}{2}\theta_1\tan\frac{1}{2}\theta_2 = 0.$$

Hence, from (i),

$$(x-a)(c-a)^2 + y\tan\frac{1}{2}\theta_1\{(c-a)^2 + (c+a)^2\} - (x+a)(c+a)^2\tan^2\frac{1}{2}\theta_1 = 0.$$

Hence the equation of the envelope is

$$(x^2 - a^2)(c^2 - a^2)^2 + (c^2 + a^2)^2y^2 = 0,$$

the envelope is therefore an ellipse of which the given circle is the auxiliary circle.

48. Let the equation of the ellipse be $x^2/a^2 + y^2/b^2 - 1 = 0$, and let the two fixed points be (x_1, y_1) and (x_2, y_2).

Let θ_1, θ_2, θ_3 be the eccentric angles of the points A, B, C respectively. Then, since AC goes through (x_1, y_1) and BC through (x_2, y_2), we have

$$\frac{x_1}{a}\cos\frac{1}{2}(\theta_1+\theta_3)+\frac{y_1}{b}\sin\frac{1}{2}(\theta_1+\theta_3)=\cos\frac{1}{2}(\theta_1-\theta_3),$$

and
$$\frac{x_2}{a}\cos\frac{1}{2}(\theta_2+\theta_3)+\frac{y_2}{b}\sin\frac{1}{2}(\theta_2+\theta_3)=\cos\frac{1}{2}(\theta_2-\theta_3).$$

Hence

$$\frac{x_1}{a}-1-\left(\frac{x_1}{a}+1\right)\tan\frac{1}{2}\theta_1\tan\frac{1}{2}\theta_3+\frac{y_1}{b}\left(\tan\frac{1}{2}\theta_1+\tan\frac{1}{2}\theta_3\right)=0,$$

and
$$\frac{x_2}{a}-1-\left(\frac{x_2}{a}+1\right)\tan\frac{1}{2}\theta_2\tan\frac{1}{2}\theta_3 +\frac{y_2}{b}\left(\tan\frac{1}{2}\theta_2+\tan\frac{1}{2}\theta_3\right)=0.$$

Eliminating $\tan\frac{1}{2}\theta_3$, we have

$$\left\{\frac{x_1}{a}-1+\frac{y_1}{b}\tan\frac{1}{2}\theta_1\right\}\left\{\left(\frac{x_2}{a}+1\right)\tan\frac{1}{2}\theta_2-\frac{y_2}{b}\right\} =\left\{\frac{x_2}{a}-1+\frac{y_2}{b}\tan\frac{1}{2}\theta_2\right\}\left\{\left(\frac{x_1}{a}+1\right)\tan\frac{1}{2}\theta_1-\frac{y_1}{b}\right\},$$

whence $$P+Q\tan\frac{1}{2}\theta_1+R\tan\frac{1}{2}\theta_2+S\tan\frac{1}{2}\theta_1\tan\frac{1}{2}\theta_2=0\ldots\ldots\ldots\text{(i)},$$

where P, Q, R, S are known constants.

Now the equation of AB is

$$\frac{x}{a}\cos\frac{1}{2}(\theta_1+\theta_2)+\frac{y}{b}\sin\frac{1}{2}(\theta_1+\theta_2)=\cos\frac{1}{2}(\theta_1-\theta_2),$$

which may be written

$$\frac{x}{a}-1-\left(\frac{x}{a}+1\right)\tan\frac{1}{2}\theta_1\tan\frac{1}{2}\theta_2+\frac{y}{b}\left(\tan\frac{1}{2}\theta_1+\tan\frac{1}{2}\theta_2\right)=0,$$

or from (i)

$$\left(\frac{x}{a}-1\right)\left(R+S\tan\frac{1}{2}\theta_1\right)+\left(\frac{x}{a}+1\right)\left(P+Q\tan\frac{1}{2}\theta_1\right)\tan\frac{1}{2}\theta_1 +\frac{y}{b}\left\{\tan\frac{1}{2}\theta_1\left(R+S\tan\frac{1}{2}\theta_1\right)-P-Q\tan\frac{1}{2}\theta_1\right\}=0,$$

whence it is obvious that the envelope is a *conic*.

To prove that the two conics have double contact with one another, it is only necessary to take the point C on the ellipse indefinitely near one of the extremities of the chord through the two fixed points, when it will be seen that the envelope has double contact with the ellipse, the chord of contact being the line joining the fixed points.

Or thus:

Project the line joining the two fixed points to infinity and the conic into a circle. Then, inscribed in a circle we have a triangle two of whose sides pass through fixed points at infinity, that is are drawn in fixed directions; and it is now obvious that the envelope of the third side is a *concentric circle*. Thus the envelope has double contact with the original circle, the chord of contact being the line joining the fixed points. Hence the original envelope must have been a conic having double contact with the original conic.

49. Let the points A, B, C be on the ellipse

$$x^2/a^2+y^2/b^2-1=0,$$

and let the sides AB, AC touch the ellipse

$$Ax^2+By^2=1.$$

Then, if α, β, γ be the eccentric angles of A, B, C respectively, the equation of AB is

$$\frac{x}{a}\cos\frac{1}{2}(\alpha+\beta)+\frac{y}{b}\sin\frac{1}{2}(\alpha+\beta)=\cos\frac{1}{2}(\alpha-\beta).$$

This line touches $\qquad Ax^2+By^2-1=0$;

$$\therefore \frac{1}{a^2A}\cos^2\frac{1}{2}(\alpha+\beta)+\frac{1}{b^2B}\sin^2\frac{1}{2}(\alpha+\beta)=\cos^2\frac{1}{2}(\alpha-\beta),$$

or
$$\frac{1}{a^2A}+\frac{1}{b^2B}-1+\left(\frac{1}{a^2A}-\frac{1}{b^2B}\right)\cos(\alpha+\beta)=\cos(\alpha-\beta).$$

Put
$$L=\frac{1}{a^2A}+\frac{1}{b^2B}-1 \text{ and } M=\frac{1}{a^2A}-\frac{1}{b^2B};$$

then we have

$$\cos\alpha\cos\beta(1-M)+\sin\alpha\sin\beta(1+M)-L=0.$$

We have similarly, since AC touches the conic

$$Ax^2+By^2-1=0,$$

$$\cos\alpha\cos\gamma(1-M)+\sin\alpha\sin\gamma(1+M)-L=0.$$

Hence

$$\frac{\cos\alpha}{L(1+M)\cos\frac{1}{2}(\beta+\gamma)}=\frac{\sin\alpha}{L(1-M)\sin\frac{1}{2}(\beta+\gamma)}=\frac{1}{(1-M^2)\cos\frac{1}{2}(\beta-\gamma)};$$

$$\therefore L^2(1+M)^2\cos^2\frac{1}{2}(\beta+\gamma)+L^2(1-M)^2\sin^2\frac{1}{2}(\beta+\gamma)$$
$$=(1-M^2)^2\cos^2\frac{1}{2}(\alpha-\beta)\ \ldots\ldots\ldots\ldots(\text{i}).$$

Now the equation of BC is

$$\frac{x}{a}\cos\frac{1}{2}(\beta+\gamma)+\frac{y}{b}\sin\frac{1}{2}(\beta+\gamma)=\cos\frac{1}{2}(\beta-\gamma),$$

and (i) shews that BC always touches the conic

$$\frac{x^2}{a^2L^2(1+M)^2}+\frac{y^2}{b^2L^2(1-M)^2}=\frac{1}{(1-M^2)^2}.$$

50. Let the equation of the conic be the general tangential equation of the second degree, and let

$$l_1x+m_1y+1=0, \text{ \&c.}$$

be the three given tangents.

Then we have

$$al_1^2+2hl_1m_1+bm_1^2+2gl_1+2fm_1+c=0 \qquad\text{(i)},$$

$$al_2^2+2hl_2m_2+bm_2^2+2gl_2+2fm_2+c=0 \qquad\text{(ii)},$$

$$al_3^2+2hl_3m_3+bm_3^2+2gl_3+2fm_3+c=0 \qquad\text{(iii)}.$$

Also, if (x, y) be the centre of the conic, we have [see 37]

$$g-cx=0 \qquad\text{(iv)},$$

$$f-cy=0 \qquad\text{(v)}.$$

The square of the radius of the director-circle is constant, and equal to k^2 suppose. Hence, from Art. 241, we have

$$k^2c^2=g^2+f^2-c(a+b);$$

or, using (iv) and (v),

$$a+b-gx-fy+k^2c=0 \qquad\text{(vi)}.$$

Eliminating a, h, b, g, f, c from the equations (i), (ii),..., (vi) we have the equation of the required locus, namely

$$\begin{vmatrix} 1, & 0, & 1, & -x, & -y, & k^2 \\ 0, & 0, & 0, & 1, & 0, & -x \\ 0, & 0, & 0, & 0, & 1, & -y \\ l_1^2, & l_1m_1, & m_1^2, & 2l_1, & 2m_1, & 1 \\ l_2^2, & l_2m_2, & m_2^2, & 2l_2, & 2m_2, & 1 \\ l_3^2, & l_3m_3, & m_3^2, & 2l_3, & 2m_3, & 1 \end{vmatrix}=0.$$

Multiply the fourth column by x and the fifth by y, and add the sum to the sixth; we then have

$$\begin{vmatrix} 1, & 0, & 1, & -x, & -y, & k^2-x^2-y^2 \\ 0, & 0, & 0, & 1, & 0, & 0 \\ 0, & 0, & 0, & 0, & 1, & 0 \\ l_1^2, & l_1m_1, & m_1^2, & 2l_1, & 2m_1, & 2l_1x+2m_1y+1 \\ l_2^2, & l_2m_2, & m_2^2, & 2l_2, & 2m_2, & 2l_2x+2m_2y+1 \\ l_3^2, & l_3m_3, & m_3^2, & 2l_3, & 2m_3, & 2l_3x+2m_3y+1 \end{vmatrix}=0,$$

that is

$$\begin{vmatrix} 1, & 0, & 1, & k^2-x^2-y^2 \\ l_1^2, & l_1m_1, & m_1^2, & 2l_1x+2m_1y+1 \\ l_2^2, & l_2m_2, & m_2^2, & 2l_2x+2m_2y+1 \\ l_3^2, & l_3m_3, & m_3^2, & 2l_3x+2m_3y+1 \end{vmatrix}=0,$$

from which it is obvious that the locus of the centres of the conics is a circle for any particular value of k, and that the circles are concentric for different values of k.

Or thus:—

The general tangential equation of a conic which touches three given straight lines is

$$\lambda_1S_1+\lambda_2S_2+\lambda_3S_3=0,$$

where λ_1, λ_2, λ_3 are arbitrary and $S_1=0$, $S_2=0$, $S_3=0$ are the tangential equations of any three conics which touch the lines.

Now from Art. 241 we see that the equation of the director-circle of a conic is of the first degree in a, h, b, &c. It therefore follows that if $C_1=0$, $C_2=0$, $C_3=0$ be the director-circles of the conics $S_1=0$, $S_2=0$, $S_3=0$ respectively, the equation of the director-circle of

$$\lambda_1S_1+\lambda_2S_2+\lambda_3S_3=0$$

will be

$$\lambda_1C_1+\lambda_2C_2+\lambda_3C_3=0.$$

Now it is easy to prove that a circle will cut the three circles $C_1=0$, $C_2=0$, $C_3=0$ orthogonally, and that this circle will also cut orthogonally any one of the systems of circles given by

$$\lambda_1C_1+\lambda_2C_2+\lambda_3C_3=0.$$

We therefore have the following theorem:—*The director-circles of all conics which touch three given straight lines are cut orthogonally by the same circle.*

It follows from the above that all those director-circles which are of given radius have their centres (which are also the centres of the corresponding conics) at a given distance from the centre of the orthogonal circle: this proves the proposition.

CHAPTER XIII.

Page 285.

1. In order that a point may be on the bisector of an angle it is sufficient and necessary that the perpendiculars from the point on the line bounding the angle should be equal. Hence for any point on the bisector of A we have $\beta=\gamma$; and similarly the equations of the other bisectors are $\gamma=\alpha$ and $\alpha=\beta$.

2. At A', the middle point of BC, it is obvious that $b\beta=c\gamma$; and the ratio $\beta:\gamma$ is clearly the same for all points on AA'. Hence at any point on AA' we have $b\beta=c\gamma$.

The equations of the three medians of the triangle are therefore

$$b\beta-c\gamma=0,\ c\gamma-a\alpha=0 \text{ and } a\alpha-b\beta=0.$$

3. At B' we have $\beta=0$ and $c\gamma-a\alpha=0$; also at C' we have $\gamma=0$ and $b\beta-a\alpha=0$.

Hence the line whose equation is $b\beta+c\gamma-a\alpha=0$ passes through B' and through C'.

Similarly the equations of $C'A'$ and $A'B'$ are respectively

$$a\alpha-b\beta+c\gamma=0 \text{ and } a\alpha+b\beta-c\gamma=0.$$

4. The trilinear co-ordinates of the centre of the in-circle are proportional to 1, 1, 1; also the co-ordinates of the centre of the circum-circle are proportional to $\cos A$, $\cos B$, $\cos C$.

Hence the equation of the line joining these two points is

$$\begin{vmatrix} \alpha, & \beta, & \gamma \\ 1, & 1, & 1 \\ \cos A, & \cos B, & \cos C \end{vmatrix}=0.$$

5. The co-ordinates of the four centres are given by

$$\alpha_1=\beta_1=\gamma_1=\frac{2\Delta}{a+b+c},$$

$$-\alpha_2=\beta_2=\gamma_2=\frac{2\Delta}{-a+b+c},$$

$$\alpha_3 = -\beta_3 = \gamma_3 = \frac{2\Delta}{a-b+c},$$

and
$$\alpha_4 = \beta_4 = -\gamma_4 = \frac{2\Delta}{a+b-c}.$$

The co-ordinates of the middle point of the line joining $(\alpha_1, \beta_1, \gamma_1)$ and $(\alpha_2, \beta_2, \gamma_2)$ are $\frac{1}{2}\left(\frac{2\Delta}{a+b+c} - \frac{2\Delta}{-a+b+c}\right)$,

$$\frac{1}{2}\left(\frac{2\Delta}{a+b+c} + \frac{2\Delta}{-a+b+c}\right), \frac{1}{2}\left(\frac{2\Delta}{a+b+c} + \frac{2\Delta}{-a+b+c}\right),$$

which are *proportional* to $-a$, $b+c$, $b+c$.

The co-ordinates of the middle point of the line joining $(\alpha_3, \beta_3, \gamma_3)$ and $(\alpha_4, \beta_4, \gamma_4)$ are

$$\frac{1}{2}\left(\frac{2\Delta}{a-b+c} + \frac{2\Delta}{a+b-c}\right), \frac{1}{2}\left(-\frac{2\Delta}{a-b+c} + \frac{2\Delta}{a+b-c}\right),$$
$$\frac{1}{2}\left(\frac{2\Delta}{a-b+c} - \frac{2\Delta}{a+b-c}\right),$$

which are *proportional* to a, $c-b$, $b-c$.

Hence the six middle points are

$$(-a, b+c, b+c), (c+a, -b, c+a),$$
$$(a+b, a+b, -c), (a, c-b, b-c),$$
$$(c-a, b, a-c) \text{ and } (b-a, a-b, c).$$

It is now easy to verify that these six points are all on

$$a\beta\gamma + b\gamma\alpha + c\alpha\beta = 0.$$

6. Let O be $(\alpha_1, \beta_1, \gamma_1)$. Then the equations of AOA', BOB', COC' are respectively

$$\frac{\beta}{\beta_1} - \frac{\gamma}{\gamma_1} = 0,$$
$$\frac{\gamma}{\gamma_1} - \frac{\alpha}{\alpha_1} = 0 \text{ and } \frac{\alpha}{\alpha_1} - \frac{\beta}{\beta_1} = 0.$$

Hence B' is the point $\beta = 0$, $\frac{\gamma}{\gamma_1} - \frac{\alpha}{\alpha_1} = 0$,

and C' is the point $\gamma = 0$, $\frac{\beta}{\beta_1} - \frac{\alpha}{\alpha_1} = 0$;

and both these points are clearly on the

$$\frac{\beta}{\beta_1} + \frac{\gamma}{\gamma_1} - \frac{\alpha}{\alpha_1} = 0.$$

Thus the equations of $B'C'$, $C'A'$ and $A'B'$ are respectively

$$\frac{\beta}{\beta_1}+\frac{\gamma}{\gamma_1}-\frac{\alpha}{\alpha_1}=0,\ \frac{\gamma}{\gamma_1}+\frac{\alpha}{\alpha_1}-\frac{\beta}{\beta_1}=0,$$

and
$$\frac{\alpha}{\alpha_1}+\frac{\beta}{\beta_1}-\frac{\gamma}{\gamma_1}=0.$$

Now P is the point of intersection of $\alpha=0$ and

$$\frac{\beta}{\beta_1}+\frac{\gamma}{\gamma_1}-\frac{\alpha}{\alpha_1}=0,$$

or of $\alpha=0$ and
$$\frac{\beta}{\beta_1}+\frac{\gamma}{\gamma_1}=0;$$

hence the line $\frac{\alpha}{\alpha_1}+\frac{\beta}{\beta_1}+\frac{\gamma}{\gamma_1}=0$ goes through P, and the symmetry of the equation shews that it also goes through Q and R. Thus P, Q, R are all on the line

$$\frac{\alpha}{\alpha_1}+\frac{\beta}{\beta_1}+\frac{\gamma}{\gamma_1}=0.$$

Again, since Q is the point of intersection of $\beta=0$ and $\frac{\alpha}{\alpha_1}+\frac{\gamma}{\gamma_1}=0$,

the equation of BQ is
$$\frac{\alpha}{\alpha_1}+\frac{\gamma}{\gamma_1}=0;$$

so also the equation of CR is
$$\frac{\alpha}{\alpha_1}+\frac{\beta}{\beta_1}=0;$$

and the equation of AA' is
$$\frac{\beta}{\beta_1}=\frac{\gamma}{\gamma_1}.$$

These three lines obviously meet in the point P' given by

$$-\frac{\alpha}{\alpha_1}=\frac{\beta}{\beta_1}=\frac{\gamma}{\gamma_1}.$$

7. At the point P it is easy to prove that $\beta=a\cos C+\frac{a}{2}\sin C$, and $\gamma=a\cos B+\frac{a}{2}\sin B$. Hence the equation of AP is

$$\frac{\beta}{2\cos C+\sin C}=\frac{\gamma}{2\cos B+\sin B},$$

or
$$\beta\,(2\cos B+\sin B)=\gamma\,(2\cos C+\sin C)\ldots\ldots\ldots\ldots\ldots\ldots\text{(i)}.$$

The equations of BQ and CR are similarly

$$\gamma\,(2\cos C+\sin C)=\alpha\,(2\cos A+\sin A)\ldots\ldots\ldots\ldots\ldots\ldots\text{(ii)},$$

and
$$\alpha\,(2\cos A+\sin A)=\beta\,(2\cos B+\sin B)\ \ldots\ldots\ldots\ldots\ldots\text{(iii)},$$

and it is obvious that the three lines whose equations are (i), (ii), (iii) meet in a point.

8. Let $l\alpha+m\beta+n\gamma=0$ be the equation of the straight line; then the lengths of the perpendiculars on this line from

$$\left(\frac{2\Delta}{a},\ 0,\ 0\right),\ \left(0,\ \frac{2\Delta}{b},\ 0\right),\ \text{and}\ \left(0,\ 0,\ \frac{2\Delta}{c}\right)$$

are proportional to

$$\frac{2\Delta l}{a},\ \frac{2\Delta m}{b},\ \text{and}\ \frac{2\Delta n}{c}\ \text{[Art. 257]}.$$

Hence we have $\dfrac{l}{ap}=\dfrac{m}{bq}=\dfrac{n}{cr}$.

The equation of the line may therefore be written in the form

$$ap\alpha+bq\beta+cr\gamma=0.$$

9. Let ABC, $A'B'C'$ be two triangles such that AA', BB', CC' meet in some point O. Take ABC for the triangle of reference, and let O be $(f,\ g,\ h)$.

Then the equation of $AA'O$ is $\dfrac{\beta}{g}=\dfrac{\gamma}{h}$, and therefore A' may be taken to be $(\alpha_1,\ g,\ h)$. So also B' and C' may be taken to be $(f,\ \beta_1,\ h)$ and $(f,\ g,\ \gamma_1)$ respectively. The equation of $B'C'$ is therefore

$$\begin{vmatrix} \alpha, & \beta, & \gamma \\ f, & \beta_1, & h \\ f, & g, & \gamma_1 \end{vmatrix}=0.$$

Hence $B'C'$ meets BC where

$$\alpha=0\ \text{and}\ \frac{\beta}{g-\beta_1}+\frac{\gamma}{h-\gamma_1}=0.$$

Similarly the point of intersection of CA and $C'A'$ is given by

$$\beta=0\ \text{and}\ \frac{\alpha}{f-\alpha_1}+\frac{\gamma}{h-\gamma_1}=0,$$

and the point of intersection of AB and $A'B'$ is given by

$$\gamma=0\ \text{and}\ \frac{\alpha}{f-\alpha_1}+\frac{\beta}{g-\beta_1}=0.$$

The three points of intersection of corresponding sides are therefore on the straight line whose equation is

$$\frac{\alpha}{f-\alpha_1}+\frac{\beta}{g-\beta_1}+\frac{\gamma}{h-\gamma_1}=0.$$

Pages 309—314.

1. The perpendicular from the centre of a conic on any tangent cannot be less than the semi-minor axis.

Hence, if $(\alpha_0, \beta_0, \gamma_0)$ be the centre of any conic inscribed in the triangle of reference, and ρ be the length of the semi-minor axis, and r be the radius of the inscribed circle, we have

$$a\alpha_0 + b\beta_0 + c\gamma_0 > (a + b + c)\,\rho\,;$$

also
$$a\alpha_0 + b\beta_0 c\gamma_0 = 2\Delta = (a + b + c)\,r\,;$$

$$\therefore \quad r > \rho.$$

2. Let $(\alpha_1, \beta_1, \gamma_1)$, &c. be the angular points of the triangle; then, as in Art. 6,

$$2 \times \text{area of triangle} = \sin C \begin{vmatrix} \alpha_1 \operatorname{cosec} C, & \alpha_1 \operatorname{cosec} C, & 1 \\ \alpha_2 \operatorname{cosec} C, & \beta_2 \operatorname{cosec} C, & 1 \\ \alpha_3 \operatorname{cosec} C, & \beta_3 \operatorname{cosec} C, & 1 \end{vmatrix}$$

$$= \frac{1}{2\Delta ab \sin C} \begin{vmatrix} a\alpha_1, & b\beta_1, & 2\Delta \\ a\alpha_2, & b\beta_2, & 2\Delta \\ a\alpha_3, & b\beta_3, & 2\Delta \end{vmatrix}$$

$$= \frac{1}{(2\Delta)^2} \begin{vmatrix} a\alpha_1, & b\beta_1, & c\gamma_1 \\ a\alpha_2, & b\beta_2, & c\gamma_2 \\ a\alpha_3, & b\beta_3, & c\gamma_3 \end{vmatrix},$$

by subtracting the sum of the two first columns from the third

$$= \frac{abc}{(2\Delta)^2} \begin{vmatrix} \alpha_1, & \beta_1, & \gamma_1 \\ \alpha_2, & \beta_2, & \gamma_2 \\ \alpha_3, & \beta_3, & \gamma_3 \end{vmatrix}.$$

3. The equations of the conics which have a common self-polar triangle may be taken to be

$$u_1\alpha^2 + v_1\beta^2 + w_1\gamma^2 = 0, \text{ \&c.}$$

The four points of intersection of the first two are given by

$$\frac{\alpha}{\pm\sqrt{(v_1w_2 - v_2w_1)}} = \frac{\beta}{\pm\sqrt{(w_1u_2 - w_2u_1)}} = \frac{\gamma}{\pm\sqrt{(u_1v_2 - u_2v_1)}}.$$

The four points of intersection of the other two are given by

$$\frac{\alpha}{\pm\sqrt{(v_3w_4 - v_4w_3)}} = \frac{\beta}{\pm\sqrt{(w_3u_4 - w_4u_3)}} = \frac{\gamma}{\pm\sqrt{(u_3v_4 - u_4v_3)}}.$$

The eight points will all lie on the conic whose equation is

$$L\alpha^2 + M\beta^2 + N\gamma^2 = 0,$$

provided $L(v_1w_2 - v_2w_1) + M(w_1u_2 - w_2u_1) + N(u_1v_2 - u_2v_1) = 0,$

and $L(v_3w_4 - v_4w_3) + M(w_3u_4 - w_4u_3) + N(u_3v_4 - u_4v_3) = 0.$

Hence the eight points are all on the conic whose equation is

$$\begin{vmatrix} \alpha^2, & \beta^2, & \gamma^2 \\ v_1w_2 - v_2w_1, & w_1u_2 - w_2u_1, & u_1v_2 - u_2v_1 \\ v_3w_4 - v_4w_3, & w_3u_4 - w_4u_3, & u_3v_4 - u_4v_3 \end{vmatrix} = 0.$$

4. Refer the conics to their common self-polar triangle; then their equations will be

$$u_1\alpha^2 + v_1\beta^2 + w_1\gamma^2 = 0 \text{ and } u_2\alpha^2 + v_2\beta^2 + w_2\gamma^2 = 0.$$

The line $$l\alpha + m\beta + n\gamma = 0$$

will touch both conics provided

$$\frac{l^2}{u_1} + \frac{m^2}{v_1} + \frac{n^2}{w_1} = 0 \text{ and } \frac{l^2}{u_2} + \frac{m^2}{v_2} + \frac{n^2}{w_2} = 0.$$

Hence the equations of the common tangents are of the form

$$l\alpha \pm m\beta \pm n\gamma = 0.$$

The points where the four common tangents touch the two conics are given by

$$\frac{u_1\alpha}{l} = \frac{v_1\beta}{\pm m} = \frac{w_1\gamma}{\pm n} \text{ and } \frac{u_2\alpha}{l} = \frac{v_2\beta}{\pm m} = \frac{w_2\gamma}{\pm n}.$$

These eight points are all on the conic whose equation is

$$\begin{vmatrix} \alpha^2, & \beta^2, & \gamma^2 \\ \frac{l^2}{u_1^2}, & \frac{m^2}{v_1^2}, & \frac{n^2}{w_1^2} \\ \frac{l^2}{u_2^2}, & \frac{m^2}{v_2^2}, & \frac{n^2}{w_2^2} \end{vmatrix} = 0.$$

5. Refer the conics to their common self-polar triangle; then their equations will be

$$u_1\alpha^2 + v_1\beta^2 + w_1\gamma^2 = 0 \text{ and } u_2\alpha^2 + v_2\beta^2 + w_2\gamma^2 = 0.$$

The co-ordinates of their four common points are given by

$$\frac{\alpha}{\sqrt{(v_1w_2 - v_2w_1)}} = \frac{\beta}{\pm\sqrt{(w_1u_2 - w_2u_1)}} = \frac{\gamma}{\pm\sqrt{(u_1v_2 - u_2v_1)}},$$

which are of the form $\pm f, \pm g, \pm h.$

Hence the equations of the eight tangents are of the form

$$\pm u_1f\alpha \pm v_1g\beta \pm w_1h\gamma = 0 \text{ and } \pm u_2f\alpha \pm v_2g\beta \pm w_2h\gamma = 0 \quad \text{.........(i).}$$

The eight tangents will all touch the conic

$$L\alpha^2 + M\beta^2 + N\gamma^2 = 0$$

provided

$$\frac{u_1^2f^2}{L}+\frac{v_1^2g^2}{M}+\frac{w_1^2h^2}{N}=0,$$

and

$$\frac{u_2^2f^2}{L}+\frac{v_2^2g^2}{M}+\frac{w_2^2h^2}{N}=0.$$

Hence $\frac{f^2}{L}\Big/(v_1^2w_2^2-v_2^2w_1^2)=\frac{g^2}{M}\Big/(w_1^2u_2^2-w_2^2u_1^2)=\frac{h^2}{N}\Big/(u_1^2v_2^2-u_2^2v_1^2).$

Hence the eight lines (i) touch the conic

$$\frac{f^2\alpha^2}{v_1^2w_2^2-v_2^2w_1^2}+\frac{g^2\beta^2}{w_1^2u_2^2-w_2^2u_1^2}+\frac{h^2\gamma^2}{u_1^2v_2^2-u_2^2v_1^2}=0,$$

that is

$$\frac{\alpha^2}{v_1w_2+v_2w_1}+\frac{\beta^2}{w_1u_2+w_2u_1}+\frac{\gamma^2}{u_1v_2+u_2v_1}=0.$$

6. Take the triangle formed by the diagonals for the triangle of reference; then the equations of the four sides may be taken to be

$$l\alpha\pm m\beta\pm n\gamma=0.$$

Hence the equation of the lines through A and through the extremities of the opposite diagonal is

$$m^2\beta^2-n^2\gamma^2=0.$$

Now the equation of any other pair of lines through A which are harmonically conjugate to

$$m^2\beta^2-n^2\gamma^2=0$$

may be put in the form

$$m^2\beta^2+n^2\gamma^2+2\lambda\beta\gamma=0.$$

Hence the three pairs of points which satisfy the given conditions are given by

$$m^2\beta^2+n^2\gamma^2+2\lambda\beta\gamma=0,\ \alpha=0;\ n^2\gamma^2+l^2\alpha^2+2\mu\gamma\alpha=0,\ \beta=0$$

and

$$l^2\alpha^2+m^2\beta^2+2\nu\alpha\beta=0,\ \gamma=0.$$

The six points are all on the conic whose equation is

$$l^2\alpha^2+m^2\beta^2+n^2\gamma^2+2\lambda\beta\gamma+2\mu\gamma\alpha+2\nu\alpha\beta=0.$$

7. The perpendicular distance of (α, β, γ) from $-a\alpha+b\beta+c\gamma=0$ is

$$\frac{-a\alpha+b\beta+c\gamma}{\sqrt{(a^2+b^2+c^2-2bc\cos A+2ca\cos B+2ab\cos C})}=\frac{-a\alpha+b\beta+c\gamma}{2a}.$$

The perpendicular distances of (α, β, γ) from

$$a\alpha - b\beta + c\gamma = 0$$

and

$$a\alpha + b\beta - c\gamma = 0$$

are similarly $\dfrac{a\alpha - b\beta + c\gamma}{2b}$ and $\dfrac{a\alpha + b\beta - c\gamma}{2c}$.

Also the lengths of the sides of the triangle formed by the three lines whose equations are

$$-a\alpha + b\beta + c\gamma = 0, \text{ \&c.}$$

are

$$\frac{a}{2}, \frac{b}{2}, \frac{c}{2}.$$

Hence, from Art. 269, the required equation is

$$\frac{a^2}{-a\alpha + b\beta + c\gamma} + \frac{b^2}{a\alpha - b\beta + c\gamma} + \frac{c^2}{a\alpha + b\beta - c\gamma} = 0.$$

8. The equation of any circle concentric with the circumscribing circle is

$$a\beta\gamma + b\gamma\alpha + c\alpha\beta + \lambda\,(a\alpha + b\beta + c\gamma)^2 = 0.$$

Along the line through A parallel to BC we have

$$a\alpha = 2\Delta \text{ and } b\beta + c\gamma = 0;$$

hence at the points, P, P' suppose, where this line meets the circle, we have

$$a^2b\beta^2 + 2b^2\beta\Delta - 2c^2\beta\Delta - 4ac\lambda\Delta^2 = 0;$$

$$\therefore\ PA\,.\,AP' \sin^2 C = \beta_1\beta_2 = -\frac{4\Delta^2 c\lambda}{ab};$$

$$\therefore\ r^2 - R^2 = -\frac{4c\lambda\Delta^2}{ab\sin^2 C} = -abc\lambda.$$

Hence the required equation is

$$a\beta\gamma + b\gamma\alpha + c\alpha\beta - \frac{r^2 - R^2}{abc}(a\alpha + b\beta + c\gamma)^2 = 0.$$

9. The point which is at a distance ρ from $(\alpha_0, \beta_0, \gamma_0)$ on a line parallel to BC is

$$(\alpha_0,\ \beta_0 + \rho\sin C,\ \gamma_0 - \rho\sin B).$$

Hence, if this point be on the circumscribing conic whose equation is

$$l\beta\gamma + m\gamma\alpha + n\alpha\beta = 0,$$

we have

$$l\,(\beta_0 + \rho\sin C)(\gamma_0 - \rho\sin B) + m\alpha_0\,(\gamma_0 - \rho\sin B) + n\alpha_0\,(\beta_0 + \rho\sin C) = 0;$$

$$\therefore\ \rho_1\rho_2 = \frac{l\beta_0\gamma_0 + m\gamma_0\alpha_0 + n\alpha_0\beta_0}{-l\sin B\sin C}.$$

Hence, from Art. 186, Cor. I., we have

$$\frac{r_1^2}{\frac{a}{l}} = \frac{r_2^2}{\frac{b}{m}} = \frac{r_3^2}{\frac{c}{n}}.$$

The equation of the conic must therefore be

$$\frac{a}{r_1^2}\beta\gamma + \frac{b}{r_2^2}\gamma\alpha + \frac{c}{r_3^2}\alpha\beta = 0.$$

10. Take ABC for the triangle of reference, and let the equation of the conic be

$$l\beta\gamma + m\gamma\alpha + n\alpha\beta = 0.$$

The equations of $B'C'$, $C'A'$, $A'B'$ will then be respectively

$$m\gamma + n\beta = 0, \quad n\alpha + l\gamma = 0 \text{ and } l\beta + m\alpha = 0.$$

The equations AA', BB', CC' are easily seen to be respectively

$$\frac{\beta}{m} - \frac{\gamma}{n} = 0, \quad \frac{\gamma}{n} - \frac{\alpha}{l} = 0 \text{ and } \frac{\alpha}{l} - \frac{\beta}{m} = 0,$$

from which it is obvious that AA', BB', CC' meet in a point.

Again the point D is the point of intersection of $\alpha = 0$, $m\gamma + n\beta = 0$.

Hence D is on the line whose equation is

$$\frac{\alpha}{l} + \frac{\beta}{m} + \frac{\gamma}{n} = 0;$$

and it is obvious that E and F are also on this straight line.

11. Take ABC for the triangle of reference, and let P be (f, g, h). Then the equations of AA', BB', CC' are

$$\frac{\beta}{g} = \frac{\gamma}{h}, \quad \frac{\gamma}{h} = \frac{\alpha}{f} \text{ and } \frac{\alpha}{f} = \frac{\beta}{g}$$

respectively. Hence the equations of $B'C'$, $C'A'$, $A'B'$ are respectively

$$-\frac{\alpha}{f} + \frac{\beta}{g} + \frac{\gamma}{h} = 0, \quad \frac{\alpha}{f} - \frac{\beta}{g} + \frac{\gamma}{h} = 0$$

and

$$\frac{\alpha}{f} + \frac{\beta}{g} - \frac{\gamma}{h} = 0.$$

Hence the points K, L, M are on the line whose equation is

$$\frac{\alpha}{f} + \frac{\beta}{g} + \frac{\gamma}{h} = 0.$$

Now, (i), if P be on the fixed straight line

$$l\alpha + m\beta + n\gamma = 0,$$

we have

$$lf + mg + nh = 0,$$

which shews that

$$\frac{\alpha}{f}+\frac{\beta}{g}+\frac{\gamma}{h}=0$$

touches the conic

$$\sqrt{l\alpha}+\sqrt{m\beta}+\sqrt{n\gamma}=0.$$

Also, (ii), if P move on the fixed conic

$$l\beta\gamma+m\gamma\alpha+n\alpha\beta=0,$$

we have
$$\frac{l}{f}+\frac{m}{g}+\frac{n}{h}=0,$$

whence it follows that

$$\frac{\alpha}{f}+\frac{\beta}{g}+\frac{\gamma}{h}=0$$

passes through the fixed point (l, m, n).

And, (iii), if P move on the fixed conic $\alpha\beta=\lambda\gamma^2$, we have $fg=\lambda h^2$. The equation of KLM may now be written in the form

$$\frac{\alpha}{f^2}+\frac{\beta}{\lambda h^2}+\frac{\gamma}{fh}=0,$$

the envelope of which is
$$4\alpha\beta=\lambda\gamma^2.$$

12. If O be (f, g, h) and O' be (f', g', h'), A' will be $\alpha=0$, $\frac{\beta}{g}=\frac{\gamma}{h}$, and so for B' and C'; also A'' will be $\alpha=0$, $\frac{\beta}{g'}=\frac{\gamma}{h'}$, and so for B'' and C''.

Hence $B'C'$ is
$$\frac{\beta}{g}+\frac{\gamma}{h}-\frac{\alpha}{f}=0,$$

and $B''C''$ is
$$\frac{\beta}{g'}+\frac{\gamma}{h'}-\frac{\alpha}{f'}=0;$$

and therefore at P
$$\beta\left(\frac{f}{g}-\frac{f'}{g'}\right)+\gamma\left(\frac{f}{h}-\frac{f'}{h'}\right)=0.$$

Hence the equation of AP is

$$\frac{\beta}{gg'(hf'-h'f)}=\frac{\gamma}{hh'(fg'-f'g)}.$$

The equations of BQ, CR are similarly

$$\frac{\gamma}{hh'(fg'-f'g)}=\frac{\alpha}{ff'(gh'-g'h)}$$

and
$$\frac{\alpha}{ff'(gh'-g'h)}=\frac{\beta}{gg'(hf'-h'f)}.$$

The point Z is therefore given by

$$\frac{\alpha}{ff'(gh'-g'h)}=\frac{\beta}{gg'(hf'-h'f)}=\frac{\gamma}{hh'(fg'-f'g)}.$$

Now, if O, O' are both on the conic

$$\frac{\lambda}{\alpha}+\frac{\mu}{\beta}+\frac{\nu}{\gamma}=0;$$

we have
$$\frac{\lambda}{\frac{1}{gh'}-\frac{1}{g'h}}=\frac{\mu}{\frac{1}{hf'}-\frac{1}{h'f}}=\frac{\nu}{\frac{1}{fg'}-\frac{1}{f'g}}.$$

Hence Z is the fixed point (λ, μ, ν).

13. Take the triangle of reference as in Art. 258, and let the given points be $(f, \pm g, \pm h)$. Then the equation of any conic through the four points is
$$u\alpha^2+v\beta^2+w\gamma^2=0,$$
with the condition
$$uf^2+vg^2+wh^2=0.$$

Let
$$\lambda\alpha+\mu\beta+\nu\gamma=0$$
be the given straight line; then, if (ξ, η, ζ) be its pole with respect to

$$u\alpha^2+v\beta^2+w\gamma^2=0,$$

the equations

$$\lambda\alpha+\mu\beta+\nu\gamma=0 \text{ and } u\xi\alpha+v\eta\beta+w\zeta\gamma=0$$

must represent the same straight line, and therefore

$$\frac{u\xi}{\lambda}=\frac{v\eta}{\mu}=\frac{w\zeta}{\nu}.$$

Hence, from the condition

$$uf^2+vg^2+wh^2=0,$$

we have
$$\frac{f^2\lambda}{\xi}+\frac{g^2\mu}{\eta}+\frac{h^2\nu}{\zeta}=0.$$

Thus the locus of (ξ, η, ζ) is a conic circumscribing the triangle of reference.

14. Take the triangle of reference as in Art. 259, and let the equations of the given straight line be

$$l\alpha\pm m\beta\pm n\gamma=0.$$

Then the equation of any conic touching the given lines is

$$u\alpha^2+v\beta^2+w\gamma^2=0,$$

with the condition
$$\frac{l^2}{u}+\frac{m^2}{v}+\frac{n^2}{w}=0.$$

Let (f, g, h) be the given point; then its polar with respect to

$$u\alpha^2+v\beta^2+w\gamma^2=0 \text{ is } uf\alpha+vg\beta+wh\gamma=0.$$

And the condition $$\frac{l^2}{u}+\frac{m^2}{v}+\frac{n^2}{w}=0$$

shews that $$uf\alpha+vg\beta+wh\gamma=0$$

touches the conic

$$\sqrt{l^2f\alpha}+\sqrt{m^2g\beta}+\sqrt{n^2h\gamma}=0.$$

Thus the envelope of the polar of (f, g, h) for all possible valnes of u, v, w is a conic which touches the diagonals of the quadrilateral.

15. The equation of any conic of the system may be taken to be

$$u\alpha^2+v\beta^2+w\gamma^2=0,$$

with the condition

$$\frac{l^2}{u}+\frac{m^2}{v}+\frac{n^2}{w}=0.$$

The tangent at $(\alpha', \beta', \gamma')$ is

$$u\alpha'\alpha+v\beta'\beta+w\gamma'\gamma=0.$$

The condition that this tangent may be parallel to the fixed line

$$l\alpha+m\beta+n\gamma=0$$

is $$u\alpha'(mc-nb)+v\beta'(na-lc)+w\gamma'(lb-ma)=0 \text{ ....(i).}$$

But, since $(\alpha', \beta', \gamma')$ is on the conic, we have

$$u\alpha'^2+v\beta'^2+w\gamma'^2=0 \text{ ....(ii).}$$

From (i) and (ii) we have

$$\frac{u\alpha'}{\gamma'(na-lc)-\beta'(lb-ma)}=\dots$$

Substitute for u, v and w in the condition

$$\frac{l^2}{u}+\frac{m^2}{v}+\frac{n^2}{w}=0;$$

and we obtain the equation of the required locus, namely

$$\Sigma l^2a\,\{\alpha(lb-ma)-\gamma(mc-nb)\}\,\{\beta(mc-nb)-\alpha(na-lc)\}=0.$$

Thus the required locus is a cubic curve which passes through the angular points of the triangle of reference.

16. Let r_1, r_2 be the semi-axes of an ellipse inscribed in the triangle of reference, and let $(\alpha_0, \beta_0, \gamma_0)$ be the centre of the ellipse. Then, if the sides of the triangle make angles θ_1, θ_2, θ_3 with the axis r_2, we have

$$\alpha_0^2 = r_1^2 \cos^2\theta_1 + r_2^2 \sin^2\theta_1,$$

$$\beta_0^2 = r_1^2 \cos^2\theta_2 + r_2^2 \sin^2\theta_2,$$

and
$$\gamma_0^2 = r_1^2 \cos^2\theta_3 + r_2^2 \sin^2\theta_3.$$

Hence $\Sigma\sqrt{(\alpha_0^2 - r_2^2)} \,.\, \sin(\theta_2 - \theta_3) = \sqrt{(r_1^2 - r_2^2)}\, \Sigma \sin(\theta_2 - \theta_3)\cos\theta_1 = 0.$

Hence, as
$$\sin(\theta_2 \sim \theta_3) = \sin A,$$

we have the following quadratic in r^2 whose roots are the squares of the axes of the conic,

$$a\sqrt{(\alpha_0^2 - r^2)} + b\sqrt{(\beta_0^2 - r^2)} + c\sqrt{(\gamma_0^2 - r^2)} = 0.$$

The above equation when rationalised becomes

$$\Sigma a^4 (\alpha_0^2 - r^2)^2 - 2\Sigma b^2c^2 (\beta_0^2 - r^2)(\gamma_0^2 - r^2) = 0,$$

or $\{2\Sigma b^2c^2\beta_0^2\gamma_0^2 - \Sigma a^4\alpha_0^4\} - 4r^2\, abc\Sigma\alpha_0^2\, a\cos A + r^4\{2\Sigma b^2c^2 - \Sigma a^4\} = 0,$

whence it follows that

$$r_1^2 + r_2^2 = \frac{abc}{4\Delta^2}\Sigma\alpha_0^2\, a\cos A,$$

and
$$r_1^2 r_2^2 = \frac{1}{16\Delta^2}(a\alpha_0 + b\beta_0 + c\gamma_0)(-a\alpha_0 + b\beta_0 + c\gamma_0)$$
$$(a\alpha_0 - b\beta_0 + c\gamma_0)(a\alpha_0 + b\beta_0 - c\gamma_0).$$

In the case before us the centre of the conic is

$$(R\cos A,\ R\cos B,\ R\cos C);$$

and it will be found that

$$r_1^2 + r_2^2 = R^2(1 - 4\cos A\cos B\cos C),$$

and
$$r_1^2 r_2^2 = 4R^4\cos^2 A\cos^2 B\cos^2 C.$$

Now
$$d^2 = R^2(1 - 8\cos A\cos B\cos C).$$

Hence
$$r_1^2 + r_2^2 = \frac{1}{2}(R^2 + d^2)$$

and
$$r_1^2 r_2^2 = \frac{1}{16}(R^2 - d^2)^2,$$

so that $2r_1$ and $2r_2$ are equal to $R + d$ and $R - d$.

17. Let AOA', BOB', COC' be diameters of the conic; then, if the centre be within the triangle DEF, it is clear that A', B', C' will all be outside the triangle ABC. Now, if the curve be an hyperbola, two at least of the points A, B, C must be on the same branch of the curve; let then B and C be on the same branch, then the arc from B to C is convex to O, and therefore A' cannot be on the curve. It therefore follows that a conic through A, B and C must be an ellipse if the centre be within the triangle DEF.

Next suppose that the centre O is in the angle diametrically opposite to D, and as before let AOA', BOB', COC' be diameters of the conic; then it is clear that B' and C' are within the angle BAC.

Now if a conic through A, B, C is an hyperbola with centre O, B and C cannot be on the same branch, for the arc of an hyperbola being everywhere convex to the centre, the line OA would cut the arc in some point X between O and BC, and there would be three points A, X, A' on the same straight line. Then, since B and C are on different branches, it follows that A and B or else A and C are on the same branch. Suppose A and B are on the same branch, then C' must also be on this branch, since C and C' are on opposite branches; but it is impossible that A, B, and C' should be on the same branch of an hyperbola for C' is within the angle BAC.

Or thus:

The general equation of the circumscribing conic is

$$\lambda yz + \mu zx + \nu xy = 0.$$

The conic will be an ellipse, a parabola or an hyperbola according as the ratios $x : y : z$ given by

$$\lambda yz + \mu zx + \nu xy = 0, \quad ax + by + cz = 0$$

are imaginary, equal or real; that is according as the roots of

$$a\lambda yz - (\mu z + \nu x)(by + cz) = 0$$

are imaginary, equal or real.

Hence the necessary and sufficient condition that the conic may be an ellipse is

$$(b\mu + c\nu - a\lambda)^2 - 4bc\mu\nu < 0,$$

that is $$a^2\lambda^2 + b^2\mu^2 + c^2\nu^2 - 2bc\mu\nu - 2ca\nu\lambda - 2ab\lambda\mu < 0 \qquad \text{(i)}.$$

Now (ξ, η, ζ), the centre of the conic, is given by

$$\frac{\mu\zeta + \nu\eta}{a} = \frac{\nu\xi + \lambda\zeta}{b} = \frac{\lambda\eta + \mu\xi}{c},$$

whence $$\frac{\lambda}{\xi(-a\xi + b\eta + c\zeta)} = \frac{\mu}{\eta(a\xi - b\eta + c\zeta)} = \frac{\nu}{\zeta(a\xi + b\eta - c\zeta)}.$$

Hence, from (i), the necessary and sufficient condition that the conic may be an ellipse is that

$$\Sigma a^2\xi^2(-a\xi + b\eta + c\zeta)^2 - 2\Sigma bc\eta\zeta(a\xi - b\eta + c\zeta)(a\xi + b\eta - c\zeta) < 0,$$

or that

$$(a\xi + b\eta + c\zeta)(-a\xi + b\eta + c\zeta)(a\xi - b\eta + c\zeta)(a\xi + b\eta - c\zeta) > 0,$$

or $$(2\Delta - 2a\xi)(2\Delta - 2b\eta)(2\Delta - 2c\zeta) > 0.$$

Now within the triangle DEF all the factors on the left are positive; and within either of the angles vertically opposite to one of the angles of the triangle DEF two of the factors are negative and the other factor is positive. This proves the theorem.

18. The conic $$u\alpha^2+v\beta^2+w\gamma^2=0$$
is a parabola if it touches
$$a\alpha+b\beta+c\gamma=0,$$
the condition for which is
$$a^2/u+b^2/v+c^2/w=0.$$

But this condition shews that the parabola also touches the lines
$$-a\alpha+b\beta+c\gamma=0,\ \ a\alpha-b\beta+c\gamma=0 \ \text{ and } \ a\alpha+b\beta-c\gamma=0.$$

The focus of the parabola is therefore on the circle circumscribing the triangle formed by these tangents, and this circle is clearly the nine-point circle of the triangle of reference.

19. Let (α, β, γ) and $(\alpha', \beta', \gamma')$ be a pair of foci of any conic touching the four lines $l\alpha\pm m\beta\pm n\gamma=0$, both foci being real or both imaginary; then since the product of the perpendiculars from the foci of a given conic on any tangent is constant, we have
$$\frac{l\alpha+m\beta+n\gamma}{P_1}\cdot\frac{l\alpha'+m\beta'+n\gamma'}{P_1}=\text{constant}=\lambda.$$

Hence
$$l\alpha'+m\beta'+n\gamma'=\frac{\lambda P_1^2}{l\alpha+m\beta+n\gamma},$$
$$l\alpha'-m\beta'-n\gamma'=\frac{\lambda P_2^2}{l\alpha-m\beta-n\gamma},$$
$$-l\alpha'+m\beta\ -n\gamma'=\frac{\lambda P_3^2}{-l\alpha+m\beta-n\gamma},$$
and
$$-l\alpha'-m\beta'+n\gamma'=\frac{\lambda P_4^2}{-l\alpha-m\beta+n\gamma}.$$

Hence, by addition,
$$\frac{P_1^2}{l\alpha+m\beta+n\gamma}+\frac{P_2^2}{l\alpha-m\beta-n\gamma}+\frac{P_3^2}{-l\alpha+m\beta-n\gamma}+\frac{P_4^2}{-l\alpha-m\beta+n\gamma}=0.$$

20. Let (ξ, η, ζ) be one focus; then the other will be
$$\left(\frac{1}{\xi},\ \frac{1}{\eta},\ \frac{1}{\zeta}\right).$$

The equation of the line joining the foci is given by
$$\begin{vmatrix} \alpha, & \beta, & \gamma \\ \xi, & \eta, & \zeta \\ \frac{1}{\xi}, & \frac{1}{\eta}, & \frac{1}{\zeta} \end{vmatrix}=0.$$

By supposition the line joining the foci passes through (f, g, h). Hence we have

$$\begin{vmatrix} f, & g, & h \\ \xi, & \eta, & \zeta \\ \frac{1}{\xi}, & \frac{1}{\eta}, & \frac{1}{\zeta} \end{vmatrix} = 0.$$

Hence the locus of (ξ, η, ζ) is the cubic

$$f\alpha(\beta^2-\gamma^2)+g\beta(\gamma^2-\alpha^2)+h\gamma(\alpha^2-\beta^2)=0.$$

21. Let the line on which the centre moves be

$$l\alpha+m\beta+n\gamma=0.$$

Let (ξ, η, ζ) be one focus; then the co-ordinates of the other will be

$$\frac{\lambda}{\xi}, \frac{\lambda}{\eta}, \frac{\lambda}{\zeta},$$

where λ is given by

$$2\Delta=\frac{a\lambda}{\xi}+\frac{b\lambda}{\eta}+\frac{c\lambda}{\zeta}.$$

If $(\alpha_0, \beta_0, \gamma_0)$ be the centre of the conic, we have

$$2\alpha_0=\xi+2\Delta/\xi\left(\frac{a}{\xi}+\frac{b}{\eta}+\frac{c}{\zeta}\right),$$

$$2\beta_0=\eta+2\Delta/\eta\left(\frac{a}{\xi}+\frac{b}{\eta}+\frac{c}{\zeta}\right),$$

and

$$2\gamma_0=\zeta+2\Delta/\zeta\left(\frac{a}{\xi}+\frac{b}{\eta}+\frac{c}{\zeta}\right).$$

Hence

$$l\xi+m\eta+n\zeta+2\Delta\left(\frac{l}{\xi}+\frac{m}{\eta}+\frac{n}{\zeta}\right)\Big/\left(\frac{a}{\xi}+\frac{b}{\eta}+\frac{c}{\zeta}\right)=0,$$

or

$$(l\xi+m\eta+n\zeta)(a\eta\zeta+b\zeta\xi+c\xi\eta)+(a\xi+b\eta+c\zeta)(l\eta\zeta+m\zeta\xi+n\xi\eta)=0.$$

22. The equation of a rectangular hyperbola with respect to which the triangle of reference is self-polar is

$$u\alpha^2+v\beta^2+w\gamma^2=0,$$

with the condition

$$u+v+w=0 \quad\text{..................................(i).}$$

The centre is given by

$$\frac{u\alpha}{a}=\frac{v\beta}{b}=\frac{w\gamma}{c}.$$

Substituting for u, v, w in (i), we have the equation of the locus of the centres, namely

$$\frac{a}{\alpha}+\frac{b}{\beta}+\frac{c}{\gamma}=0.$$

23. The equation of any one of the conics is

$$\sqrt{\lambda\alpha}+\sqrt{\mu\beta}+\sqrt{\nu\gamma}=0,$$

or in the rationalised form

$$\lambda^2\alpha^2+\mu^2\beta^2+\nu^2\gamma^2-2\mu\nu\beta\gamma-2\nu\lambda\gamma\alpha-2\lambda\mu\alpha\beta=0 \quad\ldots\ldots\ldots\ldots\text{(i)},$$

with the condition

$$\lambda^2+\mu^2+\nu^2+2\mu\nu\cos A+2\nu\lambda\cos B+2\lambda\mu\cos C=0=\ldots\quad\ldots\text{(ii)}.$$

Now the centre of (i) is given by

$$\frac{\lambda(\lambda\alpha-\mu\beta-\nu\gamma)}{a}=\frac{\mu(-\lambda\alpha+\mu\beta-\nu\gamma)}{b}=\frac{\nu(-\lambda\alpha-\mu\beta+\nu\gamma)}{c},$$

from which it follows that

$$\frac{\lambda}{a(a\alpha-b\beta-c\gamma)}=\frac{\mu}{b(-a\alpha+b\beta-c\gamma)}=\frac{\nu}{c(-a\alpha-b\beta+c\gamma)}.$$

Substituting in (ii) we have the equation of the locus of the centre, namely

$$a^2(a\alpha-b\beta-c\gamma)^2+b^2(-a\alpha+b\beta-c\gamma)^2+c^2(-a\alpha-b\beta+c\gamma)^2$$
$$+2bc(-a\alpha+b\beta-c\gamma)(-a\alpha-b\beta+c\gamma)\cos A$$
$$+2ca(-a\alpha-b\beta+c\gamma)(a\alpha-b\beta-c\gamma)\cos B$$
$$+2ab(a\alpha-b\beta-c\gamma)(-a\alpha+b\beta-c\gamma)\cos C=0,$$

which reduces to

$$\alpha^2\sin 2A+\beta^2\sin 2B+\gamma^2\sin 2C=0.$$

24. The equation of the inscribed circle is

$$\sqrt{(s-a)\,a\alpha}+\sqrt{(s-b)\,b\beta}+\sqrt{(s-c)\,c\gamma}=0,$$

or $$\Sigma(s-a)^2a^2\alpha^2-2\Sigma(s-b)(s-c)\,bc\beta\gamma=0.$$

Hence the equation of any other circle is included in

$$\Sigma(s-a)^2a^2\alpha^2-2\Sigma(s-b)(s-c)\,bc\beta\gamma+(\lambda\alpha+\mu\beta+\nu\gamma)(a\alpha+b\beta+c\gamma)=0.$$

The nine-point circle goes through the points

$$\alpha=0,\ b\beta-c\gamma=0;\quad \beta=0,\ c\gamma-a\alpha=0;\quad\text{and}\quad \gamma=0,\ a\alpha-b\beta=0.$$

Hence λ, μ, ν are given by

$$(s-b)^2+(s-c)^2-2(s-b)(s-c)+2\left(\frac{\mu}{b}+\frac{\nu}{c}\right)=0,$$

or $$(b-c)^2+2\left(\frac{\mu}{b}+\frac{\nu}{c}\right)=0,$$

and two similar equations.

Hence we have

$$\frac{\lambda}{a(a-b)(a-c)}=\frac{\mu}{b(b-c)(b-a)}=\frac{\nu}{b(c-a)(c-b)}.$$

The equation of the radical axis of the inscribed and nine-point circles is therefore

$$\frac{a\alpha}{b-c}+\frac{b\beta}{c-a}+\frac{c\gamma}{a-b}=0 \quad\ldots\ldots\ldots\ldots\ldots\ldots\ldots\ldots\text{(i)}.$$

It follows at once from Art. 278 that the radical axis of the nine-point circle and the inscribed circle *touches* the inscribed circle: the two circles must therefore touch one another.

Again the equation of the circle which touches BC externally and CA, AB internally is found as in Art. 279 to be

$$\sqrt{-as\alpha}+\sqrt{b(s-c)\beta}+\sqrt{c(s-b)\gamma}=0.$$

The common radical axis of this inscribed circle and the nine-point circle is found as above to be

$$\frac{a\alpha}{b-c}+\frac{b\beta}{c+a}+\frac{c\gamma}{-a-b}=0 \ldots\ldots\ldots\ldots\ldots\ldots\ldots\ldots\text{(ii)}.$$

It follows from Art. 278 that (ii) touches the escribed circle; from which it follows that the nine-point circle must touch the escribed circle.

25. We have found in the preceding example that the equations of the four tangents are

$$\frac{a\alpha}{b-c}+\frac{b\beta}{c-a}+\frac{c\gamma}{a-b}=0 \quad\ldots\ldots\ldots\ldots\ldots\ldots\ldots\ldots\text{(i)},$$

$$\frac{a\alpha}{b-c}+\frac{b\beta}{c+a}+\frac{c\gamma}{-a-b}=0 \quad\ldots\ldots\ldots\ldots\ldots\ldots\ldots\ldots\text{(ii)},$$

$$\frac{a\alpha}{-b-c}+\frac{b\beta}{c-a}+\frac{c\gamma}{a+b}=0 \ldots\ldots\ldots\ldots\ldots\ldots\ldots\ldots\text{(iii)},$$

and

$$\frac{a\alpha}{b+c}+\frac{b\beta}{-c-a}+\frac{c\gamma}{a-b}=0 \ldots\ldots\ldots\ldots\ldots\ldots\ldots\ldots\text{(iv)}.$$

The line joining the intersections of (i), (ii) and of (iii), (iv) is easily found to be

$$\frac{b\beta}{c^2-a^2}+\frac{c\gamma}{a^2-b^2}=0 \ldots\ldots\ldots\ldots\ldots\ldots\ldots\ldots\text{(v)}.$$

The equations of the other diagonals are respectively

$$\frac{c\gamma}{a^2-b^2}+\frac{a\alpha}{b^2-c^2}=0 \ldots\ldots\ldots\ldots\ldots\ldots\ldots\ldots\text{(vi)},$$

and

$$\frac{a\alpha}{b^2-c^2}+\frac{b\beta}{c^2-a^2}=0 \quad\ldots\ldots\ldots\ldots\ldots\ldots\ldots\ldots\text{(vii)}.$$

Now it is clear that each of the lines given by (v), (vi) and (vii) passes through one of the angular points of the triangle of reference.

The lines joining corresponding angular points of the triangle of reference and of the triangle formed by the lines (v), (vi) and (vii) are given by

$$\frac{b\beta}{c^2-a^2}=\frac{c\gamma}{a^2-b^2},\quad \frac{c\gamma}{a^2-b^2}=\frac{a\alpha}{b^2-c^2},\quad \text{and}\quad \frac{a\alpha}{b^2-c^2}=\frac{b\beta}{c^2-a^2},$$

and it is easy to shew that these lines are all parallel to

$$\alpha\cos A+\beta\cos B+\gamma\cos C=0,$$

which [Art. 286] is the radical axis of the nine-point circle and the circumscribing circle.

26. Take ABC for the triangle of reference, and let the conic be given by the general equation.

Then the equations of $B'C'$, $C'A'$ and $A'B'$ are respectively

$$u\alpha+w'\beta+v'\gamma=0,$$

$$w'\alpha+v\beta+u'\gamma=0,$$

and $$v'\alpha+u'\beta+w\gamma=0.$$

Hence the equations of AA', BB', CC' are respectively

$$\frac{\beta}{u'v'-ww'}=\frac{\gamma}{w'u'-vv'},\quad \frac{\gamma}{v'w'-uu'}=\frac{\alpha}{u'v'-ww'},\quad \frac{\alpha}{u'w'-vv'}=\frac{\beta}{v'w'-uu'}.$$

Hence AA', BB', CC' meet in the point given by

$$\alpha\,(uu'-v'w')=\beta\,(vv'-w'u')=\gamma\,(ww'-u'v').$$

27. The conic given by the general equation cuts $\alpha=0$ where

$$v\beta^2+w\gamma^2+2u'\beta\gamma=0.$$

Hence the condition that the conic should pass through the middle point of BC is

$$vc^2+wb^2+2u'bc=0 \;\ldots\ldots\ldots\ldots\ldots\ldots \text{(i)}.$$

Also, if $b\beta-c\gamma$ be one factor of $v\beta^2+w\gamma^2+2u'\beta\gamma$, the other factor must be

$$\frac{v\beta}{b}-\frac{w\gamma}{c}.$$

Hence the equation of the line through A and the other point of intersection is

$$v\beta/b-w\gamma/c=0.$$

Hence, if the conic given by the general equation of the second degree pass through the middle points of all three sides of the triangle of reference, and cut those sides again in A', B', C' respectively, the lines AA', BB', CC' all meet in the point

$$\frac{u\alpha}{a}=\frac{v\beta}{b}=\frac{w\gamma}{c}.$$

This point is on the circumscribing circle provided

$$u+v+w=0 \;\ldots\ldots\ldots\ldots\ldots\ldots \text{(ii)}.$$

Now if the conic be a rectangular hyperbola we have

$$u+v+w-2u'\cos A-2v'\cos B-2w'\cos C=0,$$

which reduces to $$u+v+w=0,$$

since we have $vc^2+wb^2+2u'bc=0$ and two similar equations.

28. Let the equations of the conics, referred to their common self-polar triangle, be

$$u_1\alpha^2+v_1\beta^2+w_1\gamma^2=0 \text{ and } u_2\alpha^2+v_2\beta^2+w_2\gamma^2=0.$$

Let the equation of the given straight line be $l\alpha+m\beta+n\gamma=0$.

Then the polars of (f, g, h) with respect to the two conics are

$$u_1\alpha f+v_1\beta g+w_1\gamma h=0,$$

and $$u_2\alpha f+v_2\beta g+w_2\gamma h=0.$$

But, since (f, g, h) is on the given line, we have

$$lf+mg+nh=0.$$

Eliminating f, g, h from the last three equations, we have the equation of the required locus, namely

$$\begin{vmatrix} u_1\alpha, & v_1\beta, & w_1\gamma \\ u_2\alpha, & v_2\beta, & w_2\gamma \\ l\ , & m\ , & n \end{vmatrix}=0,$$

which clearly is the equation of a conic circumscribing the triangle of reference.

29. The equations of the two conics may be taken to be

$$\alpha^2-2\lambda\beta\gamma=0 \text{ and } \alpha^2-2\mu\beta\gamma=0.$$

The polar of $(\alpha', \beta', \gamma')$ with respect to the first conic is

$$\alpha\alpha'-\lambda\beta\gamma'-\lambda\gamma\beta'=0.$$

Hence, if this polar touch the second conic, we have

$$\mu\alpha'^2-2\lambda^2\beta'\gamma'=0.$$

Hence the equation of the required locus is

$$\mu\alpha^2-2\lambda^2\beta\gamma=0,$$

which proves the proposition.

30. Take one of the triangles for the triangle of reference, and let $(\alpha_1, \beta_1, \gamma_1)$, &c. be the angular points of the other triangle.

Let the conic on which the six points lie be

$$\frac{\lambda}{\alpha}+\frac{\mu}{\beta}+\frac{\nu}{\gamma}=0 \quad\text{..................................(i).}$$

Then we have

$$\frac{\lambda}{\alpha_1}+\frac{\mu}{\beta_1}+\frac{\nu}{\gamma_1}=0 \text{..............................(ii)},$$

and two similar equations.

The equation of the line joining the points $(\alpha_2, \beta_2, \gamma_2)$ and $(\alpha_3, \beta_3, \gamma_3)$ is easily seen to be

$$\frac{\lambda\alpha}{\alpha_2\alpha_3}+\frac{\mu\beta}{\beta_2\beta_3}+\frac{\nu\gamma}{\gamma_2\gamma_3}=0 \text{ .... (iii)}.$$

Now the condition that the line (iii) may touch the conic

$$\lambda\sqrt{\frac{\alpha}{\alpha_1\alpha_2\alpha_3}}+\mu\sqrt{\frac{\beta}{\beta_1\beta_2\beta_3}}+\nu\sqrt{\frac{\gamma}{\gamma_1\gamma_2\gamma_3}}=0 \text{...........(iv)},$$

is, from Art. 278,

$$\frac{\lambda}{\alpha_1}+\frac{\mu}{\beta_1}+\frac{\nu}{\gamma_1}=0,$$

and this condition is satisfied since $(\alpha_1, \beta_1, \gamma_1)$ is on the conic (i).

The symmetry of the above result shews that the conic (iv) touches all six sides of the two triangles.

31. Take one of the triangles for the triangle of reference, and let $(\alpha_1, \beta_1, \gamma_1)$, &c. be the angular points of the other triangle.

Let the conic with respect to which the triangles are self-polar be

$$u\alpha^2+v\beta^2+w\gamma^2=0 \text{................................(i)}.$$

Then the equations of the sides of the second triangle are

$$u\alpha_1\alpha+v\beta_1\beta+w\gamma_1\gamma=0 \text{ .... (ii)},$$

and two similar equations.

Also, since the second triangle is self-polar with respect to (i), we have

$$u\alpha_2\alpha_3+v\beta_2\beta_3+w\gamma_2\gamma_3=0 \text{...........................(iii)},$$

and two similar equations.

Now the conditions (iii) shew that the lines (ii) all touch the conic whose equation is

$$\sqrt{u^2\alpha_1\alpha_2\alpha_3\alpha}+\sqrt{v^2\beta_1\beta_2\beta_3\beta}+\sqrt{w^2\gamma_1\gamma_2\gamma_3\gamma}=0 \text{ ....(iv)}.$$

The conditions (iii) also shew that the points $(\alpha_1, \beta_1, \gamma_1)$, &c. are all on the conic whose equation is

$$u\alpha_1\alpha_2\alpha_3/\alpha+v\beta_1\beta_2\beta_3/\beta+w\gamma_1\gamma_2\gamma_3/\gamma=0 \text{ .... (v)}.$$

This proves the proposition, since the conics (iv) and (v) are respectively inscribed in and circumscribed to the triangle of reference.

32. Let ABC be the triangle whose angular points are on the conic S and which is self-polar with respect to the conic Σ.

Take any other point A' on S, and let its polar with respect to Σ cut S in the points B' and C'. Let the polar of B' with respect to Σ, which we know will pass through A', cut $B'C'$ in the point D. Then the triangles ABC and $A'B'D$ are both self-polar with respect to Σ; and therefore, by the preceding question, the six points A, B, C, A', B', D are on the same conic. But five of the points, namely A, B, C, A', B' are on the conic S, and only one conic will pass through five points, therefore D must also be on S, so that C' and D must coincide. This proves the proposition, since A' is *any* point on S.

33. The asymptotes of the conic $u\alpha^2+v\beta^2+w\gamma^2=0$ are parallel to

$$u(b\beta+c\gamma)^2+va^2\beta^2+wa^2\gamma^2=0.$$

The angle between these lines is [Art. 44]

$$\tan^{-1}\frac{2\sqrt{\{u^2b^2c^2-(ub^2+va^2)(uc^2+wa^2)\}}\sin A}{ub^2+va^2+uc^2+wa^2-2ubc\cos A}$$

$$=\tan^{-1}\frac{2\sqrt{\{-vwa^2-wub^2-uvc^2\}}\sin A}{a(u+v+w)}.$$

Hence, if the conic be one of a system of similar conics, we have

$$k^2(u+v+w)^2+2vwa^2+2wub^2+2uvc^2=0 \quad \ldots\ldots\ldots\ldots\text{(i)},$$

where k is some constant.

Now the centre of the conic is given by

$$\frac{u\alpha}{a}=\frac{v\beta}{b}=\frac{w\gamma}{c}.$$

Substituting for u, v, w in (i) we have the equation of the locus of the centre, namely

$$k^2\left(\frac{a}{\alpha}+\frac{b}{\beta}+\frac{c}{\gamma}\right)^2+2abc\left(\frac{a}{\beta\gamma}+\frac{b}{\gamma\alpha}+\frac{c}{\alpha\beta}\right)=0,$$

or

$$k^2(a\beta\gamma+b\gamma\alpha+c\alpha\beta)^2+2abc\alpha\beta\gamma(a\alpha+b\beta+c\gamma)=0.$$

The locus is therefore a curve of the fourth degree &c.

34. This is the converse of Brianchon's Theorem, and may be proved thus:—A conic will touch the *five* lines AB', $B'C$, CA', $A'B$ and BC'; let the other tangent to this conic from the point C' cut BA' in the point X. Then by Brianchon's Theorem, BB', CC' and $A'X$ meet in a point; but BB', CC' and $A'A$ meet in a point; hence X must coincide with A, and therefore the conic touches $C'A$.

35. Let the equation of the conic be

$$\sqrt{\lambda\alpha}+\sqrt{\mu\beta}+\sqrt{\nu\gamma}=0.$$

The points of contact are $\alpha=0$, $\mu\beta=\nu\gamma$; &c.

Hence the normals at the points of contact are

$$(\mu\cos C-\nu\cos B)\,\alpha+\mu\beta-\nu\gamma=0,$$

$$(\nu\cos A-\lambda\cos C)\,\beta+\nu\gamma-\lambda\alpha=0,$$

$$(\lambda\cos B-\mu\cos A)\,\gamma+\lambda\alpha-\mu\beta=0.$$

If these lines meet in a point, the locus of the point will be found by eliminating λ, μ, ν from the three equations: the equation is

$$(\alpha\cos C+\beta)(\beta\cos A+\gamma)(\gamma\cos B+\alpha)=(\alpha\cos B+\gamma)(\beta\cos C+\alpha)(\gamma\cos A+\beta).$$

The locus is therefore a cubic curve; and by considering the three conics which have two of the sides of the triangle for asymptotes we see that the points at infinity on the locus are in directions perpendicular to the sides of the triangle.

36. Take the diagonal-triangle of the quadrilateral for the triangle of reference, and let the equations of the four lines be

$$l\alpha\pm m\beta\pm n\gamma=0.$$

The equation of the conic will be

$$u\alpha^2+v\beta^2+w\gamma^2=0,$$

with the condition

$$\frac{l^2}{u}+\frac{m^2}{v}+\frac{n^2}{w}=0 \quad \text{.................................} \quad \text{(i)}.$$

Let

$$\lambda\alpha+\mu\beta+\nu\gamma=0$$

be any other tangent to the conic; then we have

$$\frac{\lambda^2}{u}+\frac{\mu^2}{v}+\frac{\nu^2}{w}=0 \quad \text{.................................} \quad \text{(ii)}.$$

Now the three pairs of opposite angular points of the quadrilateral are

$$\alpha=0,\ m\beta\pm n\gamma=0;\ \text{\&c.}$$

The actual co-ordinates of one pair of points are therefore

$$0,\ \frac{2\Delta n}{bn+cm},\quad \frac{2\Delta m}{bn+cm};\ \text{ and }\ 0,\ \frac{2\Delta n}{bn-cm},\quad \frac{2\Delta m}{cm-bn};$$

and so for the others.

Hence the product of the perpendiculars from these three pairs of points on $\lambda\alpha+\mu\beta+\nu\gamma=0$ are

$$\frac{4\Delta^2}{P^2}\cdot\frac{\mu^2n^2-\nu^2m^2}{b^2n^2-c^2m^2},\quad \frac{4\Delta^2}{P^2}\cdot\frac{\nu^2l^2-\lambda^2n^2}{c^2l^2-a^2n^2}\ \text{ and }\ \frac{4\Delta^2}{P^2}\cdot\frac{\lambda^2m^2-\mu^2l^2}{a^2m^2-b^2l^2}.$$

But, from (i) and (ii),

$$\frac{1}{u(m^2\nu^2-n^2\mu^2)}=\frac{1}{v(n^2\lambda^2-l^2\nu^2)}=\frac{1}{w(l^2\mu^2-m^2\lambda^2)},$$

whence it follows that the ratios of the three products of perpendiculars is independent of λ, μ and ν.

37. Take ABC for the triangle of reference, and let the conic be given by the general equation. Then the equation of the polar of A is

$$u\alpha+w'\beta+v'\gamma=0.$$

Hence A' is given by

$$\alpha=0=w'\beta+v'\gamma,$$

whence it follows that A', B' and C' are on a straight line, namely on the line whose equation is

$$\frac{\alpha}{u'}+\frac{\beta}{v'}+\frac{\gamma}{w'}=0.$$

Since $A'B'C'$ is a straight line, it follows that AA', BB', CC' are the diagonals of the quadrilateral formed by the sides of the triangle and the line $A'B'C'$. The proposition is now reduced to a particular case of Art. 299 (5); for AA' is a limiting form of a conic touching the four lines, and the circle on AA' as diameter is the director-circle of this limiting conic; and so for the other diagonals.

38. The equation of any conic which touches BC at its middle point is

$$\sqrt{l\alpha}+\sqrt{b\beta}+\sqrt{c\gamma}=0.$$

The condition that the conic should touch

$$a\alpha+b\beta+c\gamma=0$$

is [Art. 278]

$$\frac{l}{a}+2=0 \qquad\qquad \text{(i)}.$$

The condition that the line

$$\lambda\alpha+\mu\beta+\nu\gamma=0$$

should touch is

$$\frac{l}{\lambda}+\frac{b}{\mu}+\frac{c}{\nu}=0 \qquad\qquad \text{(ii)}.$$

But, if p, q, r be the perpendiculars from the angular points of the triangle on the line

$$\lambda\alpha+\mu\beta+\nu\gamma=0,$$

we have

$$\frac{p}{\frac{\lambda}{a}}=\frac{q}{\frac{\mu}{b}}=\frac{r}{\frac{\nu}{c}}.$$

Hence, from (ii),

$$\frac{l}{ap}+\frac{b}{bq}+\frac{c}{cr}=0;$$

$\therefore$, from (i)

$$\frac{1}{q}+\frac{1}{r}=\frac{2}{p}.$$

39. The condition that

$$ap\alpha+bq\beta+cr\gamma=0$$

should touch

$$a\beta\gamma+b\gamma\alpha+c\alpha\beta=0$$

is [Art. 276] $$\sqrt{a^2p}+\sqrt{b^2q}+\sqrt{c^2r}=0,$$
that is $$a\sqrt{p}+b\sqrt{q}+c\sqrt{r}=0.$$

If A', B', C' be respectively the middle points of the sides BC, CA, AB of the triangle of reference, and if p', q', r' be the perpendiculars from A', B', C' on any tangent to the nine-point circle, it follows from the above that

$$a\sqrt{p'}+b\sqrt{q'}+c\sqrt{r'}=0.$$

But, if p, q, r be the perpendiculars from A, B, C on the same tangent to the nine-point circle, we have

$$2p'=q+r,\ 2q'=r+p \text{ and } 2r'=p+q\,;$$
$$\therefore \quad a\sqrt{(q+r)}+b\sqrt{(r+p)}+c\sqrt{(p+q)}=0.$$

40. From question 16, the sum of the squares of the axes of a conic inscribed in the triangle of reference is equal to

$$\frac{abc}{\Delta^2}\{\alpha_0{}^2a\cos A+\beta_0{}^2b\cos B+\gamma_0{}^2c\cos C\},$$

where $(\alpha_0, \beta_0, \gamma_0)$ is the centre of the conic.

Hence if the sum of the squares of the axes is constant and equal to k^2, the locus of the centre of the conic is

$$\alpha^2a\cos A+\beta^2b\cos B+\gamma^2c\cos C=\frac{k^2}{abc}(a\alpha+b\beta+c\gamma)^2.$$

The locus is therefore [Art. 285] a circle concentric with the self-polar circle.

41. From 40, the theorem will be true for *all* conics whose director-circles are equal if it be true for any one of them, for the locus of the centres of the director-circles is a circle concentric with the self-polar circle. Let ABC be the triangle, and let O be the orthocentre which we know is the centre of the self-polar circle. Let OA cut BC in A'. Then, if P be any point on BC, the line AP is a limiting form of an inscribed conic, and the circle on AP as diameter is its director-circle; also $OA\,.\,OA'$ is equal to the square of the tangent to this circle from O. But $OA\,.\,OA'$ is also equal to the square of the radius of the self-polar circle, and therefore the self-polar circle cuts orthogonally the circle on AP as diameter. This proves the theorem, since P is *any* point on BC.

42. In the figure to Art. 60,

$$\{BODS\}=\{AOCR\}=\{QSPR\}=-1.$$

Hence, from Ex. 6, page 59, any circle through O and S cuts orthogonally the circle on BD as diameter; and so for the other diagonals.

Hence the circle ROS cuts orthogonally the circles whose diameters are BD, AC and QP.

43. All conics which circumscribe the same quadrilateral are cut in involution by any straight line [Ex. 2, Art. 320]. If the line *touch* two of the conics, the points of contact will be double points of the involution, and the theorem follows from Art. 68.

44. This is the reciprocal of 43.

45. Refer the conics to their common self-polar triangle, and let their equations be

$$u_1\alpha^2+v_1\beta^2+w_1\gamma^2=0 \text{ and } u_2\alpha^2+v_2\beta^2+w_2\gamma^2=0.$$

The equations of the pairs of tangents from (ξ, η, ζ) to the two conics are

$$(u_1\alpha^2+v_1\beta^2+w_1\gamma^2)(u_1\xi^2+v_1\eta^2+w_1\zeta^2)-(u_1\xi\alpha+v_1\eta\beta+w_1\zeta\gamma)^2=0$$

and

$$(u_2\alpha^2+v_2\beta^2+w_2\gamma^2)(u_2\xi^2+v_2\eta^2+w_2\zeta^2)-(u_2\xi\alpha+v_2\eta\beta+w_2\zeta\gamma)^2=0.$$

If the two pairs of tangents are conjugate pairs of a harmonic pencil, the lines from A to the points where they cut BC will be conjugate pairs of a harmonic pencil, and the equations of these pairs are

$$(v_1\beta^2+w_1\gamma^2)(u_1\xi^2+v_1\eta^2+w_1\zeta^2)-(v_1\eta\beta+w_1\zeta\gamma)^2=0,$$

and

$$(v_2\beta^2+w_2\gamma^2)(u_2\xi^2+v_2\eta^2+w_2\zeta^2)-(v_2\eta\beta+w_2\zeta\gamma)^2=0.$$

The condition that these should be conjugate pairs of a harmonic pencil is

$$v_1w_2(u_1\xi^2+w_1\zeta^2)(u_2\xi^2+v_2\eta^2)+w_1v_2(u_2\xi^2+w_2\zeta^2)(u_1\xi^2+v_1\eta^2)-2v_1w_1v_2w_2\eta^2\zeta^2=0,$$

whence

$$u_1u_2(v_1w_2+v_2w_1)\xi^2+v_1v_2(w_1u_2+w_2u_1)\eta^2+w_1w_2(u_1v_2+u_2v_1)\zeta^2=0.$$

The locus of (ξ, η, ζ) is therefore a conic.

From the point of contact of either of the conics with a common tangent, three out of the four tangents are coincident and the four will therefore form a harmonic pencil. Hence the locus passes through the eight points of contact of the common tangents of the conics.

46. The equations of the circles can be taken to be

$$x^2+y^2-2ax-c^2=0 \text{ and } x^2+y^2+2ax-c^2=0.$$

The tangents from (x', y') to the first circle cut $x=0$ where

$$(y^2-c^2)(x'^2+y'^2-2ax'-c^2)-(yy'-ax'-c^2)^2=0,$$

or

$$y^2(x'^2-2ax'-c^2)+2yy'(ax'+c^2)-(c^2+a^2)x'^2-c^2y'^2=0.$$

The tangents from (x', y') to the second circle cut $x=0$ where

$$y^2(x'^2+2ax'-c^2)+2yy'(c^2-ax')-(c^2+a^2)x'^2-c^2y'^2=0.$$

Hence if the pairs of tangents are conjugate rays of a harmonic pencil we have

$$(x'^2-c^2)\{(c^2+a^2)x'^2+c^2y'^2\}+y'^2(c^4-a^2x'^2)=0,$$

whence the equation of the required locus is

$$(c^2+a^2)\,x^2+(c^2-a^2)\,y^2-c^2\,(c^2+a^2)=0.$$

Hence the locus is an ellipse of $c^2>a^2$, and two parallel straight lines if $c^2=a^2$. This proves the proposition.

47. Let the equations of the two conics be

$$u_1\alpha^2+v_1\beta^2+w_1\gamma^2=0 \text{ and } u_2\alpha^2+v_2\beta^2+w_2\gamma^2=0.$$

The co-ordinates of any point on the first conic can be taken to be

$$\sqrt{-\frac{w_1}{u_1}}\cos\theta,\quad \sqrt{-\frac{w_1}{v_1}}\sin\theta,\quad -1.$$

Hence the equations of the tangent at the three points θ, ϕ, ψ will be

$$\alpha\sqrt{u_1}\cos\theta+\beta\sqrt{v_1}\sin\theta+\gamma\sqrt{-w_1}=0 \;\ldots\ldots\ldots\ldots\ldots\ldots \text{(i)},$$

$$\alpha\sqrt{u_1}\cos\phi+\beta\sqrt{v_1}\sin\phi+\gamma\sqrt{-w_1}=0 \;\ldots\ldots\ldots\ldots\ldots\ldots \text{(ii)},$$

$$\alpha\sqrt{u_1}\cos\psi+\beta\sqrt{v_1}\sin\psi+\gamma\sqrt{-w_1}=0 \;\ldots\ldots\ldots\ldots\ldots \text{(iii)}.$$

The points of intersection of these lines in pairs are given by

$$\frac{\alpha\sqrt{u_1}}{\cos\frac{1}{2}(\phi+\psi)}=\frac{\beta\sqrt{v_1}}{\sin\frac{1}{2}(\phi+\psi)}=\frac{-\gamma\sqrt{-w_1}}{\cos\frac{1}{2}(\phi-\psi)},\ \&c.$$

Hence, if the point of intersection of (iii) and (i), and also the point of intersection of (i) and (ii) be on the second conic, we have

$$\frac{u_2}{u_1}\cos^2\frac{1}{2}(\psi+\theta)+\frac{v_2}{v_1}\sin^2\frac{1}{2}(\psi+\theta)-\frac{w_2}{w_1}\cos^2\frac{1}{2}(\psi-\theta)=0,$$

or

$$\left(\frac{u_2}{u_1}-\frac{v_2}{v_1}-\frac{w_2}{w_1}\right)\cos\psi\cos\theta+\left(\frac{v_2}{v_1}-\frac{w_2}{w_1}-\frac{u_2}{u_1}\right)\sin\psi\sin\theta-\left(\frac{w_2}{w_1}-\frac{u_2}{u_1}-\frac{v_2}{v_1}\right)=0,$$

or $$L\cos\psi\cos\theta+M\sin\psi\sin\theta-N=0,$$

and also $$L\cos\phi\cos\theta+M\sin\phi\sin\theta-N=0.$$

Whence we have

$$\frac{L\cos\theta}{\cos\frac{1}{2}(\phi+\psi)}=\frac{M\sin\theta}{\sin\frac{1}{2}(\phi+\psi)}=\frac{N}{\cos\frac{1}{2}(\phi-\psi)};$$

$$\therefore\ \frac{1}{L^2}\cos^2\frac{1}{2}(\phi+\psi)+\frac{1}{M^2}\sin^2\frac{1}{2}(\phi+\psi)-\frac{1}{N^2}\cos^2\frac{1}{2}(\phi-\psi)=0.$$

Hence at the point of intersection of (i) and (ii) we have

$$\frac{u_1}{L^2}\alpha^2+\frac{v_1}{M^2}\beta^2+\frac{w_1}{N^2}\gamma^2=0,$$

where

$$L=\frac{u_2}{u_1}-\frac{v_2}{v_1}-\frac{w_2}{w_1},\qquad M=\frac{v_2}{v_1}-\frac{w_2}{w_1}-\frac{u_2}{u_1}\quad\text{and}\quad N=\frac{w_2}{w_1}-\frac{u_2}{u_1}-\frac{v_2}{v_1}.$$

48. Refer the conic to their common self-conjugate triangle; then their equations will be

$$u_1\alpha^2+v_1\beta^2+w_1\gamma^2=0 \text{ and } u_2\alpha^2+v_2\beta^2+w_2\gamma^2=0.$$

Let (f, g, h) be any point P on the first conic; then the equation of the tangents PQ, PR from P to the second conic are

$$(u_2\alpha^2+v_2\beta^2+w_2\gamma^2)(u_2f^2+v_2g^2+w_1h^2)-(u_2f\alpha+v_2g\beta+w_2h\gamma)^2=0 \ldots \text{(i)}.$$

But the equation of PQ, PR is included in

$$\lambda(u_1\alpha^2+v_1\beta^2+w_1\gamma^2)-(u_1f\alpha+v_1g\beta+w_1h\gamma)(L\alpha+M\beta+N\gamma)=0 \ldots \text{(ii)}.$$

Comparing coefficients of $\beta\gamma$, $\gamma\alpha$ and $\alpha\beta$ in (i) and (ii) we have

$$Mw_1h+Nv_1g=\mu v_2w_2gh,$$

$$Nu_1f+Lw_1h=\mu w_2u_2hf,$$

and

$$Lv_1g+Nu_1f=\mu u_2v_2fg,$$

Hence

$$\frac{M}{v_1g}+\frac{N}{w_1h}=\mu\cdot\frac{v_2w_2}{v_1w_1}, \text{ \&c.};$$

$$\therefore \frac{L}{u_1f}=\mu\left\{-\frac{v_2w_2}{v_1w_1}+\frac{w_2u_2}{w_1u_1}+\frac{u_2v_2}{u_1v_1}\right\}, \text{ \&c.}$$

But

$$u_1f^2+v_1g^2+w_1h^2=0.$$

Hence

$$\Sigma\frac{L^2}{u_1\left\{-\frac{v_2w_2}{v_1w_1}+\frac{w_2u_2}{w_1u_1}+\frac{u_2v_2}{u_1v_1}\right\}^2}=0,$$

which shews that

$$L\alpha+M\beta+N\gamma=0$$

touches the conic

$$\Sigma u_1\left\{-\frac{v_2w_2}{v_1w_1}+\frac{w_2u_2}{w_1u_1}+\frac{u_2v_2}{u_1v_1}\right\}^2\alpha^2=0.$$

49. Take the triangle on which the angular points lie for the triangle of reference; and let the two fixed points P, Q be

$$(f_1, g_1, h_1) \text{ and } (f_2, g_2, h_2).$$

Let D, E, F be points on BC, CA, AB respectively. *Suppose that P is on DF and Q on DE.*

Then if $l\alpha+m\beta+n\gamma=0$ be the equation of FE, it is easily seen that the equations of FD, EQ will be respectively

$$h_1(l\alpha+m\beta)=\gamma(lf_1+mg_1),$$

and

$$g_2(l\alpha+n\gamma)=\beta(lf_2+nh_2).$$

But, since FP and EQ meet on BC, we have

$$h_1 g_2 mn = (lf_1 + mg_1)(lf_2 + nh_2),$$

and the envelope of $l\alpha + m\beta + n\gamma = 0$ with the above conditions is clearly a conic.

50. The equation of the two tangents from A to the given conic is

$$u(ux^2 + vy^2 + wz^2 + 2u'yz + 2v'zx + 2w'xy) - (ux + w'y + v'z)^2 = 0.$$

These meet BC in points given by

$$x = 0,\ (uv - w'^2)\,y^2 + (uw - v'^2)\,z^2 + 2(uu' - v'w')\,yz = 0,$$

that is
$$x = 0,\ Wy^2 + Vz^2 - 2U'yz = 0,$$

where U, V, W, &c. are the co-factors of u, v, w, &c. in

$$\begin{vmatrix} u, & w', & v' \\ w', & v, & u' \\ v', & u', & w \end{vmatrix}.$$

Hence these points, and by symmetry the corresponding points on CA and AB, are on the conic

$$VWx^2 + WUy^2 + UVz^2 - 2UU'yz - 2VV'zx - 2WW'xy = 0.$$

This conic intersects $\Sigma\sqrt{xU'} = 0$, that is

$$U'^2x^2 + V'^2y^2 + W'^2z^2 - 2V'W'yz - 2W'U'zx - 2U'V'xy = 0$$

in the same four points as the conic

$$\Sigma(VW - U'^2)\,x^2 + 2\Sigma(V'W' - UU')\,yz = 0.$$

But the latter conic is the original conic since

$$VW - U'^2 = u\Delta,\ \text{\&c.}$$

CHAPTER XIV.

Pages 336—338.

1. See Art. 148 (7).

2. Conics through four given points have a common self-polar triangle; and, from Art. 306, if the conics be reciprocated with respect to a vertex of this self-polar triangle, the reciprocal conics will be concentric.

3. Reciprocate with respect to any point the following known theorem:—'Four circles can be drawn so as to touch three given straight lines, and the reciprocal of the radius of one of the circles is equal to the sum of the reciprocals of the radii of the other three; also the centres of the circles lie two and two on lines through the angular points of the triangle formed by the three given straight lines.'

4. Let the two conics be

$$u_1\alpha^2+v_1\beta^2+w_1\gamma^2=0\text{..............................(i)},$$

$$u_2\alpha^2+v_2\beta^2+w_2\gamma^2=0\text{..............................(ii)}$$

the polar of the point $(\alpha', \beta', \gamma')$ with respect to the first conic is

$$u_1\alpha'\alpha+v_1\beta'\beta+w_1\gamma'\gamma=0\text{..........................(iii)}.$$

If $(\alpha', \beta', \gamma')$ is on (ii), we have

$$u_2\alpha'^2+v_2\beta'^2+w_2\gamma'^2=0\text{..........................(iv)}.$$

Now (iv) shews that (iii) touches the conic

$$\frac{u_1^2}{u_2}\alpha^2+\frac{v_1^2}{v_2}\beta^2+\frac{w_1^2}{w_2}\gamma^2=0\text{..........................(v)}.$$

Hence (v) is the equation of the reciprocal of (ii) with respect to (i). Similarly the equation of the reciprocal of (i) with respect to (ii) is

$$\frac{u_2^2}{u_1}\alpha^2+\frac{v_2^2}{v_1}\beta^2+\frac{w_2^2}{w_1}\gamma^2=0\text{ ....(vi)}.$$

It is clear that the conics given by (i), (ii), (v) and (vi) have a common self-polar triangle.

5. Let the conics L_1 and U be referred to their common self-polar triangle, and let their equations be respectively

$$l_1\alpha^2+m_1\beta^2+n_1\gamma^2=0 \quad\text{(i)}$$

and

$$u_1\alpha^2+v_1\beta^2+w_1\gamma^2=0 \quad\text{(ii)}.$$

Then, by the preceding question, the equation of L_2 will be

$$\frac{u_1^2}{l_1}\alpha^2+\frac{v_1^2}{m_1}\beta^2+\frac{w_1^2}{n_1}\gamma^2=0 \quad\text{(iii)}.$$

M_1 is the reciprocal of (i) with respect to (iii), and its equation is therefore

$$\frac{u_1^4}{l_1^3}\alpha^2+\frac{v_1^4}{m_1^3}\beta^2+\frac{w_1^4}{n_1^3}\gamma^2=0 \quad\text{(iv)}.$$

M_2 is the reciprocal of (iii) with respect to (i), and its equation is therefore

$$\frac{l_1^3}{u_1^2}\alpha^2+\frac{m_1^3}{v_1^2}\beta^2+\frac{n_1^3}{w_1^2}\gamma^2=0 \quad\text{(v)}.$$

It is now clear that (iv) and (v) are reciprocals with respect to (ii).

6. Let the pencil be cut by any straight line in the points A, A'; B, B'; C, C'; &c. Let P be the vertex of the pencil, and let APA' and BPB' be right angles. Then, if PO be perpendicular to ABC; it is clear that $AO \,.\, OA'=BO \,.\, OB'=PO^2$, so that O is centre of the involution. It therefore follows that, if C, C' are any other pair of conjugate points, $CO \,.\, OC'=PO^2$, and therefore CP, $C'P$ are at right angles.

7. Let V be the middle point of AA' and also of BB'; and let O be the centre of the involution. Let $AV=VA'=a$; $BV=VB'=b$; and $VO=x$.

Then $x^2-a^2=x^2-b^2$, from which it follows that x is *infinite*, since a and b are unequal.

Now let c, c' be any other pair of conjugate points, and let $CV=c$, $VC'=c'$; then we have

$$x^2-a^2=(x+c)(x-c'),$$

which gives a *finite* value for x, unless $c=c'$.

8. Reciprocate with respect to any point; then we have to prove the following theorem:—

The points in which any straight line cuts a system of conics through four given points are in involution.

This is proved by projection in Art. 320, Ex. 2. It may however be proved thus:—

Take the line for axis of x, and for origin the centre of the involution determined by the two points in which the line cuts any two conics of the system $S_1=0$ and $S_2=0$. Then any other conic is $S_1+\lambda S_2=0$, or

$$a_1x^2+2h_1xy+b_1y^2+2g_1x+2f_1y+c_1 \\ +\lambda(a_2x^2+2h_2xy+b_2y^2+2g_2x+2f_2y+c_2)=0.$$

This conic cuts $y=0$ in points such that $x_1x_2=(c_1+\lambda c_2)/(a_1+\lambda a_2)$. But we know that $c_1/a_1=c_2/a_2$, whence it follows that $(c_1+\lambda c_2)/(a_1+\lambda a_2)$ is independent of λ.

Let O be either of the points of intersection of the director-circles of two of the conics; then two pairs of conjugate rays of a pencil in involution are at right angles, and therefore from 6, every pair is at right angles.

Hence O is on the director-circle of every conic of the system.

9. This is a particular case of the preceding, since the line joining the centres of similitude of two circles is the limiting form of a conic which touches their four common tangents.

10. Let AB, $A'B'$ be the two given straight lines, and let P, P' be any corresponding points of division.

Then it is clear that

$$\{APB\infty\}=\{A'P'B'\infty\},$$

from which it follows that AB, $A'B'$, AA', BB', PP' and the line at infinity all touch a conic. Hence PP' touches the *parabola* determined by the four tangents AB, $A'B'$, AA' and BB'.

11. Let A, B, C be three fixed points and P any fourth point on OA; and let A', B', C' and P' be the corresponding points on OA'.

Then $$\{ABCP\}=\{A'B'C'P'\}.$$

Hence if $OA=a$, $OB=b$, $OC=c$, $OP=x$, $OA'=a'$, $OB'=b'$, $OC'=c'$ and $OP'=y$, we have

$$\frac{(b-a)(x-c)}{(x-a)(b-c)}=\frac{(b'-a')(y-c')}{(y-a')(b'-c')},$$

which is of the form

$$lxy+mx+ny+n=0.$$

Hence the locus of Q is a conic whose asymptotes are parallel to OA, OA'.

12. Project the two common points into the circular points at infinity; there the proposition becomes:—

The radical axes of three circles meet in a point, and any line through this point is cut by the three circles in six points in involution.

13. Let ABC, $A'B'C'$ be the triangles, and let P, Q, R be the points of intersection of BC and $B'C'$, CA and $C'A'$, AB and $A'B'$ respectively.

In any plane through PQR draw arcs of circles through PQ and QR each containing angles of 60^0, and let V be one of the points of intersection of these circles. Then, if V be the vertex and the plane of projection be parallel to $VPQR$, two of the angles of each triangle will be projected into angles of 60^0: the triangles will therefore be projected into equilateral triangles.

14. Let P, P'; Q, Q'; and R, R' be the lines bounding the three angles; and let X, X' be the other pair of straight lines through the points of intersection of P, P', Q, Q'. Let R, R' meet X, X' in the points r, r', and let the line through r and r' meet P, P' in p, p' and Q, Q' in q, q'. Then r, r'; p, p'; and q, q' are in involution [Art. 320, Ex. 2]. Hence, if in any plane through rr', circles be described on pp', qq' and rr' respectively as diameters, these circles will have a common point, V suppose.

Now take V for vertex, and a plane parallel to Vrr' for the plane of projection, and the three angles will all be projected into right angles.

15. Take any point P on the curve, and find the line PD which is such that $P\{ABCD\}$ is harmonic. Then, if D be on the curve, and if the circle on BD as diameter cut the conic again in X, the pencil

$$X\{ABCD\}=P\{ABCD\}.$$

And, as the pencil $X\{ABCD\}$ is harmonic and XD, XB are at right angles, it follows that XA and XC make equal angles with XB.

16. Take the given triangle for the triangle of reference, and let O be (f, g, h). Then, if the equation of $OA'B'C'P$ be

$$l\alpha+m\beta+n\gamma=0 \quad\text{.............................(i)},$$

the equation of BB' will be

$$l\alpha+n\gamma=0.$$

Hence [Art. 56] the equation of BP is

$$l\alpha-n\gamma=0 \quad\text{.................................(ii)}.$$

Also since $l\alpha+m\beta+n\gamma=0$ goes through the point O, we have

$$lf+mg+nh=0 \quad\text{..............................(iii)}.$$

Eliminating l, m, n from the equations (i), (ii) and (iii), we have the equation of the locus of P, namely

$$\begin{vmatrix} \alpha, & \beta, & \gamma \\ \alpha, & 0, & -\gamma \\ f, & g, & h \end{vmatrix}=0.$$

Thus the locus of P is a conic which passes through A, B, C and O.

17. Let $S_1=0$ and $S_2=0$ be any two conics of the system; then the equations of any two others will be of the forms $S_1-\lambda_1S_2=0$ and $S_1-\lambda_2S_2=0$. Also, it is easy to see that if $\alpha_1=0$ and $\alpha_2=0$ be the equations of the polars of any point with respect to the first two conics, the polars of the same point with respect to the other two conics will be $\alpha_1-\lambda_1\alpha_2=0$ and $\alpha_1-\lambda_2\alpha_2=0$. These four polars clearly pass through the same point, and the cross ratio of the pencil formed by them is from Art. 56 equal to λ_2/λ_1, and is therefore constant.

18. Let A, B be the two *fixed* vertices about which the angles turn. Let PAP', QAQ', RAR', SAS' be any four positions of one angle, and PBP', QBQ', RBR', SBS' be the four corresponding positions of the other angle.

Then since $PAP' = QAQ' = RAR' = SAS'$, and $PBP' = QBQ' = RBR' = SBS'$, it follows that

$$A\{PQRS\} = A\{P'Q'R'S'\},$$

and

$$B\{PQRS\} = B\{P'Q'R'S'\}.$$

But, since A, B, P, Q, R, S are on a fixed conic, we have

$$A\{PQRS\} = B\{PQRS\};$$

$$\therefore A\{P'Q'R'S'\} = B\{P'Q'R'S'\},$$

which shews that the six points A, B, P', Q', R', S' are on a conic. But five points are sufficient to determine a conic; hence the conic through A, B and any three of the points P', Q', R', S'... will pass through every other point.

19. Let $ABCD...PQR$ be the polygon, and let AB, BC,..., QR pass respectively through the fixed points a, b,..., q. Then, if any four possible positions of the polygon be taken, as in the figure to Art. 323, Ex. 4, we have

$$\{AA'A''A'''\} = a\{AA'A''A'''\} = a\{BB'B''B'''\} = \{BB'B''B'''\},$$

$$\{BB'B''B'''\} = b\{BB'B''B'''\} = b\{CC'C''C'''\} = \{CC'C''C'''\}.$$

....

Hence $\{AA'A''A'''\} = \{RR'R''R'''\}$. Therefore AA', RR', AR, $A'R'$, $A''R''$, $A'''R'''$ all touch a conic, and therefore the conic which touches AA', RR' and any three positions of AR will touch AR in every other position.

20. This is the reciprocal of Ex. 9, Art. 323.

CAMBRIDGE: PRINTED BY C. J. CLAY, M.A. AND SONS, AT THE UNIVERSITY PRESS.

April, 1888.

A Catalogue

OF

Educational Books

PUBLISHED BY

Macmillan & Co.

BEDFORD STREET, STRAND, LONDON.

CONTENTS.

29 AND 30, BEDFORD STREET, COVENT GARDEN,
LONDON, W.C., *April*, 1888.

CLASSICS.

ELEMENTARY CLASSICS.

18mo, Eighteenpence each.

THIS SERIES FALLS INTO TWO CLASSES—

(1) First Reading Books for Beginners, provided not only with **Introductions and Notes**, but with **Vocabularies**, and in some cases with **Exercises** based upon the Text.

(2) Stepping-stones to the study of particular authors, intended for more advanced students who are beginning to read such authors as Terence, Plato, the Attic Dramatists, and the harder parts of Cicero, Horace, Virgil, and Thucydides.

These are provided with Introductions and Notes, **but no Vocabulary.** The Publishers have been led to provide the more strictly Elementary Books with Vocabularies by the representations of many teachers, who hold that beginners do not understand the use of a Dictionary, and of others who, in the case of middle-class schools where the cost of books is a serious consideration, advocate the Vocabulary system on grounds of economy. It is hoped that the two parts of the Series, fitting into one another, may together fulfil all the requirements of Elementary and Preparatory Schools, and the Lower Forms of Public Schools.

The following Elementary Books, with Introductions, Notes, and Vocabularies, and in some cases with Exercises, are either ready or in preparation:—

Aeschylus.—PROMETHEUS VINCTUS. Edited by Rev. H. M. STEPHENSON, M.A.

Arrian.—SELECTIONS. Edited for the use of Schools, with Introduction, Notes, Vocabulary, and Exercises, by JOHN BOND, M.A., and A. S. WALPOLE, M.A.

Aulus Gellius, Stories from. Edited, with Notes and Vocabulary, by Rev. G. H. NALL, M.A., Assistant Master in Westminster School. [*In the Press.*

Cæsar.—THE HELVETIAN WAR. Being Selections from Book I. of the "De Bello Gallico." Adapted for the use of Beginners. With Notes, Exercises, and Vocabulary, by W. WELCH, M.A., and C. G. DUFFIELD, M.A.

THE INVASION OF BRITAIN. Being Selections from Books IV. and V. of the "De Bello Gallico." Adapted for the use of Beginners. With Notes, Vocabulary, and Exercises, by W. WELCH, M.A., and C. G. DUFFIELD, M.A.

THE GALLIC WAR. BOOK I. Edited by A. S. WALPOLE, M.A.

THE GALLIC WAR. BOOKS II. AND III. Edited by the Rev. W. G. RUTHERFORD, M.A., LL.D., Head-Master of Westminster School.

THE GALLIC WAR. BOOK IV. Edited by CLEMENT BRYANS, M.A., Assistant-Master at Dulwich College.

THE GALLIC WAR. SCENES FROM BOOKS V. AND VI. Edited by C. COLBECK, M.A., Assistant-Master at Harrow; formerly Fellow of Trinity College, Cambridge.

THE GALLIC WAR. BOOKS V. AND VI. (separately). By the same Editor. Book V. *ready*. Book VI. *in preparation*.

THE GALLIC WAR. BOOK VII. Edited by JOHN BOND, M.A., and A. S. WALPOLE, M.A.

Cicero.—DE SENECTUTE. Edited by E. S. SHUCKBURGH, M.A., late Fellow of Emmanuel College, Cambridge.

DE AMICITIA. By the same Editor.

STORIES OF ROMAN HISTORY. Adapted for the Use of Beginners. With Notes, Vocabulary, and Exercises, by the Rev. G. E. JEANS, M.A., Fellow of Hertford College, Oxford, and A. V. JONES, M.A.; Assistant-Masters at Haileybury College.

Eutropius.—Adapted for the Use of Beginners. With Notes, Vocabulary, and Exercises, by WILLIAM WELCH, M.A., and C. G. DUFFIELD, M.A., Assistant-Masters at Surrey County School, Cranleigh.

Homer.—ILIAD. BOOK I. Edited by Rev. JOHN BOND, M.A., and A. S. WALPOLE, M.A.

Homer.—ILIAD. BOOK XVIII. THE ARMS OF ACHILLES. Edited by S. R. JAMES, M.A., Assistant-Master at Eton College.

ODYSSEY. BOOK I. Edited by Rev. JOHN BOND, M.A. and A. S. WALPOLE, M.A.

Horace.—ODES. BOOKS I.—IV. Edited by T. E. PAGE, M.A., late Fellow of St. John's College, Cambridge; Assistant-Master at the Charterhouse. Each 1*s.* 6*d.*

Latin Accidence and Exercises Arranged for Beginners. By WILLIAM WELCH, M.A., and C. G. DUFFIELD, M.A., Assistant Masters at Surrey County School, Cranleigh.

Livy.—BOOK I. Edited by H. M. STEPHENSON, M.A., late Head Master of St. Peter's School, York.

THE HANNIBALIAN WAR. Being part of the XXI. AND XXII. BOOKS OF LIVY, adapted for the use of beginners, by G. C. MACAULAY, M.A., late Fellow of Trinity College, Cambridge.

THE SIEGE OF SYRACUSE. Being part of the XXIV. AND XXV. BOOKS OF LIVY, adapted for the use of beginners. With Notes, Vocabulary, and Exercises, by GEORGE RICHARDS, M.A., and A. S. WALPOLE, M.A.

LEGENDS OF EARLY ROME. Adapted for the use of beginners. With Notes, Exercises, and Vocabulary, by HERBERT WILKINSON, M.A. [*In preparation.*

Lucian.—EXTRACTS FROM LUCIAN. Edited, with Notes, Exercises, and Vocabulary, by Rev. JOHN BOND, M.A., and A. S. WALPOLE, M.A.

Nepos.—SELECTIONS ILLUSTRATIVE OF GREEK AND ROMAN HISTORY. Edited for the use of beginners with Notes, Vocabulary and Exercises, by G. S. FARNELL, M.A.

Ovid.—SELECTIONS. Edited by E. S. SHUCKBURGH, M.A. late Fellow and Assistant-Tutor of Emmanuel College, Cambridge.

EASY SELECTIONS FROM OVID IN ELEGIAC VERSE. Arranged for the use of Beginners with Notes, Vocabulary, and Exercises, by HERBERT WILKINSON, M.A.

STORIES FROM THE METAMORPHOSES. Edited for the Use of Schools. With Notes, Exercises, and Vocabulary. By J. BOND, M.A., and A. S. WALPOLE, M.A.

Phædrus.—SELECT FABLES. Adapted for the Use of Beginners. With Notes, Exercises, and Vocabularies, by A. S. WALPOLE, M.A.

Thucydides.—THE RISE OF THE ATHENIAN EMPIRE. BOOK I. cc. LXXXIX. — CXVII. AND CXXVIII. — CXXXVIII. Edited with Notes, Vocabulary and Exercises, by F. H. COLSON, M.A., Senior Classical Master at Bradford Grammar School; Fellow of St. John's College, Cambridge.

Virgil.—ÆNEID. BOOK I. Edited by A. S. WALPOLE, M.A.

ÆNEID. BOOK IV. Edited by Rev. H. M. STEPHENSON, M.A. [*In the press.*

Virgil.—ÆNEID. BOOK V. Edited by Rev. A. CALVERT, M.A., late Fellow of St. John's College, Cambridge.

ÆNEID. BOOK VI. Edited by T. E. PAGE, M.A.

ÆNEID. BOOK IX. Edited by Rev. H. M. STEPHENSON, M.A.

GEORGICS. BOOK I. Edited by C. BRYANS, M.A. [*In preparation.*

SELECTIONS. Edited by E. S. SHUCKBURGH, M.A.

Xenophon.—ANABASIS. BOOK I. Edited by A. S. WALPOLE, M.A.

ANABASIS. BOOK I. Chaps. I.—VIII. for the use of Beginners, with Titles to the Sections, Notes, Vocabulary, and Exercises, by E. A. WELLS, M.A., Assistant Master in Durham School.

ANABASIS. BOOK II. Edited by A. S. WALPOLE, M.A. [*In the press.*

ANABASIS. BOOK IV. THE RETREAT OF THE TEN THOUSAND. Edited for the use of Beginners, with Notes, Vocabulary, and Exercises, by Rev. E. D. STONE, M.A., formerly Assistant-Master at Eton. [*In preparation.*

SELECTIONS FROM THE CYROPÆDIA. Edited, with Notes, Vocabulary, and Exercises, by A. H. COOKE, M.A., Fellow and Lecturer of King's College, Cambridge.

The following more advanced Books, with Introductions and Notes, **but no Vocabulary,** are either ready, or in preparation:—

Cicero.—SELECT LETTERS. Edited by Rev. G. E. JEANS, M.A., Fellow of Hertford College, Oxford, and Assistant-Master at Haileybury College.

Euripides.—HECUBA. Edited by Rev. JOHN BOND, M.A. and A. S. WALPOLE, M.A.

Herodotus.—SELECTIONS FROM BOOKS VII. AND VIII., THE EXPEDITION OF XERXES. Edited by A. H. COOKE, M.A., Fellow and Lecturer of King's College, Cambridge.

Horace. — SELECTIONS FROM THE SATIRES AND EPISTLES. Edited by Rev. W. J. V. BAKER, M.A., Fellow of St. John's College, Cambridge.

SELECT EPODES AND ARS POETICA. Edited by H. A. DALTON, M.A., formerly Senior Student of Christchurch; Assistant-Master in Winchester College.

Plato.—EUTHYPHRO AND MENEXENUS. Edited by C. E. GRAVES, M.A., Classical Lecturer and late Fellow of St. John's College, Cambridge.

Terence.—SCENES FROM THE ANDRIA. Edited by F. W. CORNISH, M.A., Assistant-Master at Eton College.

The Greek Elegiac Poets.— FROM CALLINUS TO CALLIMACHUS. Selected and Edited by Rev. HERBERT KYNASTON, D.D., Principal of Cheltenham College, and formerly Fellow of St. John's College, Cambridge.

Thucydides.—BOOK IV. CHS. I.—XLI. THE CAPTURE OF SPHACTERIA. Edited by C. E. GRAVES, M.A.

Virgil.—GEORGICS. BOOK II. Edited by Rev. J. H. SKRINE, M.A., late Fellow of Merton College, Oxford; Warden of Trinity College, Glenalmond.

*** *Other Volumes to follow.*

CLASSICAL SERIES FOR COLLEGES AND SCHOOLS.

Fcap. 8vo.

Being select portions of Greek and Latin authors, edited with Introductions and Notes, for the use of Middle and Upper forms of Schools, or of candidates for Public Examinations at the Universities and elsewhere.

Attic Orators.—Selections from ANTIPHON, ANDOKIDES, LYSIAS, ISOKRATES, AND ISAEOS. Edited by R. C. JEBB, M.A., LL.D., Litt.D., Professor of Greek in the University of Glasgow. [*New Edition in the press.*

Æschines.— IN CTESIPHONTEM. Edited by Rev. T. GWATKIN, M.A., late Fellow of St. John's College, Cambridge. [*In the press.*

Æschylus. — PERSÆ. Edited by A. O. PRICKARD, M.A. Fellow and Tutor of New College, Oxford. With Map. 3*s*. 6*d*.
SEVEN AGAINST THEBES. Edited by A. W. VERRALL, M.A. School Edition prepared by Rev. M. A. BAYFIELD, M.A. [*In the press.*

Andocides.—DE MYSTERIIS. Edited by W. J. HICKIE, M.A., formerly Assistant-Master in Denstone College. 2*s*. 6*d*.

Cæsar.—THE GALLIC WAR. Edited, after Kraner, by Rev. JOHN BOND, M.A., and A. S. WALPOLE, M.A. With Maps. 6*s*.

Catullus.—SELECT POEMS. Edited by F. P. SIMPSON, B.A., late Scholar of Balliol College, Oxford. New and Revised Edition. 5*s*. The Text of this Edition is carefully adapted to School use.

Cicero.—THE CATILINE ORATIONS. From the German of KARL HALM. Edited, with Additions, by A. S. WILKINS, M.A., LL.D., Professor of Latin at the Owens College, Manchester, Examiner of Classics to the University of London. New Edition, 3*s*. 6*d*.

Cicero.—PRO LEGE MANILIA. Edited, after HALM, by Professor A. S. WILKINS, M.A., LL.D. 2s. 6d.

THE SECOND PHILIPPIC ORATION. From the German of KARL HALM. Edited, with Corrections and Additions, by JOHN E. B. MAYOR, Professor of Latin in the University of Cambridge, and Fellow of St. John's College. New Edition, revised. 5s.

PRO ROSCIO AMERINO. Edited, after HALM, by E. H. DONKIN, M.A., late Scholar of Lincoln College, Oxford; Assistant-Master at Sherborne School. 4s. 6d.

PRO P. SESTIO. Edited by Rev. H. A. HOLDEN, M.A., LL.D., late Fellow of Trinity College, Cambridge; and late Classical Examiner to the University of London. 5s.

Demosthenes.—DE CORONA. Edited by B. DRAKE, M.A., late Fellow of King's College, Cambridge. New and revised Edition. 4s. 6d.

ADVERSUS LEPTINEM. Edited by Rev. J. R. KING, M.A. Fellow and Tutor of Oriel College, Oxford. 4s. 6d.

THE FIRST PHILIPPIC. Edited, after C. REHDANTZ, by Rev. T. GWATKIN, M.A., late Fellow of St. John's College, Cambridge. 2s. 6d.

IN MIDIAM. Edited by Prof. A. S. WILKINS, LL.D., and HERMAN HAGER, Ph.D., of the Owens College, Manchester. [*In preparation.*

Euripides.—HIPPOLYTUS. Edited by J. P. MAHAFFY, M.A., Fellow and Professor of Ancient History in Trinity College, Dublin, and J. B. BURY, Fellow of Trinity College, Dublin. 3s. 6d.

MEDEA. Edited by A. W. VERRALL, M.A., Fellow and Lecturer of Trinity College, Cambridge. 3s. 6d.

IPHIGENIA IN TAURIS. Edited by E. B. ENGLAND, M.A., Lecturer at the Owens College, Manchester. 4s. 6d.

Herodotus.—BOOKS V. AND VI. Edited by J. STRACHAN, M.A., Professor of Greek in the Owens College, Manchester. [*In preparation.*

BOOKS VII. AND VIII. Edited by Miss A. RAMSAY. [*In the press.*

Hesiod.—THE WORKS AND DAYS. Edited by W. T. LENDRUM, Assistant Master in Dulwich College. [*In preparation.*

Homer.—ILIAD. BOOKS I., IX., XI., XVI.—XXIV. THE STORY OF ACHILLES. Edited by the late J. H. PRATT, M.A., and WALTER LEAF, M.A., Fellows of Trinity College, Cambridge. 6s.

ODYSSEY. BOOK IX. Edited by Prof. JOHN E. B. MAYOR. 2s. 6d.

ODYSSEY. BOOKS XXI.—XXIV. THE TRIUMPH OF ODYSSEUS. Edited by S. G. HAMILTON, B.A., Fellow of Hertford College, Oxford. 3s. 6d.

Horace.—THE ODES. Edited by T. E. PAGE, M.A., formerly Fellow of St. John's College, Cambridge; Assistant-Master at the Charterhouse. 6*s.* (BOOKS I., II., III., and IV. separately, 2*s.* each.)

THE SATIRES. Edited by ARTHUR PALMER, M.A., Fellow of Trinity College, Dublin; Professor of Latin in the University of Dublin. 6*s.*

THE EPISTLES AND ARS POETICA. Edited by A S. WILKINS, M.A., LL.D., Professor of Latin in Owens College, Manchester; Examiner in Classics to the University of London. 6*s.*

Isaeos.—THE ORATIONS. Edited by WILLIAM RIDGEWAY, M.A., Fellow of Caius College, Cambridge; and Professor of Greek in the University of Cork. [*In preparation.*

Juvenal. THIRTEEN SATIRES. Edited, for the Use of Schools, by E. G. HARDY, M.A., late Fellow of Jesus College, Oxford. 5*s.*

The Text of this Edition is carefully adapted to School use.

SELECT SATIRES. Edited by Professor JOHN E. B. MAYOR. X. AND XI. 3*s.* 6*d.* XII.—XVI. 4*s.* 6*d.*

Livy.—BOOKS II. AND III. Edited by Rev. H. M. STEPHENSON, M.A. 5*s.*

BOOKS XXI. AND XXII. Edited by the Rev. W. W. CAPES, M.A. Maps. 5*s.*

BOOKS XXIII. AND XXIV. Edited by G. C. MACAULAY, M.A. With Maps. 5*s.*

THE LAST TWO KINGS OF MACEDON. EXTRACTS FROM THE FOURTH AND FIFTH DECADES OF LIVY. Selected and Edited, with Introduction and Notes, by F. H. RAWLINS, M.A., Fellow of King's College, Cambridge; and Assistant-Master at Eton. With Maps. 3*s.* 6*d.*

THE SUBJUGATION OF ITALY. SELECTIONS FROM THE FIRST DECADE. Edited by G. E. MARINDIN, M.A., formerly Assistant Master at Eton. [*In preparation.*

Lucretius. BOOKS I.—III. Edited by J. H. WARBURTON LEE, M.A., late Scholar of Corpus Christi College, Oxford, and Assistant-Master at Rossall. 4*s.* 6*d.*

Lysias.—SELECT ORATIONS. Edited by E. S. SHUCKBURGH, M.A., late Assistant-Master at Eton College, formerly Fellow and Assistant-Tutor of Emmanuel College, Cambridge. New Edition, revised. 6*s.*

Martial. — SELECT EPIGRAMS. Edited by Rev. H. M. STEPHENSON, M.A. New Edition, Revised and Enlarged. 6*s.* 6*d.*

Ovid.—FASTI. Edited by G. H. HALLAM, M.A., Fellow of St. John's College, Cambridge, and Assistant-Master at Harrow. With Maps. 5*s.*

HEROIDUM EPISTULÆ XIII. Edited by E. S. SHUCKBURGH, M.A. 4*s.* 6*d.*

Ovid.—METAMORPHOSES. BOOKS I.—III. Edited by C. SIMMONS, M.A. [*In preparation.*

METAMORPHOSES. BOOKS XIII. AND XIV. Edited by C. SIMMONS, M.A. 4*s.* 6*d.*

Plato.—MENO. Edited by E. S. THOMPSON, M.A., Fellow of Christ's College, Cambridge. [*In preparation.*

APOLOGY AND CRITO. Edited by F. J. H. JENKINSON, M.A., Fellow of Trinity College, Cambridge. [*In preparation.*

LACHES. Edited, with Introduction and Notes, by M. T. TATHAM, M.A., Balliol College, Oxford, formerly Assistant Master at Westminster School. [*In the press.*

THE REPUBLIC. BOOKS I.—V. Edited by T. H. WARREN. M.A., President of Magdalen College, Oxford. [*In the press.*

Plautus.—MILES GLORIOSUS. Edited by R. Y. TYRRELL. M.A., Fellow of Trinity College, and Regius Professor of Greek in the University of Dublin. Second Edition Revised. 5*s.*

AMPHITRUO. Edited by ARTHUR PALMER, M.A., Fellow of Trinity College and Regius Professor of Latin in the University of Dublin. [*In preparation.*

CAPTIVI. Edited by A. RHYS SMITH, late Junior Student of Christ Church, Oxford. [*In preparation.*

Pliny.—LETTERS. BOOK III. Edited by Professor JOHN E. B. MAYOR. With Life of Pliny by G. H. RENDALL, M.A. 5*s.*

LETTERS. BOOKS I. and II. Edited by J. COWAN, B.A., Assistant-Master in the Grammar School, Manchester. [*In preparation.*

Plutarch.—LIFE OF THEMISTOKLES. Edited by Rev. H. A. HOLDEN, M.A., LL.D. 5*s.*

Polybius.—THE HISTORY OF THE ACHÆAN LEAGUE AS CONTAINED IN THE REMAINS OF POLYBIUS. Edited by W. W. CAPES, M.A. 6*s.* 6*d.*

Propertius.—SELECT POEMS. Edited by Professor J. P. POSTGATE, M.A., Fellow of Trinity College, Cambridge. Second Edition, revised. 6*s.*

Sallust.—CATILINA AND JUGURTHA. Edited by C. MERIVALE, D.D., Dean of Ely. New Edition, carefully revised and enlarged, 4*s.* 6*d.* Or separately, 2*s.* 6*d.* each.

BELLUM CATULINAE. Edited by A. M. COOK, M.A., Assistant Master at St. Paul's School. 4*s.* 6*d.*

JUGURTHA. By the same Editor. [*In preparation.*

Sophocles.—ANTIGONE. Edited by Rev. JOHN BOND, M.A., and A. S. WALPOLE, M.A. [*In preparation.*

Tacitus.—AGRICOLA AND GERMANIA. Edited by A. J. CHURCH, M.A., and W. J. BRODRIBB, M.A., Translators of Tacitus. New Edition, 3*s.* 6*d.* Or separately, 2*s.* each.

THE ANNALS. BOOK VI. By the same Editors. 2*s.* 6*d.*

THE HISTORIES. BOOKS I. AND II. Edited by A. D. GODLEY, M.A. 5*s.*

Tacitus.—THE HISTORIES. BOOKS III.—V. By the same Editor. [*In preparation.*

THE ANNALS. BOOKS I. AND II. Edited by J. S. REID, M.L., Litt.D. [*In preparation.*

Terence.—HAUTON TIMORUMENOS. Edited by E. S. SHUCKBURGH, M.A. 3*s.* With Translation, 4*s.* 6*d.*

PHORMIO. Edited by Rev. JOHN BOND, M.A., and A. S WALPOLE, M.A. 4*s.* 6*d.*

Thucydides.—BOOK IV. Edited by C. E. GRAVES, M.A., Classical Lecturer, and late Fellow of St. John's College, Cambridge. 5*s.*

BOOKS III. AND V. By the same Editor. To be published separately. [*In preparation.* (*Book V. in the press.*)

BOOKS I. AND II. Edited by C. BRYANS, M.A. [*In preparation.*

BOOKS VI. AND VII. THE SICILIAN EXPEDITION. Edited by the Rev. PERCIVAL FROST, M.A., late Fellow of St. John's College, Cambridge. New Edition, revised and enlarged, with Map. 5*s.*

Tibullus.—SELECT POEMS. Edited by Professor J. P. POSTGATE, M.A. [*In preparation.*

Virgil.—ÆNEID. BOOKS II. AND III. THE NARRATIVE OF ÆNEAS. Edited by E. W. HOWSON, M.A., Fellow of King's College, Cambridge, and Assistant-Master at Harrow. 3*s.*

Xenophon.—HELLENICA, BOOKS I. AND II. Edited by H. HAILSTONE, B.A., late Scholar of Peterhouse, Cambridge. With Map. 4*s.* 6*d.*

CYROPÆDIA. BOOKS VII. AND VIII. Edited by ALFRED GOODWIN, M.A., Professor of Greek in University College, London. 5*s.*

MEMORABILIA SOCRATIS. Edited by A. R. CLUER, B.A., Balliol College, Oxford. 6*s.*

THE ANABASIS. BOOKS I.—IV. Edited by Professors W. W. GOODWIN and J. W. WHITE. Adapted to Goodwin's Greek Grammar. With a Map. 5*s.*

HIERO. Edited by Rev. H. A. HOLDEN, M.A., LL.D. 3*s.* 6*d.*

OECONOMICUS. By the same Editor. With Introduction, Explanatory Notes, Critical Appendix, and Lexicon. 6*s.*

*** *Other Volumes will follow.*

CLASSICAL LIBRARY.

(1) **Texts,** Edited with **Introductions and Notes,** for the use of Advanced Students. (2) **Commentaries and Translations.**

Æschylus.—THE EUMENIDES. The Greek Text, with Introduction, English Notes, and Verse Translation. By BERNARD DRAKE, M.A., late Fellow of King's College, Cambridge. 8vo. 5*s.*

Æschylus.—AGAMEMNON. Edited, with Introduction and Notes, by A. W. VERRALL, M.A. 8vo. [*In preparation.*

AGAMEMNON, CHOEPHORŒ, AND EUMENIDES. Edited, with Introduction and Notes, by A. O. PRICKARD, M.A., Fellow and Tutor of New College, Oxford. 8vo. [*In preparation.*

AGAMEMNO. Emendavit DAVID S. MARGOLIOUTH, Coll. Nov. Oxon. Soc. Demy 8vo. 2*s.* 6*d.*

THE "SEVEN AGAINST THEBES." Edited, with Introduction, Commentary, and Translation, by A. W. VERRALL, M.A., Fellow of Trinity College, Cambridge. 8vo. 7*s.* 6*d.*

SUPPLICES. Edited, with Introduction and Notes, by T. G. TUCKER, M.A., Professor of Classics in the University of Melbourne. 8vo. [*In preparation.*

Antoninus, Marcus Aurelius.—BOOK IV. OF THE MEDITATIONS. The Text Revised, with Translation and Notes, by HASTINGS CROSSLEY, M.A., Professor of Greek in Queen's College, Belfast. 8vo. 6*s.*

Aristotle.—THE METAPHYSICS. BOOK I. Translated by a Cambridge Graduate. 8vo. 5*s.* [*Book II. in preparation.*

THE POLITICS. Edited, after SUSEMIHL, by R. D. HICKS, M.A., Fellow of Trinity College, Cambridge. 8vo. [*In the press.*

THE POLITICS. Translated by Rev. J. E. C. WELLDON, M.A., Fellow of King's College, Cambridge, and Head-Master of Harrow School. Crown 8vo. 10*s.* 6*d.*

THE RHETORIC. Translated, with an Analysis and Critical Notes, by the same. Crown 8vo. 7*s.* 6*d.*

THE ETHICS. Translated, with an Analysis and Critical Notes, by the same. Crown 8vo. [*In preparation.*

AN INTRODUCTION TO ARISTOTLE'S RHETORIC With Analysis, Notes, and Appendices. By E. M. COPE, Fellow and Tutor of Trinity College, Cambridge. 8vo. 14*s.*

THE SOPHISTICI ELENCHI. With Translation and Notes by E. POSTE, M.A., Fellow of Oriel College, Oxford. 8vo. 8*s.* 6*d.*

Aristophanes.—THE BIRDS. Translated into English Verse, with Introduction, Notes, and Appendices, by B. H. KENNEDY, D.D., Regius Professor of Greek in the University of Cambridge. Crown 8vo. 6*s.* Help Notes to the same, for the use of Students, 1*s.* 6*d.*

Attic Orators.—FROM ANTIPHON TO ISAEOS. By R. C. JEBB, M.A., LL.D., Professor of Greek in the University of Glasgow. 2 vols. 8vo. 25*s.*

Babrius.—Edited, with Introductory Dissertations, Critical Notes, Commentary and Lexicon. By Rev. W. GUNION RUTHERFORD, M.A., LL.D., Head-Master of Westminster School. 8vo. 12*s.* 6*d.*

Cicero.—THE ACADEMICA. The Text revised and explained by J. S. REID, M.L., Litt.D., Fellow of Caius College, Cambridge. 8vo. 15*s.*

Cicero.—THE ACADEMICS. Translated by J. S. REID, M.L. 8vo. 5*s.* 6*d.*

SELECT LETTERS. After the Edition of ALBERT WATSON, M.A. Translated by G. E. JEANS, M.A., Fellow of Hertford College, Oxford, and late Assistant-Master at Haileybury. Second Edition. Revised. Crown 8vo. 10*s.* 6*d.*

Ctesias.—THE FRAGMENTS OF CTESIAS. Edited, with Introduction and Notes, by J. E. GILMORE, M.A. 8vo. (Classical Library.) [*In the press.*

(See also *Classical Series.*)

Euripides.—MEDEA. Edited, with Introduction and Notes, by A. W. VERRALL, M.A., Fellow and Lecturer of Trinity College, Cambridge. 8vo. 7*s.* 6*d.*

IPHIGENIA IN AULIS. Edited, with Introduction and Notes, by E. B. ENGLAND, M.A., Lecturer in the Owens College, Manchester. 8vo. [*In preparation.*

INTRODUCTION TO THE STUDY OF EURIPIDES. By Professor J. P. MAHAFFY. Fcap. 8vo. 1*s.* 6*d.* (*Classical Writers Series.*)

(See also *Classical Series.*)

Herodotus.—BOOKS I.—III. THE ANCIENT EMPIRES OF THE EAST. Edited, with Notes, Introductions, and Appendices, by A. H. SAYCE, Deputy-Professor of Comparative Philology, Oxford; Honorary LL.D., Dublin. Demy 8vo. 16*s.*

BOOKS IV.—IX. Edited by REGINALD W. MACAN, M.A., Lecturer in Ancient History at Brasenose College, Oxford. 8vo. [*In preparation.*

Homer.—THE ILIAD. Edited, with Introduction and Notes, by WALTER LEAF, M.A., late Fellow of Trinity College, Cambridge. 8vo. Vol. I. Books I.—XII. 14*s.* [Vol. II. *in the press.*

THE ILIAD. Translated into English Prose. By ANDREW LANG, M.A., WALTER LEAF, M.A., and ERNEST MYERS, M.A. Crown 8vo. 12*s.* 6*d.*

THE ODYSSEY. Done into English by S. H. BUTCHER, M.A., Professor of Greek in the University of Edinburgh, and ANDREW LANG, M.A., late Fellow of Merton College, Oxford. Seventh and Cheaper Edition, revised and corrected. Crown 8vo. 4*s.* 6*d.*

INTRODUCTION TO THE STUDY OF HOMER. By the Right Hon. W. E. GLADSTONE, M.P. 18mo. 1*s.* (*Literature Primers.*)

HOMERIC DICTIONARY. For Use in Schools and Colleges. Translated from the German of Dr. G. AUTENRIETH, with Additions and Corrections, by R. P. KEEP, Ph.D. With numerous Illustrations. Crown 8vo. 6*s.*

(See also *Classical Series.*)

Horace.—THE WORKS OF HORACE RENDERED INTO ENGLISH PROSE. With Introductions, Running Analysis, Notes, &c. By J. LONSDALE, M.A., and S. LEE, M.A. (*Globe Edition.*) 3*s.* 6*d.*

Horace.—STUDIES, LITERARY AND HISTORICAL, IN THE ODES OF HORACE. By A. W. VERRALL, Fellow of Trinity College, Cambridge. Demy 8vo. 8*s.* 6*d.*

(See also *Classical Series.*)

Juvenal.—THIRTEEN SATIRES OF JUVENAL. With a Commentary. By JOHN E. B. MAYOR, M.A., Professor of Latin in the University of Cambridge. Crown 8vo.

*** Vol. I. Fourth Edition, Revised and Enlarged. 10*s.* 6*d.* Vol. II. Second Edition. 10*s.* 6*d.*

*** The new matter consists of an Introduction (pp. 1—53), Additional Notes (pp. 333—466) and Index (pp. 467—526). It is also issued separately, as a Supplement to the previous edition, at 5*s.*

THIRTEEN SATIRES. Translated into English after the Text of J. E. B. MAYOR by ALEXANDER LEEPER, M.A., Warden of Trinity College, in the University of Melbourne. Crown 8vo. 3*s.* 6*d.*

(See also *Classical Series.*)

Livy.—BOOKS I.—IV. Translated by Rev. H. M. STEPHENSON, M.A., late Head-Master of St. Peter's School, York. [*In preparation.*

BOOKS XXI.—XXV. Translated by ALFRED JOHN CHURCH, M.A., of Lincoln College, Oxford, Professor of Latin, University College, London, and WILLIAM JACKSON BRODRIBB, M.A., late Fellow of St. John's College, Cambridge. Cr. 8vo. 7*s.* 6*d.*

INTRODUCTION TO THE STUDY OF LIVY. By Rev. W. W. CAPES, Reader in Ancient History at Oxford. Fcap. 8vo. 1*s.* 6*d.* (*Classical Writers Series.*)

(See also *Classical Series.*)

Martial.—BOOKS I. AND II. OF THE EPIGRAMS. Edited, with Introduction and Notes, by Professor J. E. B. MAYOR, M.A. 8vo. [*In the press.*

(See also *Classical Series.*)

Pausanias.—DESCRIPTION OF GREECE. Translated by J. G. FRAZER, M.A., Fellow of Trinity College, Cambridge. [*In preparation.*

Phrynichus.—THE NEW PHRYNICHUS; being a Revised Text of the Ecloga of the Grammarian Phrynichus. With Introduction and Commentary by Rev. W. GUNION RUTHERFORD, M.A., LL.D., Head-Master of Westminster School. 8vo. 18*s.*

Pindar.—THE EXTANT ODES OF PINDAR. Translated into English, with an Introduction and short Notes, by ERNEST MYERS, M.A., late Fellow of Wadham College, Oxford. Second Edition. Crown 8vo. 5*s.*

THE OLYMPIAN AND PYTHIAN ODES. Edited, with an Introductory Essay, Notes, and Indexes, by BASIL GILDERSLEEVE, Professor of Greek in the Johns Hopkins University, Baltimore. Crown 8vo. 7*s.* 6*d.*

Plato.—PHÆDO. Edited, with Introduction, Notes, and Appendices, by R. D. ARCHER-HIND, M.A., Fellow of Trinity College, Cambridge. 8vo. 8*s.* 6*d.*

TIMAEUS.—Edited, with Introduction, Notes, and a Translation, by the same Editor. 8vo. 16*s.*

PHÆDO. Edited, with Introduction and Notes, by W. D. GEDDES, LL.D., Principal of the University of Aberdeen. Second Edition. Demy 8vo. 8*s.* 6*d.*

PHILEBUS. Edited, with Introduction and Notes, by HENRY JACKSON, M.A., Fellow of Trinity College, Cambridge. 8vo. [*In preparation.*

THE REPUBLIC.—Edited, with Introduction and Notes, by H. C. GOODHART, M.A., Fellow of Trinity College, Cambridge. 8vo. [*In preparation.*

THE REPUBLIC OF PLATO. Translated into English, with an Analysis and Notes, by J. LL. DAVIES, M.A., and D. J. VAUGHAN, M.A. 18mo. 4*s.* 6*d.*

EUTHYPHRO, APOLOGY, CRITO, AND PHÆDO. Translated by F. J. CHURCH. 18mo. 4*s.* 6*d.*

PHÆDRUS, LYSIS, AND PROTAGORAS. Translated by Rev. J. WRIGHT, M.A. [*New edition in the press.*

(See also *Classical Series.*)

Plautus.—THE MOSTELLARIA OF PLAUTUS. With Notes, Prolegomena, and Excursus. By WILLIAM RAMSAY, M.A., formerly Professor of Humanity in the University of Glasgow. Edited by Professor GEORGE G. RAMSAY, M.A., of the University of Glasgow. 8vo. 14*s.*

(See also *Classical Series.*)

Pliny.—LETTERS TO TRAJAN. Edited, with Introductory Essays and Notes, by E. G. HARDY, M.A., late Fellow of Jesus College, Oxford. 8vo. [*In the press.*

Polybius.—THE HISTORIES. Translated, with Introduction and Notes, by E. S. SHUCKBURGH, M.A. 2 vols. Crown 8vo. [*In the press.*

Sallust.—CATILINE AND JUGURTHA. Translated, with Introductory Essays, by A. W. POLLARD, B.A. Crown 8vo. 6*s.*

THE CATILINE (separately). Crown 8vo. 3*s.*

(See also *Classical Series.*)

Sophocles.—ŒDIPUS THE KING. Translated from the Greek of Sophocles into English Verse by E. D. A. MORSHEAD, M.A., late Fellow of New College, Oxford; Assistant Master at Winchester College. Fcap. 8vo. 3*s.* 6*d.*

Studia Scenica.—Part I., Section I. Introductory Study on the Text of the Greek Dramas. The Text of SOPHOCLES' TRACHINIAE, 1–300. By DAVID S. MARGOLIOUTH, Fellow of New College, Oxford. Demy 8vo. 2*s.* 6*d.*

Tacitus.—THE ANNALS. Edited, with Introductions and Notes, by G. O. HOLBROOKE, M.A., Professor of Latin in Trinity College, Hartford, U.S.A. With Maps. 8vo. 16*s.*

Tacitus.—THE ANNALS. Translated by A. J. CHURCH, M.A., and W. J. BRODRIBB, M.A. With Notes and Maps. New Edition. Crown 8vo. 7*s*. 6*d*.

THE HISTORIES. Edited, with Introduction and Notes, by Rev. W. A. SPOONER, M.A., Fellow of New College, and H. M. SPOONER, M.A., formerly Fellow of Magdalen College, Oxford. 8vo. [*In preparation.*

THE HISTORY. Translated by A. J. CHURCH, M.A., and W. J. BRODRIBB, M.A. With Notes and a Map. Crown 8vo. 6*s*.

THE AGRICOLA AND GERMANY, WITH THE DIALOGUE ON ORATORY. Translated by A. J. CHURCH, M.A., and W. J. BRODRIBB, M.A. With Notes and Maps. New and Revised Edition. Crown 8vo. 4*s*. 6*d*.

INTRODUCTION TO THE STUDY OF TACITUS. By A. J. CHURCH, M.A. and W. J. BRODRIBB, M.A. Fcap. 8vo. 1*s*. 6*d*. (*Classical Writers Series.*)

Theocritus, Bion, and Moschus. Rendered into English Prose, with Introductory Essay, by A. LANG, M.A. Crown 8vo. 6*s*.

Virgil.—THE WORKS OF VIRGIL RENDERED INTO ENGLISH PROSE, with Notes, Introductions, Running Analysis, and an Index, by JAMES LONSDALE, M.A., and SAMUEL LEE, M.A. New Edition. Globe 8vo. 3*s*. 6*d*.

THE ÆNEID. Translated by J. W. MACKAIL, M.A., Fellow of Balliol College, Oxford. Crown 8vo. 7*s*. 6*d*.

Xenophon.—COMPLETE WORKS. Translated, with Introduction and Essays, by H. G DAKYNS, M.A., Assistant-Master in Clifton College. Four Volumes. Crown 8vo. [*In the press.*

GRAMMAR, COMPOSITION, & PHILOLOGY.

Belcher.—SHORT EXERCISES IN LATIN PROSE COMPOSITION AND EXAMINATION PAPERS IN LATIN GRAMMAR, to which is prefixed a Chapter on Analysis of Sentences. By the Rev. H. BELCHER, M.A., Rector of the High School, Dunedin, N.Z. New Edition. 18mo. 1*s*. 6*d*.

KEY TO THE ABOVE (for Teachers only). 3*s*. 6*d*.

SHORT EXERCISES IN LATIN PROSE COMPOSITION. Part II., On the Syntax of Sentences, with an Appendix, including EXERCISES IN LATIN IDIOMS, &c. 18mo. 2*s*.

KEY TO THE ABOVE (for Teachers only). 3*s*.

Blackie.—GREEK AND ENGLISH DIALOGUES FOR USE IN SCHOOLS AND COLLEGES. By JOHN STUART BLACKIE, Emeritus Professor of Greek in the University of Edinburgh. New Edition. Fcap. 8vo. 2*s*. 6*d*.

Bryans.—LATIN PROSE EXERCISES BASED UPON CAESAR'S GALLIC WAR. With a Classification of Cæsar's Chief Phrases and Grammatical Notes on Cæsar's Usages. By CLEMENT BRYANS, M.A., Assistant-Master in Dulwich College. Second Edition, Revised and Enlarged. Extra fcap. 8vo. 2*s*. 6*d*.

KEY TO THE ABOVE (for Teachers only). 3*s*. 6*d*.

GREEK PROSE EXERCISES based upon Thucydides. By the same Author. Extra fcap. 8vo. [*In preparation.*

Colson.—A FIRST GREEK READER. Stories and Legends from Greek Writers. By F. H. COLSON, M.A., Fellow of St. John's College, Cambridge, and Senior Classical Master at Bradford Grammar School. Globe 8vo. [*In the press.*

Eicke.—FIRST LESSONS IN LATIN. By K. M. EICKE, B.A., Assistant-Master in Oundle School. Globe 8vo. 2*s*.

England.—EXERCISES ON LATIN SYNTAX AND IDIOM. ARRANGED WITH REFERENCE TO ROBY'S SCHOOL LATIN GRAMMAR. By E. B. ENGLAND, M.A., Assistant Lecturer at the Owens College, Manchester. Crown 8vo. 2*s*. 6*d*. Key for Teachers only, 2*s*. 6*d*.

Goodwin.—Works by W. W. GOODWIN, LL.D., Professor of Greek in Harvard University, U.S.A.

SYNTAX OF THE MOODS AND TENSES OF THE GREEK VERB. New Edition, revised. Crown 8vo. 6*s*. 6*d*.

A GREEK GRAMMAR. New Edition, revised. Crown 8vo. 6*s*.

"It is the best Greek Grammar of its size in the English language."—ATHENÆUM.

A GREEK GRAMMAR FOR SCHOOLS. Crown 8vo. 3*s*. 6*d*.

Greenwood.—THE ELEMENTS OF GREEK GRAMMAR, including Accidence, Irregular Verbs, and Principles of Derivation and Composition; adapted to the System of Crude Forms. By J. G. GREENWOOD, Principal of Owens College, Manchester. New Edition. Crown 8vo. 5*s*. 6*d*.

Hadley and Allen.—A GREEK GRAMMAR FOR SCHOOLS AND COLLEGES. By JAMES HADLEY, late Professor in Yale College. Revised and in part Rewritten by FREDERIC DE FOREST ALLEN, Professor in Harvard College. Crown 8vo. 6*s*.

Hardy.—A LATIN READER. By H. J. HARDY, M.A., Assistant Master in Winchester College. Globe 8vo. [*In preparation.*

Hodgson.—MYTHOLOGY FOR LATIN VERSIFICATION. A brief Sketch of the Fables of the Ancients, prepared to be rendered into Latin Verse for Schools. By F. HODGSON, B.D., late Provost of Eton. New Edition, revised by F. C. HODGSON, M.A. 18mo. 3*s*.

Jackson.—FIRST STEPS TO GREEK PROSE COMPOSITION. By BLOMFIELD JACKSON, M.A., Assistant-Master in King's College School, London. New Edition, revised and enlarged. 18mo. 1*s.* 6*d.*

KEY TO FIRST STEPS (for Teachers only). 18mo. 3*s.* 6*d.*

SECOND STEPS TO GREEK PROSE COMPOSITION, with Miscellaneous Idioms, Aids to Accentuation, and Examination Papers in Greek Scholarship. 18mo. 2*s.* 6*d.*

KEY TO SECOND STEPS (for Teachers only). 18mo. 3*s.* 6*d.*

Kynaston.—EXERCISES IN THE COMPOSITION OF GREEK IAMBIC VERSE by Translations from English Dramatists. By Rev. H. KYNASTON, D.D., Principal of Cheltenham College. With Introduction, Vocabulary, &c. New Edition, revised and enlarged. Extra fcap. 8vo. 5*s.*

KEY TO THE SAME (for Teachers only). Extra fcap. 8vo. 4*s.* 6*d.*

Lupton.—Works by J. H. LUPTON, M.A., Sur-Master of St. Paul's School, and formerly Fellow of St. John's College, Cambridge.

AN INTRODUCTION TO LATIN ELEGIAC VERSE COMPOSITION. Globe 8vo. 2*s.* 6*d.*

LATIN RENDERING OF THE EXERCISES IN PART II. (XXV.-C.). Globe 8vo. 3*s.* 6*d.*

AN INTRODUCTION TO THE COMPOSITION OF LATIN LYRICS. Globe 8vo. [*In preparation.*

Mackie.—PARALLEL PASSAGES FOR TRANSLATION INTO GREEK AND ENGLISH. Carefully graduated for the use of Colleges and Schools. With Indexes. By Rev. ELLIS C. MACKIE, Classical Master at Heversham Grammar School. Globe 8vo. 4*s.* 6*d.*

Macmillan.—FIRST LATIN GRAMMAR. By M. C. MACMILLAN, M.A., late Scholar of Christ's College, Cambridge; sometime Assistant-Master in St. Paul's School. New Edition, enlarged. Fcap. 8vo. 1*s.* 6*d.*

Macmillan's Greek Course.—Edited by Rev. W. GUNION RUTHERFORD, M.A., LL.D., Head Master of Westminster School. [*In preparation.*

I.—FIRST GREEK GRAMMAR. By the Editor. 18mo. 1*s.* 6*d.* [*In preparation.*

II.—FIRST GREEK EXERCISE BOOK. By H. G. UNDERHILL.

III.—SECOND GREEK EXERCISE BOOK.

IV.—MANUAL OF GREEK ACCIDENCE.

V.—MANUAL OF GREEK SYNTAX.

VI.—ELEMENTARY GREEK COMPOSITION.

Macmillan's Latin Course. FIRST YEAR. By A. M. COOK, M.A., Assistant-Master at St. Paul's School. New Edition, revised and enlarged. Globe 8vo. 3*s.* 6*d.*

*** *The Second Part is in preparation.*

Macmillan's Shorter Latin Course. By A. M. COOK, M.A., Assistant-Master at St. Paul's School. Being an abridgment of "Macmillan's Latin Course," First Year. Globe 8vo. 1*s*. 6*d*.

Marshall.—A TABLE OF IRREGULAR GREEK VERBS, classified according to the arrangement of Curtius's Greek Grammar. By J. M. MARSHALL, M.A., Head Master of the Grammar School, Durham. New Edition. 8vo. 1*s*.

Mayor (John E. B.)—FIRST GREEK READER. Edited after KARL HALM, with Corrections and large Additions by Professor JOHN E. B. MAYOR, M.A., Fellow of St. John's College, Cambridge. New Edition, revised. Fcap. 8vo. 4*s*. 6*d*.

Mayor (Joseph B.)—GREEK FOR BEGINNERS. By the Rev. J. B. MAYOR, M.A., Professor of Classical Literature in King's College, London. Part I., with Vocabulary, 1*s*. 6*d*. Parts II. and III., with Vocabulary and Index, 3*s*. 6*d*. Complete in one Vol. fcap. 8vo. 4*s*. 6*d*.

Nixon.—PARALLEL EXTRACTS, Arranged for Translation into English and Latin, with Notes on Idioms. By J. E. NIXON, M.A., Fellow and Classical Lecturer, King's College, Cambridge. Part I.—Historical and Epistolary. New Edition, revised and enlarged. Crown 8vo. 3*s*. 6*d*.

PROSE EXTRACTS, Arranged for Translation into English and Latin, with General and Special Prefaces on Style and Idiom. I. Oratorical. II. Historical. III. Philosophical and Miscellaneous. By the same Author. Crown 8vo. 3*s*. 6*d*.

*** *Translations of Select Passages supplied by Author only.*

Peile.—A PRIMER OF PHILOLOGY. By J. PEILE, M.A., Litt. D., Master of Christ's College, Cambridge. 18mo. 1*s*.

Postgate.—PASSAGES FOR TRANSLATION INTO LATIN PROSE. With Introduction and Notes, by J. P. POSTGATE, M.A. Crown 8vo. [*In the press.*

Postgate and Vince.—A DICTIONARY OF LATIN ETYMOLOGY. By J. P. POSTGATE, M.A., and C. A. VINCE, M.A. [*In preparation.*

Potts (A. W.)—Works by ALEXANDER W. POTTS, M.A., LL.D., late Fellow of St. John's College, Cambridge; Head Master of the Fettes College, Edinburgh.

HINTS TOWARDS LATIN PROSE COMPOSITION. New Edition. Extra fcap. 8vo. 3*s*.

PASSAGES FOR TRANSLATION INTO LATIN PROSE. Edited with Notes and References to the above. New Edition. Extra fcap. 8vo. 2*s*. 6*d*.

LATIN VERSIONS OF PASSAGES FOR TRANSLATION INTO LATIN PROSE (for Teachers only). 2*s*. 6*d*.

Preston.—EXERCISES IN LATIN VERSE COMPOSITION. By Rev. G. PRESTON, M.A., Head Master of the King's School, Chester. (With Key.) Globe 8vo. [*In preparation.*

Reid.—A GRAMMAR OF TACITUS. By J. S. REID, M.L., Fellow of Caius College, Cambridge. [*In preparation.*

A GRAMMAR OF VERGIL. By the same Author. [*In preparation.*

*** *Similar Grammars to other Classical Authors will probably follow.*

Roby.—A GRAMMAR OF THE LATIN LANGUAGE, from Plautus to Suetonius. By H. J. ROBY, M.A., late Fellow of St. John's College, Cambridge. In Two Parts. Part I. Fifth Edition, containing:—Book I. Sounds. Book II. Inflexions. Book III. Word-formation. Appendices. Crown 8vo. 9*s.* Part II. Syntax, Prepositions, &c. Crown 8vo. 10*s.* 6*d.*

"Marked by the clear and practised insight of a master in his art. A book that would do honour to any country"—ATHENÆUM.

SCHOOL LATIN GRAMMAR. By the same Author. Crown 8vo. 5*s.*

Rush.—SYNTHETIC LATIN DELECTUS. A First Latin Construing Book arranged on the Principles of Grammatical Analysis. With Notes and Vocabulary. By E. RUSH, B.A. With Preface by the Rev. W. F. MOULTON, M.A., D.D. New and Enlarged Edition. Extra fcap. 8vo. 2*s.* 6*d.*

Rust.—FIRST STEPS TO LATIN PROSE COMPOSITION. By the Rev. G. RUST, M.A., of Pembroke College, Oxford, Master of the Lower School, King's College, London. New Edition. 18mo. 1*s.* 6*d.*

KEY TO THE ABOVE. By W. M. YATES, Assistant-Master in the High School, Sale. 18mo. 3*s.* 6*d.*

Rutherford.—Works by the Rev. W. GUNION RUTHERFORD, M.A., LL.D., Head-Master of Westminster School.

REX LEX. A Short Digest of the principal Relations between Latin, Greek, and Anglo-Saxon Sounds. 8vo. [*In preparation.*

THE NEW PHRYNICHUS; being a Revised Text of the Ecloga of the Grammarian Phrynichus. With Introduction and Commentary. 8vo. 18*s.* (See also Macmillan's Greek Course.)

Simpson.—LATIN PROSE AFTER THE BEST AUTHORS. By F. P. SIMPSON, B.A., late Scholar of Balliol College, Oxford. Part I. CÆSARIAN PROSE. Extra fcap. 8vo. 2*s.* 6*d.*

KEY TO THE ABOVE, for Teachers only. Extra fcap. 8vo. 5*s.*

Thring.—Works by the Rev. E. THRING, M.A., late Head-Master of Uppingham School.

A LATIN GRADUAL. A First Latin Construing Book for Beginners. New Edition, enlarged, with Coloured Sentence Maps. Fcap. 8vo. 2*s.* 6*d.*

A MANUAL OF MOOD CONSTRUCTIONS. Fcap. 8vo. 1*s.* 6*d.*

Welch and Duffield.—LATIN ACCIDENCE AND EXERCISES ARRANGED FOR BEGINNERS. By WILLIAM WELCH, M.A., and C. G. DUFFIELD, M.A., Assistant Masters at Cranleigh School. 18mo. 1*s.* 6*d.*

This book is intended as an introduction to Macmillan's *Elementary Classics*, and is the development of a plan which has been in use for some time and has been worked satisfactorily.

White.—FIRST LESSONS IN GREEK. Adapted to GOODWIN'S GREEK GRAMMAR, and designed as an introduction to the ANABASIS OF XENOPHON. By JOHN WILLIAMS WHITE, Ph.D., Assistant-Professor of Greek in Harvard University. Crown 8vo. 4*s.* 6*d.*

Wilkins and Strachan.—PASSAGES FOR TRANSLATION FROM GREEK AND LATIN. Selected and Arranged by A. S. WILKINS, M.A., Professor of Latin, and J. STRACHAN, M.A., Professor of Greek, in the Owens College, Manchester. [*In the press.*

Wright.—Works by J. WRIGHT, M.A., late Head Master of Sutton Coldfield School.

A HELP TO LATIN GRAMMAR; or, The Form and Use of Words in Latin, with Progressive Exercises. Crown 8vo. 4*s.* 6*d.*

THE SEVEN KINGS OF ROME. An Easy Narrative, abridged from the First Book of Livy by the omission of Difficult Passages; being a First Latin Reading Book, with Grammatical Notes and Vocabulary. New and revised Edition. Fcap. 8vo. 3*s.* 6*d.*

FIRST LATIN STEPS; OR, AN INTRODUCTION BY A SERIES OF EXAMPLES TO THE STUDY OF THE LATIN LANGUAGE. Crown 8vo. 3*s.*

ATTIC PRIMER. Arranged for the Use of Beginners. Extra fcap. 8vo. 2*s.* 6*d.*

A COMPLETE LATIN COURSE, comprising Rules with Examples, Exercises, both Latin and English, on each Rule, and Vocabularies. Crown 8vo. 2*s.* 6*d.*

ANTIQUITIES, ANCIENT HISTORY, AND PHILOSOPHY.

Arnold.—Works by W. T. ARNOLD, M.A.

A HANDBOOK OF LATIN EPIGRAPHY. [*In preparation.*

THE ROMAN SYSTEM OF PROVINCIAL ADMINISTRATION TO THE ACCESSION OF CONSTANTINE THE GREAT. Crown 8vo. 6*s.*

Arnold (T.)—THE SECOND PUNIC WAR. Being Chapters on THE HISTORY OF ROME. By the late THOMAS ARNOLD, D.D., formerly Head Master of Rugby School, and Regius Professor of Modern History in the University of Oxford. Edited, with Notes, by W. T. ARNOLD, M.A. With 8 Maps. Crown 8vo. 8*s.* 6*d.*

Beesly.—STORIES FROM THE HISTORY OF ROME. By Mrs. BEESLY. Fcap. 8vo. 2*s.* 6*d.*

Burn.—ROMAN LITERATURE IN RELATION TO ROMAN ART. By Rev. ROBERT BURN, M.A., Fellow of Trinity College, Cambridge. With numerous Illustrations. Extra Crown 8vo. 14*s.*

Classical Writers.—Edited by JOHN RICHARD GREEN, M.A., LL.D. Fcap. 8vo. 1*s.* 6*d.* each.

EURIPIDES. By Professor MAHAFFY.
MILTON. By the Rev. STOPFORD A. BROOKE, M.A.
LIVY. By the Rev. W. W. CAPES, M.A.
VIRGIL. By Professor NETTLESHIP, M.A.
SOPHOCLES. By Professor L. CAMPBELL, M.A.
DEMOSTHENES. By Professor S. H. BUTCHER, M.A.
TACITUS. By Professor A. J. CHURCH, M.A., and W. J. BRODRIBB, M.A.

Freeman.—Works by EDWARD A. FREEMAN, D.C.L., LL.D., Hon. Fellow of Trinity College, Oxford, Regius Professor of Modern History in the University of Oxford.

HISTORY OF ROME. (*Historical Course for Schools.*) 18mo. [*In preparation.*

A SCHOOL HISTORY OF ROME. Crown 8vo. [*In preparation.*

HISTORICAL ESSAYS. Second Series. [Greek and Roman History.] 8vo. 10*s.* 6*d.*

Fyffe.—A SCHOOL HISTORY OF GREECE. By C. A. FYFFE, M.A. Crown 8vo. [*In preparation.*

Geddes.—THE PROBLEM OF THE HOMERIC POEMS. By W. D. GEDDES, Principal of the University of Aberdeen. 8vo. 14*s.*

Gladstone.—Works by the Rt. Hon. W. E. GLADSTONE, M.P.

THE TIME AND PLACE OF HOMER. Crown 8vo. 6*s.* 6*d.*
A PRIMER OF HOMER. 18mo. 1*s.*

Gow.—A COMPANION TO SCHOOL CLASSICS. By JAMES GOW, M.A., Litt.D., Head Master of the High School, Nottingham; formerly Fellow of Trinity College, Cambridge. With Illustrations. Crown 8vo. [*In the press.*

Jackson.—A MANUAL OF GREEK PHILOSOPHY. By HENRY JACKSON, M.A., Litt.D., Fellow and Prælector in Ancient Philosophy, Trinity College, Cambridge. [*In preparation.*

Jebb.—Works by R. C. JEBB, M.A., LL.D., Professor of Greek in the University of Glasgow.

THE ATTIC ORATORS FROM ANTIPHON TO ISAEOS. 2 vols. 8vo. 25*s.*
A PRIMER OF GREEK LITERATURE. 18mo. 1*s.*

(See also *Classical Series.*)

Kiepert.—MANUAL OF ANCIENT GEOGRAPHY, Translated from the German of Dr. HEINRICH KIEPERT. Crown 8vo. 5*s.*

Mahaffy.—Works by J. P. MAHAFFY, M.A., D.D., Fellow and Professor of Ancient History in Trinity College, Dublin, and Hon. Fellow of Queen's College, Oxford.

SOCIAL LIFE IN GREECE; from Homer to Menander. Fifth Edition, revised and enlarged. Crown 8vo. 9*s.*

GREEK LIFE AND THOUGHT; from the Age of Alexander to the Roman Conquest. Crown 8vo. 12*s.* 6*d.*

RAMBLES AND STUDIES IN GREECE. With Illustrations. Third Edition, Revised and Enlarged. With Map. Crown 8vo. 10*s.* 6*d.*

A PRIMER OF GREEK ANTIQUITIES. With Illustrations. 18mo. 1*s.*

EURIPIDES. 18mo. 1*s.* 6*d.* (*Classical Writers Series.*)

Mayor (J. E. B.)—BIBLIOGRAPHICAL CLUE TO LATIN LITERATURE. Edited after HÜBNER, with large Additions by Professor JOHN E. B. MAYOR. Crown 8vo. 10*s.* 6*d.*

Newton.—ESSAYS IN ART AND ARCHÆOLOGY. By Sir CHARLES NEWTON, K.C.B., D.C.L., Professor of Archæology in University College, London, and formerly Keeper of Greek and Roman Antiquities at the British Museum. 8vo. 12*s.* 6*d.*

Ramsay.—A SCHOOL HISTORY OF ROME. By G. G. RAMSAY, M.A., Professor of Humanity in the University of Glasgow. With Maps. Crown 8vo. [*In preparation.*

Sayce.—THE ANCIENT EMPIRES OF THE EAST. By A. H. SAYCE, Deputy-Professor of Comparative Philosophy, Oxford, Hon. LL.D. Dublin. Crown 8vo. 6*s.*

Stewart.—THE TALE OF TROY. Done into English by AUBREY STEWART, M.A., late Fellow of Trinity College, Cambridge. Globe 8vo. 3*s.* 6*d.*

Wilkins.—A PRIMER OF ROMAN ANTIQUITIES. By Professor WILKINS, M.A., LL.D. Illustrated. 18mo. 1*s.*

A PRIMER OF LATIN LITERATURE. By the same Author. [*In preparation.*

MATHEMATICS.

(1) Arithmetic and Mensuration, (2) Algebra, (3) Euclid and Elementary Geometry, (4) Trigonometry, (5) Higher Mathematics.

ARITHMETIC AND MENSURATION.

Aldis.—THE GREAT GIANT ARITHMOS. A most Elementary Arithmetic for Children. By MARY STEADMAN ALDIS. With Illustrations. Globe 8vo. 2*s.* 6*d.*

Bradshaw.—EASY EXERCISES IN ARITHMETIC. By GERALD BRADSHAW, M.A., Assistant Master in Clifton College. Globe 8vo. [*In preparation.*

Brook-Smith (J.).—ARITHMETIC IN THEORY AND PRACTICE. By J. BROOK-SMITH, M.A., LL.B., St. John's College, Cambridge; Barrister-at-Law; one of the Masters of Cheltenham College. New Edition, revised. Crown 8vo. 4*s.* 6*d.*

Candler.—HELP TO ARITHMETIC. Designed for the use of Schools. By H. CANDLER, M.A., Mathematical Master of Uppingham School. Second Edition. Extra fcap. 8vo. 2*s.* 6*d.*

Dalton.—RULES AND EXAMPLES IN ARITHMETIC. By the Rev. T. DALTON, M.A., Assistant-Master in Eton College. New Edition. 18mo. 2*s.* 6*d.*
[*Answers to the Examples are appended.*

Goyen.—HIGHER ARITHMETIC AND ELEMENTARY MENSURATION. By P. GOYEN, M.A., Inspector of Schools, Dunedin, N. Z. Crown 8vo. 5*s.* [*Just ready.*

Hall and Knight.—ARITHMETICAL EXERCISES AND EXAMINATION PAPERS. By H. S. HALL, M.A., formerly Scholar of Christ's College, Cambridge; Master of the Military and Engineering Side, Clifton College; and S. R. KNIGHT, B.A., formerly Scholar of Trinity College, Cambridge, late Assistant Master at Marlborough College, Author of "Elementary Algebra," "Algebraical Exercises and Examination Papers," "Higher Algebra," &c. Globe 8vo. [*In the press.*

Lock.—Works by Rev. J. B. LOCK, M.A., Senior Fellow, Assistant Tutor, and Lecturer of Caius College, Teacher of Physics in the University of Cambridge, formerly Assistant-Master at Eton.

ARITHMETIC FOR SCHOOLS. With Answers and 1000 additional Examples for Exercise. Second Edition, revised. Globe 8vo. 4*s.* 6*d.* Or in Two Parts:—Part I. Up to and including Practice, with Answers. Globe 8vo. 2*s.* Part II. With Answers and 1000 additional Examples for Exercise. Globe 8vo. 3*s.* [*A Key is in the press.*

⁎ *The complete book and both parts can also be obtained* without answers *at the same price, though in different binding. But the edition* with answers *will always be supplied unless the other is specially asked for.*

ARITHMETIC FOR BEGINNERS. Globe 8vo. [*In the press.*

COMMERCIAL ARITHMETIC. Globe 8vo. [*In preparation.*

Pedley.—EXERCISES IN ARITHMETIC for the Use of Schools. Containing more than 7,000 original Examples. By S. PEDLEY, late of Tamworth Grammar School. Crown 8vo. 5*s.* Also in Two Parts 2*s.* 6*d.* each.

Smith.—Works by the Rev. BARNARD SMITH, M.A., late Rector of Glaston, Rutland, and Fellow and Senior Bursar of S. Peter's College, Cambridge.

ARITHMETIC AND ALGEBRA, in their Principles and Application; with numerous systematically arranged Examples taken from the Cambridge Examination Papers, with especial reference to the Ordinary Examination for the B.A. Degree. New Edition, carefully Revised. Crown 8vo. 10*s*. 6*d*.

ARITHMETIC FOR SCHOOLS. New Edition. Crown 8vo. 4*s*. 6*d*.

A KEY TO THE ARITHMETIC FOR SCHOOLS. New Edition. Crown 8vo. 8*s*. 6*d*.

EXERCISES IN ARITHMETIC. Crown 8vo, limp cloth, 2*s*. With Answers, 2*s*. 6*d*. Answers separately, 6*d*.

SCHOOL CLASS-BOOK OF ARITHMETIC. 18mo, cloth. 3*s*. Or sold separately, in Three Parts, 1*s*. each.

KEYS TO SCHOOL CLASS-BOOK OF ARITHMETIC. Parts I., II., and III., 2*s*. 6*d*. each.

SHILLING BOOK OF ARITHMETIC FOR NATIONAL AND ELEMENTARY SCHOOLS. 18mo, cloth. Or separately, Part I. 2*d*.; Part II. 3*d*.; Part III. 7*d*. Answers, 6*d*.

THE SAME, with Answers complete. 18mo, cloth. 1*s*. 6*d*.

KEY TO SHILLING BOOK OF ARITHMETIC. 18mo. 4*s*. 6*d*.

EXAMINATION PAPERS IN ARITHMETIC. 18mo. 1*s*. 6*d*. The same, with Answers, 18mo, 2*s*. Answers, 6*d*.

KEY TO EXAMINATION PAPERS IN ARITHMETIC. 18mo. 4*s*. 6*d*.

THE METRIC SYSTEM OF ARITHMETIC, ITS PRINCIPLES AND APPLICATIONS, with numerous Examples, written expressly for Standard V. in National Schools. New Edition. 18mo, cloth, sewed. 3*d*.

A CHART OF THE METRIC SYSTEM, on a Sheet, size 42 in. by 34 in. on Roller, mounted and varnished. New Edition. Price 3*s*. 6*d*.

Also a Small Chart on a Card, price 1*d*.

EASY LESSONS IN ARITHMETIC, combining Exercises in Reading, Writing, Spelling, and Dictation. Part I. for Standard I. in National Schools. Crown 8vo. 9*d*.

EXAMINATION CARDS IN ARITHMETIC. (Dedicated to Lord Sandon.) With Answers and Hints.

Standards I. and II. in box, 1*s*. Standards III., IV., and V., in boxes, 1*s*. each. Standard VI. in Two Parts, in boxes, 1*s*. each.

A and B papers, of nearly the same difficulty, are given so as to prevent copying, and the colours of the A and B papers differ in each Standard, and from those of every other Standard, so that a master or mistress can see at a glance whether the children have the proper papers.

Todhunter.—MENSURATION FOR BEGINNERS. By I. TODHUNTER, M.A., F.R.S., D.Sc., late of St. John's College, Cambridge. With Examples. New Edition. 18mo. 2*s*. 6*d*.

KEY TO MENSURATION FOR BEGINNERS. By the Rev. FR. LAWRENCE McCARTHY, Professor of Mathematics in St. Peter's College, Agra. Crown 8vo. 7*s*. 6*d*.

ALGEBRA.

Dalton.—RULES AND EXAMPLES IN ALGEBRA. By the Rev. T. DALTON, M.A., Assistant-Master of Eton College. Part I. New Edition. 18mo. 2*s.* Part II. 18mo. 2*s.* 6*d.*

*** *A Key to Part I. for Teachers only, 7s. 6d.*

Hall and Knight.—ELEMENTARY ALGEBRA FOR SCHOOLS. By H. S. HALL, M.A., formerly Scholar of Christ's College, Cambridge, Master of the Military and Engineering Side, Clifton College; and S. R. KNIGHT, B.A., formerly Scholar of Trinity College, Cambridge, late Assistant-Master at Marlborough College. Fourth Edition, Revised and Corrected. Globe 8vo, bound in maroon coloured cloth, 3*s.* 6*d.*; with Answers, bound in green coloured cloth, 4*s.* 6*d.*

ALGEBRAICAL EXERCISES AND EXAMINATION PAPERS. To accompany ELEMENTARY ALGEBRA. Second Edition, revised. Globe 8vo. 2*s.* 6*d.*

HIGHER ALGEBRA. A Sequel to "ELEMENTARY ALGEBRA FOR SCHOOLS." Second Edition. Crown 8vo. 7*s.* 6*d.*

Jones and Cheyne.—ALGEBRAICAL EXERCISES. Progressively Arranged. By the Rev. C. A. JONES, M.A., and C. H. CHEYNE, M.A., F.R.A.S., Mathematical Masters of Westminster School. New Edition. 18mo. 2*s.* 6*d.*

SOLUTIONS AND HINTS FOR THE SOLUTION OF SOME OF THE EXAMPLES IN THE ALGEBRAICAL EXERCISES OF MESSRS. JONES AND CHEYNE. By Rev. W. FAILES, M.A., Mathematical Master at Westminster School, late Scholar of Trinity College, Cambridge. Crown 8vo. 7*s.* 6*d.*

Smith (Barnard).—ARITHMETIC AND ALGEBRA, in their Principles and Application; with numerous systematically arranged Examples taken from the Cambridge Examination Papers, with especial reference to the Ordinary Examination for the B.A. Degree. By the Rev. BARNARD SMITH, M.A., late Rector of Glaston, Rutland, and Fellow and Senior Bursar of St. Peter's College, Cambridge. New Edition, carefully Revised. Crown 8vo. 10*s.* 6*d.*

Smith (Charles).—Works by CHARLES SMITH, M.A., Fellow and Tutor of Sidney Sussex College, Cambridge.

ELEMENTARY ALGEBRA. Globe 8vo. 4*s.* 6*d.*

In this work the author has endeavoured to explain the principles of Algebra in as simple a manner as possible for the benefit of beginners, bestowing great care upon the explanations and proofs of the fundamental operations and rules.

A TREATISE ON ALGEBRA. Crown 8vo. 7*s.* 6*d.*

Todhunter.—Works by I. TODHUNTER, M.A., F.R.S., D.Sc., late of St. John's College, Cambridge.

"Mr. Todhunter is chiefly known to Students of Mathematics as the author of a series of admirable mathematical text-books, which possess the rare qualities of being clear in style and absolutely free from mistakes, typographical or other."—SATURDAY REVIEW.

ALGEBRA FOR BEGINNERS. With numerous Examples. New Edition. 18mo. 2*s.* 6*d.*

KEY TO ALGEBRA FOR BEGINNERS. Crown 8vo. 6*s.* 6*d.*

ALGEBRA. For the Use of Colleges and Schools. New Edition. Crown 8vo. 7*s.* 6*d.*

KEY TO ALGEBRA FOR THE USE OF COLLEGES AND SCHOOLS. Crown 8vo. 10*s.* 6*d.*

EUCLID, & ELEMENTARY GEOMETRY.

Constable.—GEOMETRICAL EXERCISES FOR BEGINNERS. By SAMUEL CONSTABLE. Crown 8vo. 3*s.* 6*d.*

Cuthbertson.—EUCLIDIAN GEOMETRY. By FRANCIS CUTHBERTSON, M.A., LL.D., Head Mathematical Master of the City of London School. Extra fcap. 8vo. 4*s.* 6*d.*

Dodgson.—Works by CHARLES L. DODGSON, M.A., Student and late Mathematical Lecturer of Christ Church, Oxford.

EUCLID. BOOKS I. AND II. Fourth Edition, with words substituted for the Algebraical Symbols used in the First Edition. Crown 8vo. 2*s.*

*** The text of this Edition has been ascertained, by counting the words, to be *less than five-sevenths* of that contained in the ordinary editions.

EUCLID AND HIS MODERN RIVALS. Second Edition. Crown 8vo. 6*s.*

Eagles.—CONSTRUCTIVE GEOMETRY OF PLANE CURVES. By T. H. EAGLES, M.A., Instructor in Geometrical Drawing, and Lecturer in Architecture at the Royal Indian Engineering College, Cooper's Hill. With numerous Examples. Crown 8vo. 12*s.*

Hall and Stevens.—A TEXT BOOK OF EUCLID'S ELEMENTS. Including alternative Proofs, together with additional Theorems and Exercises, classified and arranged. By H. S. HALL, M.A., formerly Scholar of Christ's College, Cambridge, and F. H. STEVENS, M.A., formerly Scholar of Queen's College, Oxford: Masters of the Military and Engineering Side, Clifton College. Globe 8vo. Part I., containing Books I. and II. 2*s.* Books I.—VI. complete. [*In the press.*

Halsted.—THE ELEMENTS OF GEOMETRY. By GEORGE BRUCE HALSTED, Professor of Pure and Applied Mathematics in the University of Texas. 8vo. 12*s.* 6*d.*

Kitchener.—A GEOMETRICAL NOTE-BOOK, containing Easy Problems in Geometrical Drawing preparatory to the Study of Geometry. For the Use of Schools. By F. E. KITCHENER, M.A., Head-Master of the Grammar School, Newcastle, Staffordshire. New Edition. 4to. 2*s.*

Lock.—EUCLID FOR BEGINNERS.—By Rev. J. B. LOCK, M.A. [*In preparation.*

Mault.—NATURAL GEOMETRY: an Introduction to the Logical Study of Mathematics. For Schools and Technical Classes. With Explanatory Models, based upon the Tachymetrical works of Ed. Lagout. By A. MAULT. 18mo. 1*s.*

Models to Illustrate the above, in Box, 12*s.* 6*d.*

Millar.—ELEMENTS OF DESCRIPTIVE GEOMETRY. By J. B. MILLAR, M.E., Civil Engineer, Lecturer on Engineering in the Victoria University, Manchester. Second Edition. Cr. 8vo. 6*s.*

Snowball.— HE ELEMENTS OF PLANE AND SPHERICAL TRIGONOMETRY. By J. C. SNOWBALL, M.A. Fourteenth Edition. Crown 8vo. 7*s.* 6*d.*

Syllabus of Plane Geometry (corresponding to Euclid, Books I.—VI.). Prepared by the Association for the Improvement of Geometrical Teaching. New Edition. Crown 8vo. 1*s.*

Todhunter.—THE ELEMENTS OF EUCLID. For the Use of Colleges and Schools. By I. TODHUNTER, M.A., F.R.S., D.Sc., of St. John's College, Cambridge. New Edition. 18mo. 3*s* 6*d.*

KEY TO EXERCISES IN EUCLID. Crown 8vo. 6*s.* 6*d.*

Wilson (J. M.).—ELEMENTARY GEOMETRY. BOOKS I.—V. Containing the Subjects of Euclid's first Six Books. Following the Syllabus of the Geometrical Association. By the Rev. J. M. WILSON, M.A., Head Master of Clifton College. New Edition. Extra fcap. 8vo. 4*s.* 6*d.*

TRIGONOMETRY.

Beasley.—AN ELEMENTARY TREATISE ON PLANE TRIGONOMETRY. With Examples. By R. D. BEASLEY, M.A. Ninth Edition, revised and enlarged. Crown 8vo. 3*s.* 6*d.*

Lock.—Works by Rev. J. B. LOCK, M.A., Senior Fellow, Assistant Tutor and Lecturer of Caius College, Teacher of Physics in the University of Cambridge; formerly Assistant-Master at Eton.

TRIGONOMETRY FOR BEGINNERS, as far as the Solution of Triangles. Globe 8vo. 2*s.* 6*d.*

ELEMENTARY TRIGONOMETRY. Fifth Edition (in this edition the chapter on logarithms has been carefully revised). Globe 8vo. 4*s.* 6*d.* [*A Key is in the press.*

Mr. E. J. ROUTH, D.Sc., F.R.S., writes:—"It is an able treatise. It takes the difficulties of the subject one at a time, and so leads the young student easily along."

HIGHER TRIGONOMETRY. Fifth Edition. Globe 8vo. 4*s.* 6*d.*

Both Parts complete in One Volume. Globe 8vo. 7*s.* 6*d.*

(See also under *Arithmetic*, *Higher Mathematics*, and *Euclid.*)

M'Clelland and Preston.—A TREATISE ON SPHERICAL TRIGONOMETRY. With numerous Examples. By WILLIAM J. M'CLELLAND, Sch.B.A., Principal of the Incorporated Society's School, Santry, Dublin, and THOMAS PRESTON, Sch.B.A. In Two Parts. Crown 8vo. Part I. To the End of Solution of Triangles, 4*s.* 6*d.* Part II., 5*s.*

Todhunter.—Works by I. TODHUNTER, M.A., F.R.S., D.Sc., late of St. John's College, Cambridge.

TRIGONOMETRY FOR BEGINNERS. With numerous Examples. New Edition. 18mo. 2*s*. 6*d*.

KEY TO TRIGONOMETRY FOR BEGINNERS. Cr. 8vo. 8*s*. 6*d*.

PLANE TRIGONOMETRY. For Schools and Colleges. New Edition. Crown 8vo. 5*s*.

KEY TO PLANE TRIGONOMETRY. Crown 8vo. 10*s*. 6*d*.

A TREATISE ON SPHERICAL TRIGONOMETRY. New Edition, enlarged. Crown 8vo. 4*s*. 6*d*.

(See also under *Arithmetic and Mensuration*, *Algebra*, and *Higher Mathematics*.)

HIGHER MATHEMATICS.

Airy.—Works by Sir G. B. AIRY, K.C.B., formerly Astronomer-Royal.

ELEMENTARY TREATISE ON PARTIAL DIFFERENTIAL EQUATIONS. Designed for the Use of Students in the Universities. With Diagrams. Second Edition. Crown 8vo. 5*s*. 6*d*.

ON THE ALGEBRAICAL AND NUMERICAL THEORY OF ERRORS OF OBSERVATIONS AND THE COMBINATION OF OBSERVATIONS. Second Edition, revised. Crown 8vo. 6*s*. 6*d*.

Alexander (T.).—ELEMENTARY APPLIED MECHANICS. Being the simpler and more practical Cases of Stress and Strain wrought out individually from first principles by means of Elementary Mathematics. By T. ALEXANDER, C.E., Professor of Civil Engineering in the Imperial College of Engineering, Tokei, Japan. Part I. Crown 8vo. 4*s*. 6*d*.

Alexander and Thomson.—ELEMENTARY APPLIED MECHANICS. By THOMAS ALEXANDER, C.E., Professor of Engineering in the Imperial College of Engineering, Tokei, Japan; and ARTHUR WATSON THOMSON, C.E., B.Sc., Professor of Engineering at the Royal College, Cirencester. Part II. TRANSVERSE STRESS. Crown 8vo. 10*s*. 6*d*.

Army Preliminary Examination, 1882-1887, Specimens of Papers set at the. With answers to the Mathematical Questions. Subjects: Arithmetic, Algebra, Euclid, Geometrical Drawing, Geography, French, English Dictation. Cr. 8vo. 3*s*. 6*d*.

Boole.—THE CALCULUS OF FINITE DIFFERENCES. By G. BOOLE, D.C.L., F.R.S., late Professor of Mathematics in the Queen's University, Ireland. Third Edition, revised by J. F. MOULTON. Crown 8vo. 10*s*. 6*d*.

Cambridge Senate-House Problems and Riders, with Solutions:—

1875—PROBLEMS AND RIDERS. By A. G. GREENHILL, M.A. Crown 8vo. 8*s*. 6*d*.

1878—SOLUTIONS OF SENATE-HOUSE PROBLEMS. By the Mathematical Moderators and Examiners. Edited by J. W. L. GLAISHER, M.A., Fellow of Trinity College, Cambridge. 12*s*.

Carll.—A TREATISE ON THE CALCULUS OF VARIATIONS. Arranged with the purpose of Introducing, as well as Illustrating, its Principles to the Reader by means of Problems, and Designed to present in all Important Particulars a Complete View of the Present State of the Science. By LEWIS BUFFETT CARLL, A.M. Demy 8vo. 21*s.*

Cheyne.—AN ELEMENTARY TREATISE ON THE PLANETARY THEORY. By C. H. H. CHEYNE, M.A., F.R.A.S. With a Collection of Problems. Third Edition. Edited by Rev. A. FREEMAN, M.A., F.R.A.S. Crown 8vo. 7*s.* 6*d.*

Christie.—A COLLECTION OF ELEMENTARY TEST-QUESTIONS IN PURE AND MIXED MATHEMATICS; with Answers and Appendices on Synthetic Division, and on the Solution of Numerical Equations by Horner's Method. By JAMES R. CHRISTIE, F.R.S., Royal Military Academy, Woolwich. Crown 8vo. 8*s.* 6*d.*

Clausius.—MECHANICAL THEORY OF HEAT. By R CLAUSIUS. Translated by WALTER R. BROWNE, M.A., late Fellow of Trinity College, Cambridge. Crown 8vo. 10*s.* 6*d.*

Clifford.—THE ELEMENTS OF DYNAMIC. An Introduction to the Study of Motion and Rest in Solid and Fluid Bodies. By W. K. CLIFFORD, F.R.S., late Professor of Applied Mathematics and Mechanics at University College, London. Part I.—KINEMATIC. Crown 8vo. Books I—III. 7*s.* 6*d.*; Book IV. and Appendix 6*s.*

Cockshott and Walters.—GEOMETRICAL CONICS. An Elementary Treatise. Drawn up in accordance with the Syllabus issued by the Society for the Improvement of Geometrical Teaching. By A. COCKSHOTT, M.A., formerly Fellow and Assistant-Tutor of Trinity College, Cambridge, and Assistant-Master at Eton; and Rev. F. B. WALTERS, M.A., Fellow of Queens' College, Cambridge, and Principal of King William's College, Isle of Man. With Diagrams. Crown 8vo.

[*In the press.*

Cotterill.—APPLIED MECHANICS: an Elementary General Introduction to the Theory of Structures and Machines. By JAMES H. COTTERILL, F.R.S., Associate Member of the Council of the Institution of Naval Architects, Associate Member of the Institution of Civil Engineers, Professor of Applied Mechanics in the Royal Naval College, Greenwich. Medium 8vo. 18*s.*

Day (R. E.)—ELECTRIC LIGHT ARITHMETIC. By R. E. DAY, M.A., Evening Lecturer in Experimental Physics at King's College, London. Pott 8vo. 2*s.*

Drew.—GEOMETRICAL TREATISE ON CONIC SECTIONS. By W. H. DREW, M.A., St. John's College, Cambridge. New Edition, enlarged. Crown 8vo. 5*s.*

Dyer.—EXERCISES IN ANALYTICAL GEOMETRY. Compiled and arranged by J. M. DYER, M.A., Senior Mathematical Master in the Classical Department of Cheltenham College. With Illustrations. Crown 8vo. 4*s*. 6*d*.

Eagles.—CONSTRUCTIVE GEOMETRY OF PLANE CURVES. By T. H. EAGLES, M.A., Instructor in Geometrical Drawing, and Lecturer in Architecture at the Royal Indian Engineering College, Cooper's Hill. With numerous Examples. Crown 8vo. 12*s*.

Edgar (J. H.) and Pritchard (G. S.).—NOTE-BOOK ON PRACTICAL SOLID OR DESCRIPTIVE GEOMETRY. Containing Problems with help for Solutions. By J. H. EDGAR, M.A., Lecturer on Mechanical Drawing at the Royal School of Mines, and G. S. PRITCHARD. Fourth Edition, revised by ARTHUR MEEZE. Globe 8vo. 4*s*. 6*d*.

Edwards.—THE DIFFERENTIAL CALCULUS. With Applications and numerous Examples. An Elementary Treatise by JOSEPH EDWARDS, M.A., formerly Fellow of Sidney Sussex College, Cambridge. Crown 8vo. 10*s*. 6*d*.

Ferrers.—Works by the Rev. N. M. FERRERS, M.A., Master of Gonville and Caius College, Cambridge.

AN ELEMENTARY TREATISE ON TRILINEAR CO-ORDINATES, the Method of Reciprocal Polars, and the Theory of Projectors. New Edition, revised. Crown 8vo. 6*s*. 6*d*.

AN ELEMENTARY TREATISE ON SPHERICAL HARMONICS, AND SUBJECTS CONNECTED WITH THEM. Crown 8vo. 7*s*. 6*d*.

Forsyth.—A TREATISE ON DIFFERENTIAL EQUATIONS. By ANDREW RUSSELL FORSYTH, M.A., F.R.S., Fellow and Assistant Tutor of Trinity College, Cambridge. 8vo. 14*s*.

Frost.—Works by PERCIVAL FROST, M.A., D.Sc., formerly Fellow of St. John's College, Cambridge; Mathematical Lecturer at King's College.

AN ELEMENTARY TREATISE ON CURVE TRACING. 8vo. 12*s*.

SOLID GEOMETRY. Third Edition. Demy 8vo. 16*s*.

HINTS FOR THE SOLUTION OF PROBLEMS in the Third Edition of SOLID GEOMETRY. 8vo. 8*s*. 6*d*.

Greaves.—A TREATISE ON ELEMENTARY STATICS. By JOHN GREAVES, M.A., Fellow and Mathematical Lecturer of Christ's College, Cambridge. Crown 8vo. 6*s*. 6*d*.

STATICS FOR BEGINNERS. By the Same Author. [*In preparation.*

Greenhill.—DIFFERENTIAL AND INTEGRAL CALCULUS. With Applications. By A. G. GREENHILL, M.A., Professor of Mathematics to the Senior Class of Artillery Officers, Woolwich, and Examiner in Mathematics to the University of London. Crown 8vo. 7*s*. 6*d*.

Hemming.—AN ELEMENTARY TREATISE ON THE DIFFERENTIAL AND INTEGRAL CALCULUS, for the Use of Colleges and Schools. By G. W. HEMMING, M.A., Fellow of St. John's College, Cambridge. Second Edition, with Corrections and Additions. 8vo. 9*s.*

Ibbetson.—THE MATHEMATICAL THEORY OF PERFECTLY ELASTIC SOLIDS, with a short account of Viscous Fluids. An Elementary Treatise. By WILLIAM JOHN IBBETSON, M.A., Fellow of the Royal Astronomical Society, and of the Cambridge Philosophical Society, Member of the London Mathematical Society, late Senior Scholar of Clare College, Cambridge. 8vo. 21*s.*

Jellett (John H.).—A TREATISE ON THE THEORY OF FRICTION. By JOHN H. JELLETT, B.D., late Provost of Trinity College, Dublin; President of the Royal Irish Academy. 8vo. 8*s.* 6*d.*

Johnson.—Works by WILLIAM WOOLSEY JOHNSON, Professor of Mathematics at the U.S. Naval Academy, Annopolis, Maryland.

INTEGRAL CALCULUS, an Elementary Treatise on the; Founded on the Method of Rates or Fluxions. Demy 8vo. 9*s.*

CURVE-TRACING IN CARTESIAN CO-ORDINATES. Crown 8vo. 4*s.* 6*d.*

Jones.—EXAMPLES IN PHYSICS. By D. E. JONES, B.Sc., Lecturer in Physics in University College, Aberystwyth. Fcap. 8vo. [*In the press.*

Kelland and Tait.—INTRODUCTION TO QUATERNIONS, with numerous examples. By P. KELLAND, M.A., F.R.S., and P. G. TAIT, M.A., Professors in the Department of Mathematics in the University of Edinburgh. Second Edition. Crown 8vo. 7*s.* 6*d.*

Kempe.—HOW TO DRAW A STRAIGHT LINE: a Lecture on Linkages. By A. B. KEMPE. With Illustrations. Crown 8vo. 1*s.* 6*d.* (*Nature Series.*)

Kennedy.—THE MECHANICS OF MACHINERY. By A. B. W. KENNEDY, F.R.S., M.Inst.C.E., Professor of Engineering and Mechanical Technology in University College, London. With Illustrations. Crown 8vo. 12*s.* 6*d.*

Knox.—DIFFERENTIAL CALCULUS FOR BEGINNERS. By ALEXANDER KNOX. Fcap. 8vo. 3*s.* 6*d.*

Lock.—Works by the Rev. J. B. LOCK, M.A., Author of "Trigonometry," "Arithmetic for Schools," &c., and Teacher of Physics in the University of Cambridge.

HIGHER TRIGONOMETRY. Fifth Edition. Globe 8vo. 4*s.* 6*d.*

DYNAMICS FOR BEGINNERS. Globe 8vo. 3*s.* 6*d.*

STATICS FOR BEGINNERS. Globe 8vo. [*In the press.*

(See also under *Arithmetic*, *Euclid*, and *Trigonometry*.)

Lupton.—CHEMICAL ARITHMETIC. With 1,200 Examples. By SYDNEY LUPTON, M.A., F.C.S., F.I.C., formerly Assistant Master in Harrow School. Second Edition. Fcap. 8vo. 4*s.* 6*d.*

Macfarlane.—PHYSICAL ARITHMETIC. By ALEXANDER MACFARLANE, M.A., D.Sc., F.R.S.E., Examiner in Mathematics to the University of Edinburgh. Crown 8vo. *7s. 6d.*

MacGregor.—KINEMATICS AND DYNAMICS. An Elementary Treatise. By JAMES GORDON MACGREGOR, M.A., D.Sc., Fellow of the Royal Societies of Edinburgh and of Canada Munro Professor of Physics in Dalhousie College, Halifax, Nova Scotia. With Illustrations. Crown 8vo. *10s. 6d.*

Merriman.—A TEXT BOOK OF THE METHOD OF LEAST SQUARES. By MANSFIELD MERRIMAN, Professor of Civil Engineering at Lehigh University, Member of the American Philosophical Society, American Association for the Advancement of Science, &c. Demy 8vo. *8s. 6d.*

Millar.—ELEMENTS OF DESCRIPTIVE GEOMETRY. By J.B. MILLAR, C.E., Assistant Lecturer in Engineering in Owens College, Manchester. Second Edition. Crown 8vo. *6s.*

Milne.—Works by the Rev. JOHN J. MILNE, M.A., Private Tutor, late Scholar, of St. John's College, Cambridge, &c., &c., formerly Second Master of Heversham Grammar School.

WEEKLY PROBLEM PAPERS. With Notes intended for the use of students preparing for Mathematical Scholarships, and for the Junior Members of the Universities who are reading for Mathematical Honours. Pott 8vo. *4s. 6d.*

SOLUTIONS TO WEEKLY PROBLEM PAPERS. Crown 8vo. *10s. 6d.*

COMPANION TO "WEEKLY PROBLEM PAPERS." Crown 8vo. *10s. 6d.*

Muir.—A TREATISE ON THE THEORY OF DETERMINANTS. With graduated sets of Examples. For use in Colleges and Schools. By THOS. MUIR, M.A., F.R.S.E., Mathematical Master in the High School of Glasgow. Crown 8vo. *7s. 6d.*

Parkinson.—AN ELEMENTARY TREATISE ON MECHANICS. For the Use of the Junior Classes at the University and the Higher Classes in Schools. By S. PARKINSON, D.D., F.R.S., Tutor and Prælector of St. John's College, Cambridge. With a Collection of Examples. Sixth Edition, revised. Crown 8vo. *9s. 6d.*

Pirie.—LESSONS ON RIGID DYNAMICS. By the Rev. G. PIRIE, M.A., late Fellow and Tutor of Queen's College, Cambridge; Professor of Mathematics in the University of Aberdeen. Crown 8vo. *6s.*

Puckle.—AN ELEMENTARY TREATISE ON CONIC SECTIONS AND ALGEBRAIC GEOMETRY. With Numerous Examples and Hints for their Solution; especially designed for the Use of Beginners. By G. H. PUCKLE, M.A. Fifth Edition, revised and enlarged. Crown 8vo. *7s. 6d.*

Reuleaux.—THE KINEMATICS OF MACHINERY. Outlines of a Theory of Machines. By Professor F. REULEAUX. Translated and Edited by Professor A. B. W. KENNEDY, F.R.S., C.E. With 450 Illustrations. Medium 8vo. 21*s.*

Rice and Johnson.—DIFFERENTIAL CALCULUS, an Elementary Treatise on the; Founded on the Method of Rates or Fluxions. By JOHN MINOT RICE, Professor of Mathematics in the United States Navy, and WILLIAM WOOLSEY JOHNSON, Professor of Mathematics at the United States Naval Academy. Third Edition, Revised and Corrected. Demy 8vo. 18*s.* Abridged Edition, 9*s.*

Robinson.—TREATISE ON MARINE SURVEYING. Prepared for the use of younger Naval Officers. With Questions for Examinations and Exercises principally from the Papers of the Royal Naval College. With the results. By Rev. JOHN L. ROBINSON, Chaplain and Instructor in the Royal Naval College, Greenwich. With Illustrations. Crown 8vo. 7*s.* 6*d.*

CONTENTS.—Symbols used in Charts and Surveying—The Construction and Use of Scales—Laying off Angles—Fixing Positions by Angles — Charts and Chart-Drawing—Instruments and Observing — Base Lines—Triangulation—Levelling—Tides and Tidal Observations—Soundings—Chronometers—Meridian Distaoces—Method of Plotting a Survey—Miscellaneous Exercises—Index.

Routh.—Works by EDWARD JOHN ROUTH, D.Sc., LL.D., F.R.S., Fellow of the University of London, Hon. Fellow of St. Peter's College, Cambridge.

A TREATISE ON THE DYNAMICS OF THE SYSTEM OF RIGID BODIES. With numerous Examples. Fourth and enlarged Edition. Two Vols. 8vo. Vol. I.—Elementary Parts. 14*s.* Vol. II.—The Advanced Parts. 14*s.*

STABILITY OF A GIVEN STATE OF MOTION, PARTICULARLY STEADY MOTION. Adams' Prize Essay for 1877. 8vo. 8*s.* 6*d.*

Smith (C.).—Works by CHARLES SMITH, M.A., Fellow and Tutor of Sidney Sussex College, Cambridge.

CONIC SECTIONS. Fourth Edition. Crown 8vo. 7*s.* 6*d.*

AN ELEMENTARY TREATISE ON SOLID GEOMETRY. Second Edition. Crown 8vo. 9*s.* 6*d.* (See also under *Algebra.*)

Tait and Steele.—A TREATISE ON DYNAMICS OF A PARTICLE. With numerous Examples. By Professor TAIT and Mr. STEELE. Fifth Edition, revised. Crown 8vo. 12*s.*

Thomson.—Works by J. J. THOMSON, Fellow of Trinity College, Cambridge, and Professor of Experimental Physics in the University.

A TREATISE ON THE MOTION OF VORTEX RINGS. An Essay to which the Adams Prize was adjudged in 1882 in the University of Cambridge. With Diagrams. 8vo. 6*s.*

APPLICATIONS OF DYNAMICS TO PHYSICS AND CHEMISTRY. Crown 8vo. [*In the press.*

Todhunter.—Works by I. TODHUNTER, M.A., F.R.S., D.Sc., late of St. John's College, Cambridge.

"Mr. Todhunter is chiefly known to students of Mathematics as the author of a series of admirable mathematical text-books, which possess the rare qualities of being clear in style and absolutely free from mistakes, typographical and other."—SATURDAY REVIEW.

MECHANICS FOR BEGINNERS. With numerous Examples. New Edition. 18mo. 4*s.* 6*d.*

KEY TO MECHANICS FOR BEGINNERS. Crown 8vo. 6*s.* 6*d.*

AN ELEMENTARY TREATISE ON THE THEORY OF EQUATIONS. New Edition, revised. Crown 8vo. 7*s.* 6*d.*

PLANE CO-ORDINATE GEOMETRY, as applied to the Straight Line and the Conic Sections. With numerous Examples. New Edition, revised and enlarged. Crown 8vo. 7*s.* 6*d.*

KEY TO PLANE CO-ORDINATE GEOMETRY. By C. W. BOURNE, M.A. Head Master of the College, Inverness. Crown 8vo. 10*s.* 6*d.*

A TREATISE ON THE DIFFERENTIAL CALCULUS. With numerous Examples. New Edition. Crown 8vo. 10*s.* 6*d.*

A KEY TO DIFFERENTIAL CALCULUS. By H. ST. J. HUNTER, M.A. Crown 8vo. 10*s.* 6*d.*

A TREATISE ON THE INTEGRAL CALCULUS AND ITS APPLICATIONS. With numerous Examples. New Edition, revised and enlarged. Crown 8vo. 10*s.* 6*d.*

EXAMPLES OF ANALYTICAL GEOMETRY OF THREE DIMENSIONS. New Edition, revised. Crown 8vo. 4*s.*

A TREATISE ON ANALYTICAL STATICS. With numerous Examples. Fifth Edition. Edited by Professor J. D. EVERETT, F.R.S. Crown 8vo. 10*s.* 6*d.*

A HISTORY OF THE MATHEMATICAL THEORY OF PROBABILITY, from the time of Pascal to that of Laplace. 8vo. 18*s.*

A HISTORY OF THE MATHEMATICAL THEORIES OF ATTRACTION, AND THE FIGURE OF THE EARTH, from the time of Newton to that of Laplace. 2 vols. 8vo. 24*s.*

AN ELEMENTARY TREATISE ON LAPLACE'S, LAME'S, AND BESSEL'S FUNCTIONS. Crown 8vo. 10*s.* 6*d.*

(See also under *Arithmetic and Mensuration*, *Algebra*, and *Trigonometry*.)

Wilson (J. M.).—SOLID GEOMETRY AND CONIC SECTIONS. With Appendices on Transversals and Harmonic Division. For the Use of Schools. By Rev. J. M. WILSON, M.A. Head Master of Clifton College. New Edition. Extra fcap. 8vo. 3*s.* 6*d.*

Woolwich Mathematical Papers, for Admission into the Royal Military Academy, Woolwich, 1880—1884 inclusive Crown 8vo. 3*s.* 6*d.*

Wolstenholme.—MATHEMATICAL PROBLEMS, on Subjects included in the First and Second Divisions of the Schedule of subjects for the Cambridge Mathematical Tripos Examination. Devised and arranged by JOSEPH WOLSTENHOLME, D.Sc., late Fellow of Christ's College, sometime Fellow of St. John's College, and Professor of Mathematics in the Royal Indian Engineering College. New Edition, greatly enlarged. 8vo. 18*s*.

EXAMPLES FOR PRACTICE IN THE USE OF SEVEN-FIGURE LOGARITHMS. By the same Author. [*In preparation.*

SCIENCE.

(1) Natural Philosophy, (2) Astronomy, (3) Chemistry, (4) Biology, (5) Medicine, (6) Anthropology, (7) Physical Geography and Geology, (8) Agriculture.

NATURAL PHILOSOPHY.

Airy.—Works by Sir G. B. AIRY, K.C.B., formerly Astronomer-Royal.

ON SOUND AND ATMOSPHERIC VIBRATIONS. With the Mathematical Elements of Music. Designed for the Use of Students in the University. Second Edition, revised and enlarged. Crown 8vo 9*s*.

A TREATISE ON MAGNETISM. Designed for the Use of Students in the University. Crown 8vo. 9*s*. 6*d*.

GRAVITATION: an Elementary Explanation of the Principal Perturbations in the Solar System. Second Edition. Crown 8vo. 7*s*. 6*d*.

Alexander (T.).—ELEMENTARY APPLIED MECHANICS. Being the simpler and more practical Cases of Stress and Strain wrought out individually from first principles by means of Elementary Mathematics. By T. ALEXANDER, C.E., Professor of Civil Engineering in the Imperial College of Engineering, Tokei, Japan. Crown 8vo. Part I. 4*s*. 6*d*.

Alexander — Thomson. — ELEMENTARY APPLIED MECHANICS. By THOMAS ALEXANDER, C.E., Professor of Engineering in the Imperial College of Engineering, Tokei, Japan; and ARTHUR WATSON THOMSON, C.E., B.Sc., Professor of Engineering at the Royal College, Cirencester. Part II. TRANSVERSE STRESS; upwards of 150 Diagrams, and 200 Examples carefully worked out. Crown 8vo. 10*s*. 6*d*.

Ball (R. S.).—EXPERIMENTAL MECHANICS. A Course of Lectures delivered at the Royal College of Science for Ireland. By Sir R. S. BALL, M.A., Astronomer Royal for Ireland. Cr. 8vo. [*New and Cheaper Edition in the press.*

Bottomley.—FOUR-FIGURE MATHEMATICAL TABLES. Comprising Logarithmic and Trigonometrical Tables, and Tables of Squares, Square Roots, and Reciprocals. By J. T. BOTTOMLEY, M.A., F.R.S.E., F.C.S., Lecturer in Natural Philosophy in the University of Glasgow. 8vo. 2*s*. 6*d*.

Chisholm.—THE SCIENCE OF WEIGHING AND MEASURING, AND THE STANDARDS OF MEASURE AND WEIGHT. By H. W. CHISHOLM, Warden of the Standards. With numerous Illustrations. Crown 8vo. 4*s*. 6*d*. (*Nature Series*).

Clausius.—MECHANICAL THEORY OF HEAT. By R. CLAUSIUS. Translated by WALTER R. BROWNE, M.A., late Fellow of Trinity College, Cambridge. Crown 8vo. 10*s*. 6*d*.

Cotterill.—APPLIED MECHANICS: an Elementary General Introduction to the Theory of Structures and Machines. By JAMES H. COTTERILL, F.R.S., Associate Member of the Council of the Institution of Naval Architects, Associate Member of the Institution of Civil Engineers, Professor of Applied Mechanics in the Royal Naval College, Greenwich. Medium 8vo. 18*s*.

Cumming.—AN INTRODUCTION TO THE THEORY OF ELECTRICITY. By LINNÆUS CUMMING, M.A., one of the Masters of Rugby School. With Illustrations. Crown 8vo. 8*s*. 6*d*.

Daniell.—A TEXT-BOOK OF THE PRINCIPLES OF PHYSICS. By ALFRED DANIELL, M.A., LL.B., D.Sc., F.R.S.E., late Lecturer on Physics in the School of Medicine, Edinburgh. With Illustrations. Second Edition. Revised and Enlarged. Medium 8vo. 21*s*.

Day.—ELECTRIC LIGHT ARITHMETIC. By R. E. DAY, M.A., Evening Lecturer in Experimental Physics at King's College, London. Pott 8vo. 2*s*.

Everett.—UNITS AND PHYSICAL CONSTANTS. By J. D. EVERETT, M.A., D.C.L., F.R.S., F.R.S.E., Professor of Natural Philosophy, Queen's College, Belfast. Second Edition Extra fcap. 8vo. 5*s*.

Gray.—ABSOLUTE MEASUREMENTS IN ELECTRICITY AND MAGNETISM. By ANDREW GRAY, M.A., F.R.S.E., Professor of Physics in the University College of North Wales. Two Vols. Crown 8vo. Vol. I. [*Immediately*.

Greaves.—STATICS FOR BEGINNERS. By JOHN GREAVES, M.A., Fellow and Mathematical Lecturer of Christ's College, Cambridge. [*In preparation*.

Grove.—A DICTIONARY OF MUSIC AND MUSICIANS. (A.D. 1450—1886). By Eminent Writers, English and Foreign. Edited by Sir GEORGE GROVE, D.C.L., Director of the Royal College of Music, &c. Demy 8vo.

Vols. I., II., and III. Price 21*s*. each.

Grove—*continued.*

Vol. I. A to IMPROMPTU. Vol. II. IMPROPERIA to PLAIN SONG. Vol. III. PLANCHE TO SUMER IS ICUMEN IN. Demy 8vo. cloth, with Illustrations in Music Type and Woodcut. Also published in Parts. Parts I. to XIV., Parts XIX—XXII., price 3*s.* 6*d.* each. Parts XV., XVI., price 7*s.* Parts XVII., XVIII., price 7*s.*

*** (Part XXII.) just published, completes the DICTIONARY OF MUSIC AND MUSICIANS as originally contemplated. But an Appendix and a full general Index are in the press.

"Dr. Grove's Dictionary will be a boon to every intelligent lover of music."—SATURDAY REVIEW.

Huxley.—INTRODUCTORY PRIMER OF SCIENCE. By T. H. HUXLEY, F.R.S., &c. 18mo. 1*s.*

Ibbetson.—THE MATHEMATICAL THEORY OF PERFECTLY ELASTIC SOLIDS, with a Short Account of Viscous Fluids. An Elementary Treatise. By WILLIAM JOHN IBBETSON, B.A., F.R.A.S., Senior Scholar of Clare College, Cambridge. 8vo. Price 21*s.*

Jones.—EXAMPLES IN PHYSICS. By D. E. JONES, B.Sc. Lecturer in Physics in University College, Aberystwith. Fcap. 8vo. [*In the press.*

Kempe.—HOW TO DRAW A STRAIGHT LINE; a Lecture on Linkages. By A. B. KEMPE. With Illustrations. Crown 8vo. 1*s.* 6*d.* (*Nature Series.*)

Kennedy.—THE MECHANICS OF MACHINERY. By A. B. W. KENNEDY, F.R.S., M.Inst.C.E., Professor of Engineering and Mechanical Technology in University College, London. With numerous Illustrations. Crown 8vo. 12*s.* 6*d.*

Lang.—EXPERIMENTAL PHYSICS. By P. R. SCOTT LANG, M.A., Professor of Mathematics in the University of St. Andrews. With Illustrations. Crown 8vo. [*In the press.*

Lock.—Works by Rev. J. B. LOCK, M.A., Senior Fellow, Assistant Tutor, and Lecturer in Mathematics and Physics, of Gonville and Caius College, Teacher of Physics in the University of Cambridge, &c.

DYNAMICS FOR BEGINNERS. Globe 8vo. 3*s.* 6*d.*

STATICS FOR BEGINNERS. Globe 8vo. [*In preparation.*

Lodge.—MODERN VIEWS OF ELECTRICITY. By OLIVER J. LODGE, F.R.S., Professor of Physics in University College, Liverpool. Illustrated. Crown 8vo. [*In preparation.*

Lupton.—NUMERICAL TABLES AND CONSTANTS IN ELEMENTARY SCIENCE. By SYDNEY LUPTON, M.A., F.C.S., F.I.C., Assistant Master at Harrow School. Extra fcap. 8vo. 2*s.* 6*d.*

Macfarlane.—PHYSICAL ARITHMETIC. By ALEXANDER MACFARLANE, D.Sc., Examiner in Mathematics in the University of Edinburgh. Crown 8vo. 7*s.* 6*d.*

Macgregor.—KINEMATICS AND DYNAMICS. An Elementary Treatise. By JAMES GORDON MACGREGOR, M.A., D. Sc., Fellow of the Royal Societies of Edinburgh and of Canada, Munro Professor of Physics in Dalhousie College, Halifax, Nova Scotia. With Illustrations. Crown 8vo. 10*s.* 6*d.*

Mayer.—SOUND: a Series of Simple, Entertaining, and Inexpensive Experiments in the Phenomena of Sound, for the Use of Students of every age. By A. M. MAYER, Professor of Physics in the Stevens Institute of Technology, &c. With numerous Illustrations. Crown 8vo. 2*s.* 6*d.* (*Nature Series.*)

Mayer and Barnard.—LIGHT: a Series of Simple, Entertaining, and Inexpensive Experiments in the Phenomena of Light, for the Use of Students of every age. By A. M. MAYER and C. BARNARD. With numerous Illustrations. Crown 8vo. 2*s.* 6*d.* (*Nature Series.*)

Newton.—PRINCIPIA. Edited by Professor Sir W. THOMSON and Professor BLACKBURNE. 4to, cloth. 31*s.* 6*d.*

THE FIRST THREE SECTIONS OF NEWTON'S PRINCIPIA. With Notes and Illustrations. Also a Collection of Problems, principally intended as Examples of Newton's Methods. By PERCIVAL FROST, M.A. Third Edition. 8vo. 12*s.*

Parkinson.—A TREATISE ON OPTICS. By S. PARKINSON, D.D., F.R.S., Tutor and Prælector of St. John's College, Cambridge. Fourth Edition, revised and enlarged. Crown 8vo. 10*s.* 6*d.*

Perry.—STEAM. AN ELEMENTARY TREATISE. By JOHN PERRY, C.E., Whitworth Scholar, Fellow of the Chemical Society, Professor of Mechanical Engineering and Applied Mechanics at the Technical College, Finsbury. With numerous Woodcuts and Numerical Examples and Exercises. 18mo. 4*s.* 6*d.*

Ramsay.—EXPERIMENTAL PROOFS OF CHEMICAL THEORY FOR BEGINNERS. By WILLIAM RAMSAY, Ph.D., Professor of Chemistry in University College, Bristol. Pott 8vo. 2*s.* 6*d.*

Rayleigh.—THE THEORY OF SOUND. By LORD RAYLEIGH, M.A., F.R.S., formerly Fellow of Trinity College, Cambridge, 8vo. Vol. I. 12*s.* 6*d.* Vol. II. 12*s.* 6*d.* [*Vol. III. in the press.*

Reuleaux.—THE KINEMATICS OF MACHINERY. Outlines of a Theory of Machines. By Professor F. REULEAUX. Translated and Edited by Professor A. B. W. KENNEDY, F.R.S., C.E. With 450 Illustrations. Medium 8vo. 21*s.*

Roscoe and Schuster.—SPECTRUM ANALYSIS. Lectures delivered in 1868 before the Society of Apothecaries of London. By Sir HENRY E. ROSCOE, LL.D., F.R.S., formerly Professor of Chemistry in the Owens College, Victoria University, Manchester. Fourth Edition, revised and considerably enlarged by the Author and by ARTHUR SCHUSTER, F.R.S., Ph.D., Professor of Applied Mathematics in the Owens College, Victoria University. With Appendices, numerous Illustrations, and Plates. Medium 8vo. 21*s.*

Shann.—AN ELEMENTARY TREATISE ON HEAT, IN RELATION TO STEAM AND THE STEAM-ENGINE. By G. SHANN, M.A. With Illustrations. Crown 8vo. 4*s*. 6*d*.

Spottiswoode.—POLARISATION OF LIGHT. By the late W. SPOTTISWOODE, F.R.S. With many Illustrations. New Edition. Crown 8vo. 3*s*. 6*d*. (*Nature Series.*)

Stewart (Balfour).—Works by BALFOUR STEWART, F.R.S., late Professor of Natural Philosophy in the Owens College, Victoria University, Manchester.

PRIMER OF PHYSICS. With numerous Illustrations. New Edition, with Questions. 18mo. 1*s*. (*Science Primers.*)

LESSONS IN ELEMENTARY PHYSICS. With numerous Illustrations and Chromolitho of the Spectra of the Sun, Stars, and Nebulæ. New Edition. Fcap. 8vo. 4*s*. 6*d*.

QUESTIONS ON BALFOUR STEWART'S ELEMENTARY LESSONS IN PHYSICS. By Prof. THOMAS H. CORE, Owens College, Manchester. Fcap. 8vo. 2*s*.

Stewart and Gee.—ELEMENTARY PRACTICAL PHYSICS, LESSONS IN. By BALFOUR STEWART, M.A., LL.D., F.R.S., and W. W. HALDANE GEE, B.Sc. Crown 8vo.

Vol. I.—GENERAL PHYSICAL PROCESSES. 6*s*.

Vol. II.—ELECTRICITY AND MAGNETISM. 7*s*. 6*d*.

Vol. III.—OPTICS, HEAT, AND SOUND. [*In the press.*

PRACTICAL PHYSICS FOR SCHOOLS AND THE JUNIOR STUDENTS OF COLLEGES. By the same Authors.

Vol. I.—ELECTRICITY AND MAGNETISM. 2*s*. 6*d*.

Stokes.—ON LIGHT. Being the Burnett Lectures, delivered in Aberdeen in 1883, 1884–1885. By GEORGE GABRIEL STOKES, M.A., P.R.S., &c., Fellow of Pembroke College, and Lucasian Professor of Mathematics in the University of Cambridge. First Course: ON THE NATURE OF LIGHT.—Second Course: ON LIGHT AS A MEANS OF INVESTIGATION.—Third Course: ON THE BENEFICIAL EFFECTS OF LIGHT. Crown 8vo. 2*s*. 6*d*. each. Also complete in one volume. 7*s*. 6*d*.

Stone.—AN ELEMENTARY TREATISE ON SOUND. By W. H. STONE, M.D. With Illustrations. 18mo. 3*s*. 6*d*.

Tait.—HEAT. By P. G. TAIT, M.A., Sec. R.S.E., formerly Fellow of St. Peter's College, Cambridge, Professor of Natural Philosophy in the University of Edinburgh. Crown 8vo. 6*s*.

Thompson.—ELEMENTARY LESSONS IN ELECTRICITY AND MAGNETISM. By SILVANUS P. THOMPSON, Principal and Professor of Physics in the Technical College, Finsbury. With Illustrations. New Edition, Revised. Twenty-Eighth Thousand. Fcap. 8vo. 4*s*. 6*d*.

Thomson, Sir W.—ELECTROSTATICS AND MAGNETISM, REPRINTS OF PAPERS ON. By Sir WILLIAM THOMSON, D.C.L., LL.D., F.R.S., F.R.S.E., Fellow of St. Peter's College, Cambridge, and Professor of Natural Philosophy in the University of Glasgow. Second Edition. Medium 8vo. 18*s.*

Thomson, J. J.—THE MOTION OF VORTEX RINGS, A TREATISE ON. An Essay to which the Adams Prize was adjudged in 1882 in the University of Cambridge. By J. J. THOMSON, Fellow of Trinity College, Cambridge, and Professor of Experimental Physics in the University. With Diagrams. 8vo. 6*s.*

APPLICATIONS OF DYNAMICS TO PHYSICS AND CHEMISTRY. By the same Author. Crown 8vo. [*In the press.*

Todhunter.—NATURAL PHILOSOPHY FOR BEGINNERS. By I. TODHUNTER, M.A., F.R.S., D.Sc.

Part I. The Properties of Solid and Fluid Bodies. 18mo. 3*s.* 6*d.*

Part II. Sound, Light, and Heat. 18mo. 3*s.* 6*d.*

Turner.—HEAT AND ELECTRICITY, A COLLECTION OF EXAMPLES ON. By H. H. TURNER, B.A., Fellow of Trinity College, Cambridge. Crown 8vo. 2*s.* 6*d.*

Wright (Lewis). — LIGHT; A COURSE OF EXPERIMENTAL OPTICS, CHIEFLY WITH THE LANTERN. By LEWIS WRIGHT. With nearly 200 Engravings and Coloured Plates. Crown 8vo. 7*s.* 6*d.*

ASTRONOMY.

Airy.—POPULAR ASTRONOMY. With Illustrations by Sir G. B. AIRY, K.C.B., formerly Astronomer-Royal. New Edition. 18mo. 4*s.* 6*d.*

Forbes.—TRANSIT OF VENUS. By G. FORBES, M.A., Professor of Natural Philosophy in the Andersonian University, Glasgow. Illustrated. Crown 8vo. 3*s.* 6*d.* (*Nature Series.*)

Godfray.—Works by HUGH GODFRAY, M.A., Mathematical Lecturer at Pembroke College, Cambridge.

A TREATISE ON ASTRONOMY, for the Use of Colleges and Schools. Fourth Edition. 8vo. 12*s.* 6*d.*

AN ELEMENTARY TREATISE ON THE LUNAR THEORY, with a Brief Sketch of the Problem up to the time of Newton. Second Edition, revised. Crown 8vo. 5*s.* 6*d.*

Lockyer.—Works by J. NORMAN LOCKYER, F.R.S.

PRIMER OF ASTRONOMY. With numerous Illustrations. New Edition. 18mo. 1*s.* (*Science Primers.*)

ELEMENTARY LESSONS IN ASTRONOMY. With Coloured Diagram of the Spectra of the Sun, Stars, and Nebulæ, and numerous Illustrations. New Edition. Fcap. 8vo. 5*s.* 6*d.*

Lockyer—*continued.*

QUESTIONS ON LOCKYER'S ELEMENTARY LESSONS IN ASTRONOMY. For the Use of Schools. By JOHN FORBES-ROBERTSON. 18mo, cloth limp. 1*s*. 6*d*.

THE CHEMISTRY OF THE SUN. With Illustrations. 8vo. 14*s*.

Newcomb.—POPULAR ASTRONOMY. By S. NEWCOMB, LL.D., Professor U.S. Naval Observatory. With 112 Illustrations and 5 Maps of the Stars. Second Edition, revised. 8vo. 18*s*.

"It is unlike anything else of its kind, and will be of more use in circulating a knowledge of Astronomy than nine-tenths of the books which have appeared on the subject of late years."—SATURDAY REVIEW.

CHEMISTRY.

Armstrong.—A MANUAL OF INORGANIC CHEMISTRY. By HENRY ARMSTRONG, Ph.D., F.R.S., Professor of Chemistry in the City and Guilds of London Technical Institute. Crown 8vo. [*In preparation.*

Cohen.—THE OWENS COLLEGE COURSE OF PRACTICAL ORGANIC CHEMISTRY. By JULIUS B. COHEN, Ph.D., F.C.S., Assistant Lecturer on Chemistry in the Owens College, Manchester. With a Preface by SIR HENRY ROSCOE, F.R.S., and C. SCHORLEMMER, F.R.S. Fcap. 8vo. 2*s*. 6*d*.

Cooke.—ELEMENTS OF CHEMICAL PHYSICS. By JOSIAH P. COOKE, Junr., Erving Professor of Chemistry and Mineralogy in Harvard University. Fourth Edition. Royal 8vo. 21*s*.

Fleischer.—A SYSTEM OF VOLUMETRIC ANALYSIS. By EMIL FLEISCHER. Translated, with Notes and Additions, from the Second German Edition by M. M. PATTISON MUIR, F.R.S.E. With Illustrations. Crown 8vo. 7*s*. 6*d*.

Frankland.—AGRICULTURAL CHEMICAL ANALYSIS, A Handbook of. By PERCY FARADAY FRANKLAND, Ph.D., B.Sc., F.C.S. Associate of the Royal School of Mines, and Demonstrator of Practical and Agricultural Chemistry in the Normal School of Science and Royal School of Mines, South Kensington Museum. Founded upon *Leitfaden für die Agriculture Chemiche Analyse*, von Dr. F. KROCKER. Crown 8vo. 7*s*. 6*d*.

Hartley.—A COURSE OF QUANTITATIVE ANALYSIS FOR STUDENTS. By W. NOEL HARTLEY, F.R.S., Professor of Chemistry, and of Applied Chemistry, Science and Art Department, Royal College of Science, Dublin. Globe 8vo. 5*s*.

Hiorns. — A TEXT-BOOK OF METALLURGY AND ASSAYING. By A. H. HIORNS. Illustrated. Globe 8vo. [*In the press.*

Jones.—Works by FRANCIS JONES, F.R.S.E., F.C.S., Chemical Master in the Grammar School, Manchester.

THE OWENS COLLEGE JUNIOR COURSE OF PRACTICAL CHEMISTRY. With Preface by Sir HENRY ROSCOE, F.R.S., and Illustrations. New Edition. 18mo. 2*s*. 6*d*.

Jones—*continued.*

QUESTIONS ON CHEMISTRY. A Series of Problems and Exercises in Inorganic and Organic Chemistry. Fcap. 8vo. 3*s.*

Landauer.—BLOWPIPE ANALYSIS. By J. LANDAUER. Authorised English Edition by J. TAYLOR and W. E. KAY, of Owens College, Manchester. Extra fcap. 8vo. 4*s.* 6*d.*

Lupton.—CHEMICAL ARITHMETIC. With 1,200 Problems. By SYDNEY LUPTON, M.A., F.C.S., F.I.C., formerly Assistant-Master at Harrow. Second Edition, Revised and Abridged. Fcap. 8vo. 4*s.* 6*d.*

Meldola.—PHOTOGRAPHIC CHEMISTRY. By RAPHAEL MELDOLA, Professor of Chemistry in the Technical College, Finsbury. Crown 8vo. [*In preparation.*

Muir.—PRACTICAL CHEMISTRY FOR MEDICAL STUDENTS. Specially arranged for the first M.B. Course. By M. M. PATTISON MUIR, F.R.S.E. Fcap. 8vo. 1*s.* 6*d.*

Muir and Wilson.—THE ELEMENTS OF THERMAL CHEMISTRY. By M. M. PATTISON MUIR, M.A., F.R.S.E., Fellow and Prælector of Chemistry in Gonville and Caius College, Cambridge; Assisted by DAVID MUIR WILSON. 8vo. 12*s.* 6*d.*

Remsen.—Works by IRA REMSEN, Professor of Chemistry in the Johns Hopkins University.

COMPOUNDS OF CARBON; or, Organic Chemistry, an Introduction to the Study of. Crown 8vo. 6*s.* 6*d.*

AN INTRODUCTION TO THE STUDY OF CHEMISTRY (INORGANIC CHEMISTRY). Crown 8vo. 6*s.* 6*d.*

THE ELEMENTS OF CHEMISTRY. A Text Book for Beginners. Fcap. 8vo. 2*s.* 6*d.*

Roscoe.—Works by Sir HENRY E. ROSCOE, F.R.S., formerly Professor of Chemistry in the Victoria University the Owens College, Manchester.

PRIMER OF CHEMISTRY. With numerous Illustrations. New Edition. With Questions. 18mo. 1*s.* (*Science Primers.*)

LESSONS IN ELEMENTARY CHEMISTRY, INORGANIC AND ORGANIC. With numerous Illustrations and Chromolitho of the Solar Spectrum, and of the Alkalies and Alkaline Earths. New Edition. Fcap. 8vo. 4*s.* 6*d.* (*See under* THORPE.)

Roscoe and Schorlemmer.—INORGANIC AND ORGANIC CHEMISTRY. A Complete Treatise on Inorganic and Organic Chemistry. By Sir HENRY E. ROSCOE, F.R.S., and Prof. C. SCHORLEMMER, F.R.S. With Illustrations. Medium 8vo.

Vols. I. and II.—INORGANIC CHEMISTRY.

Vol. I.—The Non-Metallic Elements. 21*s.* Vol. II. Part I.—Metals. 18*s.* Vol. II. Part II.—Metals. 18*s.*

Vol. III.—ORGANIC CHEMISTRY.

THE CHEMISTRY OF THE HYDROCARBONS and their Derivatives, or ORGANIC CHEMISTRY. With numerous Illustrations. Four Parts. Parts I., II., and IV. 21*s.* each. Part III. 18*s.*

Thorpe.—A SERIES OF CHEMICAL PROBLEMS, prepared with Special Reference to Sir H. E. Roscoe's Lessons in Elementary Chemistry, by T. E. THORPE, Ph.D., F.R.S., Professor of Chemistry in the Normal School of Science, South Kensington, adapted for the Preparation of Students for the Government, Science, and Society of Arts Examinations. With a Preface by Sir HENRY E. ROSCOE, F.R.S. New Edition, with Key. 18mo. 2*s.*

Thorpe and Rücker.—A TREATISE ON CHEMICAL PHYSICS. By T. E. THORPE, Ph.D., F.R.S. Professor of Chemistry in the Normal School of Science, and Professor A. W. RÜCKER. Illustrated. 8vo. [*In preparation.*

Wright.—METALS AND THEIR CHIEF INDUSTRIAL APPLICATIONS. BY C. ALDER WRIGHT, D.Sc., &c., Lecturer on Chemistry in St. Mary's Hospital Medical School. Extra fcap. 8vo. 3*s.* 6*d.*

BIOLOGY.

Allen.—ON THE COLOUR OF FLOWERS, as Illustrated in the British Flora. By GRANT ALLEN. With Illustrations. Crown 8vo. 3*s.* 6*d.* (*Nature Series.*)

Balfour.—A TREATISE ON COMPARATIVE EMBRYOLOGY. By F. M. BALFOUR, M.A., F.R.S., Fellow and Lecturer of Trinity College, Cambridge. With Illustrations. Second Edition, reprinted without alteration from the First Edition. In 2 vols. 8vo. Vol. I. 18*s.* Vol. II. 21*s.*

Balfour and Ward.—A GENERAL TEXT BOOK OF BOTANY. By ISAAC BAYLEY BALFOUR, F.R.S., Professor of Botany in the University of Edinburgh, and H. MARSHALL WARD, Fellow of Christ's College, Cambridge, and Professor of Botany in the Royal Indian Engineering College, Cooper's Hill. 8vo. [*In preparation.*

Bettany.—FIRST LESSONS IN PRACTICAL BOTANY. By G. T. BETTANY, M.A., F.L.S., formerly Lecturer in Botany at Guy's Hospital Medical School. 18mo. 1*s.*

Bower—Vines.—A COURSE OF PRACTICAL INSTRUCTION IN BOTANY. By F. O. BOWER, M.A., F.L.S., Professor of Botany in the University of Glasgow, and SYDNEY H. VINES, M.A., D.Sc., F.R.S., Fellow and Lecturer, Christ's College, Cambridge. With a Preface by W. T. THISELTON DYER, M.A., C.M.G., F.R.S., F.L.S., Director of the Royal Gardens, Kew. Crown 8vo.

Part I.—PHANEROGAMÆ—PTERIDOPHYTA. 6*s.* Part II.—BRYOPHYTA—THALLOPHYTA. 4*s.* 6*d.*

Darwin (Charles).—MEMORIAL NOTICES OF CHARLES DARWIN, F.R.S., &c. By THOMAS HENRY HUXLEY, F.R.S., G. J. ROMANES, F.R.S., ARCHIBALD GEIKIE, F.R.S., and W. T. THISELTON DYER, F.R.S. Reprinted from *Nature.* With a Portrait, engraved by C. H. JEENS. Crown 8vo. 2*s.* 6*d.* (*Nature Series.*)

Fearnley.—A MANUAL OF ELEMENTARY PRACTICAL HISTOLOGY. By WILLIAM FEARNLEY. With Illustrations. Crown 8vo. *7s. 6d.*

Flower and Gadow.—AN INTRODUCTION TO THE OSTEOLOGY OF THE MAMMALIA. By WILLIAM HENRY FLOWER, LL.D., F.R.S., Director of the Natural History Departments of the British Museum, late Hunterian Professor of Comparative Anatomy and Physiology in the Royal College of Surgeons of England. With numerous Illustrations. Third Edition. Revised with the assistance of HANS GADOW, Ph.D., M.A., Lecturer on the Advanced Morphology of Vertebrates and Strickland Curator in the University of Cambridge. Crown 8vo. *10s. 6d.*

Foster.—Works by MICHAEL FOSTER, M.D., Sec. R.S., Professor of Physiology in the University of Cambridge.

PRIMER OF PHYSIOLOGY. With numerous Illustrations. New Edition. 18mo. *1s.*

A TEXT-BOOK OF PHYSIOLOGY. With Illustrations. Fourth Edition, revised. 8vo. *21s.*

Foster and Balfour.—THE ELEMENTS OF EMBRYOLOGY. By MICHAEL FOSTER, M.A., M.D., LL.D., Sec. R.S., Professor of Physiology in the University of Cambridge, Fellow of Trinity College, Cambridge, and the late FRANCIS M. BALFOUR, M.A., LL.D., F.R.S., Fellow of Trinity College, Cambridge, and Professor of Animal Morphology in the University. Second Edition, revised. Edited by ADAM SEDGWICK, M.A., Fellow and Assistant Lecturer of Trinity College, Cambridge, and WALTER HEAPE, Demonstrator in the Morphological Laboratory of the University of Cambridge. With Illustrations. Crown 8vo. *10s. 6d.*

Foster and Langley.—A COURSE OF ELEMENTARY PRACTICAL PHYSIOLOGY. By Prof. MICHAEL FOSTER, M.D., Sec. R.S., &c., and J. N. LANGLEY, M.A., F.R.S., Fellow of Trinity College, Cambridge. Fifth Edition. Crown 8vo. *7s. 6d.*

Gamgee.—A TEXT-BOOK OF THE PHYSIOLOGICAL CHEMISTRY OF THE ANIMAL BODY. Including an Account of the Chemical Changes occurring in Disease. By A. GAMGEE, M.D., F.R.S., formerly Professor of Physiology in the Victoria University the Owens College, Manchester. 2 Vols. 8vo. With Illustrations. Vol. I. *18s.* [*Vol. II. in the press.*

Gray.—STRUCTURAL BOTANY, OR ORGANOGRAPHY ON THE BASIS OF MORPHOLOGY. To which are added the principles of Taxonomy and Phytography, and a Glossary of Botanical Terms. By Professor ASA GRAY, LL.D. 8vo. *10s. 6d.*

Hamilton.—A PRACTICAL TEXT-BOOK OF PATHOLOGY. By D. J. HAMILTON, Professor of Pathological Anatomy (Sir Erasmus Wilson Chair), University of Aberdeen. 8vo. [*In the press.*

Hooker.—Works by Sir J. D. HOOKER, K.C.S.I., C.B., M.D., F.R.S., D.C.L.

PRIMER OF BOTANY. With numerous Illustrations. New Edition. 18mo. 1*s.* (*Science Primers.*)

THE STUDENT'S FLORA OF THE BRITISH ISLANDS. Third Edition, revised. Globe 8vo. 10*s.* 6*d.*

Howes.—AN ATLAS OF PRACTICAL ELEMENTARY BIOLOGY. By G. B. HOWES, Assistant Professor of Zoology, Normal School of Science and Royal School of Mines. With a Preface by THOMAS HENRY HUXLEY, F.R.S. Royal 4to. 14*s.*

Huxley.—Works by THOMAS HENRY HUXLEY, F.R.S.

INTRODUCTORY PRIMER OF SCIENCE. 18mo. 1*s.* (*Science Primers.*)

LESSONS IN ELEMENTARY PHYSIOLOGY. With numerous Illustrations. New Edition Revised. Fcap. 8vo. 4*s.* 6*d.*

QUESTIONS ON HUXLEY'S PHYSIOLOGY FOR SCHOOLS. By T. ALCOCK, M.D. New Edition. 18mo. 1*s.* 6*d.*

Huxley and Martin.—A COURSE OF PRACTICAL INSTRUCTION IN ELEMENTARY BIOLOGY. By T. H. HUXLEY, F.R.S., LL.D., assisted by H. N. MARTIN, M.A., M.B., D.Sc., F.R.S., Fellow of Christ's College, Cambridge. New Edition, revised and extended by G. B. HOWES, Assistant Professor of Zoology, Normal School of Science, and Royal School of Mines, and D. H. SCOTT, M.A., PH.D., Assistant Professor of Botany, Normal School of Science, and Royal School of Mines. New Edition, thoroughly revised. With a Preface by T. H. HUXLEY, F.R.S. Crown 8vo. 10*s.* 6*d.*

Kane.—EUROPEAN BUTTERFLIES, A HANDBOOK OF. By W. F. DE VISMES KANE, M.A., M.R.I.A., Member of the Entomological Society of London, &c. With Copper Plate Illustrations. Crown 8vo. 10*s.* 6*d.*

A LIST OF EUROPEAN RHOPALOCERA WITH THEIR VARIETIES AND PRINCIPAL SYNONYMS. Reprinted from the *Handbook of European Butterflies.* Crown 8vo. 1*s.*

Klein.—MICRO-ORGANISMS AND DISEASE. An Introduction into the Study of Specific Micro-Organisms. By E. KLEIN, M.D., F.R.S,, Lecturer on General Anatomy and Physiology in the Medical School of St. Bartholomew's Hospital, London. With 121 Illustrations. Third Edition, Revised. Crown 8vo. 6*s.*

THE BACTERIA IN ASIATIC CHOLERA. By the Same. Crown 8vo. [*In preparation.*

Lankester.—Works by Professor E. RAY LANKESTER, F.R.S.

A TEXT BOOK OF ZOOLOGY. 8vo. [*In preparation.*

DEGENERATION: A CHAPTER IN DARWINISM. Illustrated. Crown 8vo. 2*s.* 6*d.* (*Nature Series.*)

Lubbock.—Works by SIR JOHN LUBBOCK, M.P., F.R.S., D.C.L.

THE ORIGIN AND METAMORPHOSES OF INSECTS. With numerous Illustrations. New Edition. Crown 8vo. 3*s.* 6*d.* (*Nature Series.*)

Lubbock—*continued.*

ON BRITISH WILD FLOWERS CONSIDERED IN RELATION TO INSECTS. With numerous Illustrations. New Edition. Crown 8vo. 4*s.* 6*d.* (*Nature Series*).

FLOWERS, FRUITS, AND LEAVES. With Illustrations Second Edition. Crown 8vo. 4*s.* 6*d.* (*Nature Series.*)

Martin and Moale.—ON THE DISSECTION OF VERTEBRATE ANIMALS. By Professor H. N. MARTIN and W. A. MOALE. Crown 8vo. [*In preparation.*

Mivart.—Works by ST. GEORGE MIVART, F.R.S., Lecturer on Comparative Anatomy at St. Mary's Hospital.

LESSONS IN ELEMENTARY ANATOMY. With upwards of 400 Illustrations. Fcap. 8vo. 6*s.* 6*d.*

THE COMMON FROG. Illustrated. Cr. 8vo. 3*s.* 6*d.* (*Nature Series.*)

Müller.—THE FERTILISATION OF FLOWERS. By Professor HERMANN MÜLLER. Translated and Edited by D'ARCY W. THOMPSON, B.A., Professor of Biology in University College, Dundee. With a Preface by CHARLES DARWIN, F.R.S. With numerous Illustrations. Medium 8vo. 21*s.*

Oliver.—Works by DANIEL OLIVER, F.R.S., &c., Professor of Botany in University College, London, &c.

FIRST BOOK OF INDIAN BOTANY. With numerous Illustrations. Extra fcap. 8vo. 6*s.* 6*d.*

LESSONS IN ELEMENTARY BOTANY. With nearly 200 Illustrations. New Edition. Fcap. 8vo. 4*s.* 6*d.*

Parker.—A COURSE OF INSTRUCTION IN ZOOTOMY (VERTEBRATA). By T. JEFFREY PARKER, B.Sc. London, Professor of Biology in the University of Otago, New Zealand. With Illustrations. Crown 8vo. 8*s.* 6*d.*

LESSONS IN ELEMENTARY BIOLOGY. By the same Author. With Illustrations. 8vo. [*In the press.*

Parker and Bettany.—THE MORPHOLOGY OF THE SKULL. By Professor W. K. PARKER, F.R.S., and G. T. BETTANY. Illustrated. Crown 8vo. 10*s.* 6*d.*

Romanes.—THE SCIENTIFIC EVIDENCES OF ORGANIC EVOLUTION. By GEORGE J. ROMANES, M.A., LL.D. F.R.S., Zoological Secretary of the Linnean Society. Crown 8vo. 2*s.* 6*d.* (*Nature Series.*)

Sedgwick.—A SUPPLEMENT TO F. M. BALFOUR'S TREATISE ON EMBRYOLOGY. By ADAM SEDGWICK, M.A., F.R.S., Fellow and Lecturer of Trinity College, Cambridge. 8vo. Illustrated. [*In preparation.*

Smith (W. G.)—DISEASES OF FIELD AND GARDEN CROPS, CHIEFLY SUCH AS ARE CAUSED BY FUNGI. By WORTHINGTON G. SMITH, F.L.S., M.A.I., Member of the Scientific Committee R.H.S. With 143 New Illustrations drawn and engraved from Nature by the Author. Fcap. 8vo. 4*s.* 6*d.*

Ward.—TIMBER AND ITS DISEASES. By H. MARSHALL WARD, Professor of Botany in the Royal Indian Engineering College, Cooper's Hill. Crown 8vo. Illustrated. [*In preparation.*

Wiedersheim (Prof.).—ELEMENTS OF THE COMPARATIVE ANATOMY OF VERTEBRATES. Adapted from the German of ROBERT WIEDERSHEIM, Professor of Anatomy, and Director of the Institute of Human and Comparative Anatomy in the University of Freiburg-in-Baden, by W. NEWTON PARKER, Professor of Biology in the University College of South Wales and Monmouthshire. With Additions by the Author and Translator. With Two Hundred and Seventy Woodcuts. Medium 8vo. 12*s.* 6*d.*

MEDICINE.

Brunton.—Works by T. LAUDER BRUNTON, M.D., D.Sc., F.R.C.P., F.R.S., Assistant Physician and Lecturer on Materia Medica at St. Bartholomew's Hospital; Examiner in Materia Medica in the University of London, in the Victoria University, and in the Royal College of Physicians, London; late Examiner in the University of Edinburgh.

A TEXT-BOOK OF PHARMACOLOGY, THERAPEUTICS, AND MATERIA MEDICA. Adapted to the United States Pharmacopœia, by FRANCIS H. WILLIAMS, M.D., Boston, Mass. Third Edition. Adapted to the New British Pharmacopœia, 1885. Medium 8vo. 21*s.*

TABLES OF MATERIA MEDICA: A Companion to the Materia Medica Museum. With Illustrations. New Edition Enlarged. 8vo. 10*s.* 6*d.*

Griffiths.—LESSONS ON PRESCRIPTIONS AND THE ART OF PRESCRIBING. By W. HANDSEL GRIFFITHS, PH.D., L.R.C.P.E. New Edition. Adapted to the Pharmacopœia, 1885. 18mo. 3*s.* 6*d.*

Hamilton.—A TEXT-BOOK OF PATHOLOGY. By D. J. HAMILTON, Professor of Pathological Anatomy University of Aberdeen. With Illustrations. 8vo. [*In the press.*

Klein.—MICRO-ORGANISMS AND DISEASE. An Introduction into the Study of Specific Micro-Organisms. By E. KLEIN, M.D., F.R.S., Lecturer on General Anatomy and Physiology in the Medical School of St. Bartholomew's Hospital, London. With 121 Illustrations. Third Edition, Revised. Crown 8vo 6*s.*

THE BACTERIA IN ASIATIC CHOLERA. By the Same Author. Crown 8vo. [*In preparation.*

Ziegler-Macalister.—TEXT-BOOK OF PATHOLOGICAL ANATOMY AND PATHOGENESIS. By Professor ERNST ZIEGLER of Tübingen. Translated and Edited for English Students by DONALD MACALISTER, M.A., M.D., B.Sc., F.R.C.P., Fellow and Medical Lecturer of St. John's College, Cambridge, Physician to Addenbrooke's Hospital, and Teacher of Medicine in the University. With numerous Illustrations. Medium 8vo.

Part I.—GENERAL PATHOLOGICAL ANATOMY. Second Edition. 12*s.* 6*d.*

ANTHROPOLOGY.

Flower.—FASHION IN DEFORMITY, as Illustrated in the Customs of Barbarous and Civilised Races. By Professor FLOWER, F.R.S., F.R.C.S. With Illustrations. Crown 8vo. 2*s.* 6*d.* (*Nature Series.*)

Tylor.—ANTHROPOLOGY. An Introduction to the Study of Man and Civilisation. By E. B. TYLOR, D.C.L., F.R.S. With numerous Illustrations. Crown 8vo. 7*s.* 6*d.*

PHYSICAL GEOGRAPHY & GEOLOGY.

Blanford.—THE RUDIMENTS OF PHYSICAL GEOGRAPHY FOR THE USE OF INDIAN SCHOOLS; with a Glossary of Technical Terms employed. By H. F. BLANFORD, F.R.S. New Edition, with Illustrations. Globe 8vo. 2*s.* 6*d.*

Geikie.—Works by ARCHIBALD GEIKIE, LL.D., F.R.S., Director General of the Geological Survey of Great Britain and Ireland, and Director of the Museum of Practical Geology, London, formerly Murchison Professor of Geology and Mineralogy in the University of Edinburgh, &c.

PRIMER OF PHYSICAL GEOGRAPHY. With numerous Illustrations. New Edition. With Questions. 18mo. 1*s.* (*Science Primers.*)

ELEMENTARY LESSONS IN PHYSICAL GEOGRAPHY. With numerous Illustrations. New Edition. Fcap. 8vo. 4*s.* 6*d.* QUESTIONS ON THE SAME. 1*s.* 6*d.*

PRIMER OF GEOLOGY. With numerous Illustrations. New Edition. 18mo. 1*s.* (*Science Primers.*)

CLASS BOOK OF GEOLOGY With upwards of 200 New Illustrations. Crown 8vo. 10*s.* 6*d.*

TEXT-BOOK OF GEOLOGY. With numerous Illustrations. Second Edition, Sixth Thousand, Revised and Enlarged. 8vo. 28*s.*

OUTLINES OF FIELD GEOLOGY. With Illustrations. New Edition. Extra fcap. 8vo. 3*s.* 6*d.*

THE SCENERY AND GEOLOGY OF SCOTLAND, VIEWED IN CONNEXION WITH ITS PHYSICAL GEOLOGY. With numerous Illustrations. Crown 8vo. 12*s.* 6*d.*

(See also under *History and Geography.*)

Huxley.—PHYSIOGRAPHY. An Introduction to the Study of Nature. By THOMAS HENRY HUXLEY, F.R.S. With numerous Illustrations, and Coloured Plates. New and Cheaper Edition. Crown 8vo. 6*s.*

Lockyer.—OUTLINES OF PHYSIOGRAPHY—THE MOVEMENTS OF THE EARTH. By J. NORMAN LOCKYER, F.R.S., Correspondent of the Institute of France, Foreign Member of the Academy of the Lyncei of Rome, &c., &c.; Professor of Astronomical Physics in the Normal School of Science, and Examiner in Physiography for the Science and Art Department. With Illustrations. Crown 8vo. Sewed, 1*s.* 6*d.*

Phillips.—A TREATISE ON ORE DEPOSITS. By J. ARTHUR PHILLIPS, F.R.S., V.P.G.S., F.C.S., M.Inst.C.E., Ancien Élève de l'École des Mines, Paris; Author of "A Manual of Metallurgy," "The Mining and Metallurgy of Gold and Silver," &c. With numerous Illustrations. 8vo. 25*s.*

AGRICULTURE.

Frankland.—AGRICULTURAL CHEMICAL ANALYSIS, A Handbook of. By PERCY FARADAY FRANKLAND, Ph.D., B.Sc., F.C.S., Associate of the Royal School of Mines, and Demonstrator of Practical and Agricultural Chemistry in the Normal School of Science and Royal School of Mines, South Kensington Museum. Founded upon *Leitfaden für die Agriculture Chemiche Analyse*, von Dr. F. KROCKER. Crown 8vo. 7*s.* 6*d.*

Smith (Worthington G.).—DISEASES OF FIELD AND GARDEN CROPS, CHIEFLY SUCH AS ARE CAUSED BY FUNGI. By WORTHINGTON G. SMITH, F.L.S., M.A.I., Member of the Scientific Committee of the R.H.S. With 143 Illustrations, drawn and engraved from Nature by the Author. Fcap. 8vo. 4*s.* 6*d.*

Tanner.—Works by HENRY TANNER, F.C.S., M.R.A.C., Examiner in the Principles of Agriculture under the Government Department of Science; Director of Education in the Institute of Agriculture, South Kensington, London; sometime Professor of Agricultural Science, University College, Aberystwith.

ELEMENTARY LESSONS IN THE SCIENCE OF AGRICULTURAL PRACTICE. Fcap. 8vo. 3*s.* 6*d.*

FIRST PRINCIPLES OF AGRICULTURE. 18mo. 1*s.*

THE PRINCIPLES OF AGRICULTURE. A Series of Reading Books for use in Elementary Schools. Prepared by HENRY TANNER, F.C.S., M.R.A.C. Extra fcap. 8vo.

I. The Alphabet of the Principles of Agriculture. 6*d.*

II. Further Steps in the Principles of Agriculture. 1*s.*

III. Elementary School Readings on the Principles of Agriculture for the third stage. 1*s.*

POLITICAL ECONOMY.

Cairnes.—THE CHARACTER AND LOGICAL METHOD OF POLITICAL ECONOMY. By J. E. CAIRNES, LL.D., Emeritus Professor of Political Economy in University College, London. New Edition. Crown 8vo. 6*s.*

Cossa.—GUIDE TO THE STUDY OF POLITICAL ECONOMY. By Dr. LUIGI COSSA, Professor in the University of Pavia. Translated from the Second Italian Edition. With a Preface by W. STANLEY JEVONS, F.R.S. Crown 8vo. 4*s.* 6*d.*

Fawcett (Mrs.).—Works by MILLICENT GARRETT FAWCETT:—

POLITICAL ECONOMY FOR BEGINNERS, WITH QUESTIONS. Fourth Edition. 18mo. 2*s.* 6*d.*

TALES IN POLITICAL ECONOMY. Crown 8vo. 3*s.*

Fawcett.—A MANUAL OF POLITICAL ECONOMY. By Right Hon. HENRY FAWCETT, F.R.S. Sixth Edition, revised, with a chapter on "State Socialism and the Nationalisation of the Land," and an Index. Crown 8vo. 12*s.*

AN EXPLANATORY DIGEST of the above. By CYRIL A. WATERS, B.A. Crown 8vo. 2*s.* 6*d.*

Gunton.—WEALTH AND PROGRESS: A CRITICAL EXAMINATION OF THE WAGES QUESTION AND ITS ECONOMIC RELATION TO SOCIAL REFORM. By GEORGE GUNTON. Crown 8vo. 6*s.*

Jevons.—PRIMER OF POLITICAL ECONOMY. By W. STANLEY JEVONS, LL.D., M.A., F.R.S. New Edition. 18mo. 1*s.* (*Science Primers.*)

Marshall.—THE ECONOMICS OF INDUSTRY. By A. MARSHALL, M.A., Professor of Political Economy in the University of Cambridge, and MARY P. MARSHALL, late Lecturer at Newnham Hall, Cambridge. Extra fcap. 8vo. 2*s.* 6*d.*

Marshall.—ECONOMICS. By ALFRED MARSHALL, M.A., Professor of Political Economy in the University of Cambridge. 2 vols 8vo. [*In the press.*

Sidgwick.—THE PRINCIPLES OF POLITICAL ECONOMY. By Professor HENRY SIDGWICK, M.A., LL.D., Knightbridge Professor of Moral Philosophy in the University of Cambridge &c., Author of "The Methods of Ethics." Second Edition, revised. 8vo. 16*s.*

Walker.—Works by FRANCIS A. WALKER, M.A., Ph.D., Author of "Money," "Money in its Relation to Trade," &c.

POLITICAL ECONOMY. Second Edition, revised and enlarged. 8vo. 12*s.* 6*d.*

A BRIEF TEXT-BOOK OF POLITICAL ECONOMY. Crown 8vo. 6*s.* 6*d.*

THE WAGES QUESTION. 8vo. 14*s.*

MENTAL & MORAL PHILOSOPHY.

Boole.—THE MATHEMATICAL ANALYSIS OF LOGIC Being an Essay towards a Calculus of Deductive Reasoning. By GEORGE BOOLE. 8vo. Sewed. 5*s.*

Calderwood.—HANDBOOK OF MORAL PHILOSOPHY. By the Rev. HENRY CALDERWOOD, LL.D., Professor of Moral Philosophy, University of Edinburgh. New Edition. Crown 8vo. 6*s.*

Clifford.—SEEING AND THINKING. By the late Professor W. K. CLIFFORD, F.R.S. With Diagrams. Crown 8vo. 3*s.* 6*d.* (*Nature Series.*)

Jardine.—THE ELEMENTS OF THE PSYCHOLOGY OF COGNITION. By the Rev. ROBERT JARDINE, B.D., D.Sc. (Edin.), Ex-Principal of the General Assembly's College, Calcutta. Third Edition, revised and improved. Crown 8vo. 6*s.* 6*d.*

Jevons.—Works by the late W. STANLEY JEVONS, LL.D., M.A., F.R.S.

PRIMER OF LOGIC. New Edition. 18mo. 1*s*. (*Science Primers.*)

ELEMENTARY LESSONS IN LOGIC; Deductive and Inductive, with copious Questions and Examples, and a Vocabulary of Logical Terms. New Edition. Fcap. 8vo. 3*s*. 6*d*.

THE PRINCIPLES OF SCIENCE. A Treatise on Logic and Scientific Method. New and Revised Edition. Crown 8vo. 12*s*. 6*d*.

STUDIES IN DEDUCTIVE LOGIC. Second Edition. Cr. 8vo. 6*s*.

Keynes.—FORMAL LOGIC, Studies and Exercises in. Including a Generalisation of Logical Processes in their application to Complex Inferences. By JOHN NEVILLE KEYNES, M.A., late Fellow of Pembroke College, Cambridge. Second Edition, Revised and Enlarged. Crown 8vo. 10*s*. 6*d*.

Kant—Max Müller.—CRITIQUE OF PURE REASON. By IMMANUEL KANT. In commemoration of the Centenary of its first Publication. Translated into English by F. MAX MÜLLER. With an Historical Introduction by LUDWIG NOIRÉ. 2 vols 8vo. 16*s*. each.

Volume I. HISTORICAL INTRODUCTION, by LUDWIG NOIRÉ; &c., &c.

Volume II. CRITIQUE OF PURE REASON, translated by F. MAX MÜLLER.

For the convenience of students these volumes are now sold separately.

Kant—Mahaffy and Bernard.—COMMENTARY ON KANT'S CRITIQUE. By J. P. MAHAFFY, M.A., Professor of Ancient History in the University of Dublin, and J. H. BERNARD, M.A. New and completed Edition. Crown 8vo. [*In press.*

McCosh.—PSYCHOLOGY. By JAMES McCOSH, D.D., LL.D., Litt.D., President of Princeton College, Author of "Intuitions of the Mind," "Laws of Discursive Thought," &c. Crown 8vo.

I. THE COGNITIVE POWERS. 6*s*. 6*d*.

II. THE MOTIVE POWERS. Crown 8vo. 6*s*. 6*d*.

Ray.—A TEXT-BOOK OF DEDUCTIVE LOGIC FOR THE USE OF STUDENTS. By P. K. RAY, D.Sc. (Lon. and Edin.), Professor of Logic and Philosophy, Presidency College, Calcutta. Third Edition. Globe 8vo. 4*s*. 6*d*.

The SCHOOLMASTER says:—"This work . . . is deservedly taking a place among the recognised text-books on Logic."

Sidgwick.—Works by HENRY SIDGWICK, M.A., LL.D., Knightbridge Professor of Moral Philosophy in the University of Cambridge.

THE METHODS OF ETHICS. Third Edition. 8vo. 14*s*. A Supplement to the Second Edition, containing all the important Additions and Alterations in the Third Edition. Demy 8vo. 6*s*.

OUTLINES OF THE HISTORY OF ETHICS, for English Readers. Crown 8vo. 3*s*. 6*d*.

Venn.—THE LOGIC OF CHANCE. An Essay on the Foundations and Province of the Theory of Probability, with special Reference to its Logical Bearings and its Application to Moral and Social Science. By JOHN VENN, M.A., Fellow and Lecturer in Moral Sciences in Gonville and Caius College, Cambridge, Examiner in Moral Philosophy in the University of London. Second Edition, rewritten and greatly enlarged. Crown 8vo. 10*s*. 6*d*.

SYMBOLIC LOGIC. By the same Author. Crown 8vo. 10*s*. 6*d*.

HISTORY AND GEOGRAPHY.

Arnold (T.).—THE SECOND PUNIC WAR. Being Chapters from THE HISTORY OF ROME. By THOMAS ARNOLD, D.D. Edited, with Notes, by W. T. ARNOLD, M.A. With 8 Maps. Crown 8vo. 8*s*. 6*d*.

Arnold (W. T.).—THE ROMAN SYSTEM OF PROVINCIAL ADMINISTRATION TO THE ACCESSION OF CONSTANTINE THE GREAT. By W. T. ARNOLD, M.A. Crown 8vo. 6*s*.

"Ought to prove a valuable handbook to the student of Roman history."—GUARDIAN.

Bartholomew.—THE ELEMENTARY SCHOOL ATLAS. By JOHN BARTHOLOMEW, F.R.G.S. 1*s*. [*In the press.*

This Elementary Atlas is designed to illustrate the principal textbooks on Elementary Geography.

Beesly.—STORIES FROM THE HISTORY OF ROME. By Mrs. BEESLY. Fcap. 8vo. 2*s*. 6*d*.

Bryce.—THE HOLY ROMAN EMPIRE. By JAMES BRYCE, D.C.L., Fellow of Oriel College, and Regius Professor of Civil Law in the University of Oxford. Eighth Edition. Crown 8vo. 7*s*. 6*d*.

Buckland.—OUR NATIONAL INSTITUTIONS. A Short Sketch for Schools. By ANNA BUCKLAND. With Glossary. 18mo. 1*s*.

Buckley.—A HISTORY OF ENGLAND FOR BEGINNERS. By ARABELLA B. BUCKLEY. Author of "A Short History of Natural Science," &c. With Coloured Maps, Chronological and Genealogical Tables. Globe 8vo. 3*s*.

Clarke.—CLASS-BOOK OF GEOGRAPHY. By C. B. CLARKE, M.A., F.L.S., F.G.S., F.R.S. New Edition, with Eighteen Coloured Maps. Fcap. 8vo. 3*s*.

Dicey.—LECTURES INTRODUCTORY TO THE STUDY OF THE LAW OF THE CONSTITUTION. By A. V. DICEY, B.C.L., of the Inner Temple, Barrister-at-Law; Vinerian Professor of English Law; Fellow of All Souls College, Oxford; Hon. LL.D. Glasgow. Second Edition. Demy 8vo. 12*s*. 6*d*.

Dickens's DICTIONARY OF THE UNIVERSITY OF OXFORD, 1886-7. 18mo, sewed. 1*s*.

Dickens—*continued.*

DICTIONARY OF THE UNIVERSITY OF CAMBRIDGE, 1886-7. 18mo, sewed. 1*s.*

Both books (Oxford and Cambridge) bound together in one volume. Cloth. 2*s.* 6*d.*

Freeman.—Works by EDWARD A. FREEMAN, D.C.L., LL.D., Regius Professor of Modern History in the University of Oxford, &c.

OLD ENGLISH HISTORY. With Five Coloured Maps. New Edition. Extra fcap. 8vo. 6*s.*

A SCHOOL HISTORY OF ROME. Crown 8vo. [*In preparation.*

METHODS OF HISTORICAL STUDY. A Course of Lectures. 8vo. 10*s.* 6*d.*

THE CHIEF PERIODS OF EUROPEAN HISTORY. Six Lectures read in the University of Oxford in Trinity Term, 1885. With an Essay on Greek Cities under Roman Rule. 8vo. 10*s.* 6*d.*

HISTORICAL ESSAYS. First Series. Fourth Edition. 8vo. 10*s.* 6*d.*

Contents:—The Mythical and Romantic Elements in Early English History—The Continuity of English History—The Relations between the Crown of England and Scotland—St. Thomas of Canterbury and his Biographers, &c.

HISTORICAL ESSAYS. Second Series. Second Edition, with additional Essays. 8vo. 10*s.* 6*d.*

Contents:—Ancient Greece and Mediæval Italy—Mr. Gladstone's Homer and the Homeric Ages—The Historians of Athens—The Athenian Democracy—Alexander the Great—Greece during the Macedonian Period—Mommsen's History of Rome—Lucius Cornelius Sulla—The Flavian Cæsars, &c., &c.

HISTORICAL ESSAYS. Third Series. 8vo. 12*s.*

Contents:—First Impressions of Rome—The Illyrian Emperors and their Land—Augusta Treverorum—The Goths at Ravenna—Race and Language—The Byzantine Empire—First Impressions of Athens—Mediæval and Modern Greece—The Southern Slaves—Sicilian Cycles—The Normans at Palermo.

THE GROWTH OF THE ENGLISH CONSTITUTION FROM THE EARLIEST TIMES. Fourth Edition. Crown 8vo. 5*s.*

GENERAL SKETCH OF EUROPEAN HISTORY. New Edition. Enlarged, with Maps, &c. 18mo. 3*s.* 6*d.* (Vol. I. of Historical Course for Schools.)

EUROPE. 18mo. 1*s.* (*History Primers.*)

Fyffe.—A SCHOOL HISTORY OF GREECE. By C. A. FYFFE, M.A. Crown 8vo. [*In preparation.*

Geikie.—Works by ARCHIBALD GEIKIE, F.R.S., Director-General of the Geological Survey of the United Kingdom, and Director of the Museum of Practical Geology, Jermyn Street, London; formerly Murchison Professor of Geology and Mineralogy in the University of Edinburgh.

THE TEACHING OF GEOGRAPHY. A Practical Handbook for the use of Teachers. Crown 8vo. 2*s.* Being Volume I. of a New Geographical Series Edited by ARCHIBALD GEIKIE, F.R.S.

*** The aim of this volume is to advocate the claims of geography as an educational discipline of a high order, and to show how these claims may be practically recognised by teachers.

Geikie.—Works by ARCHIBALD, *continued.*
AN ELEMENTARY GEOGRAPHY OF THE BRITISH ISLES. 18mo. 1*s.*

Green.—Works by JOHN RICHARD GREEN, M.A., LL.D., late Honorary Fellow of Jesus College, Oxford.
A SHORT HISTORY OF THE ENGLISH PEOPLE. New and Thoroughly Revised Edition. With Coloured Maps, Genealogical Tables, and Chronological Annals. Crown 8vo. 8*s.* 6*d.* 131st Thousand.
Also the same in Four Parts, with the corresponding portion of Mr. Tait's "Analysis." Crown 8vo. 3*s.* 6*d.* each. Part I. 607—1265. Part II. 1265—1540. Part III. 1540—1660. Part IV. 1660—1873.
HISTORY OF THE ENGLISH PEOPLE. In four vols. 8vo.
Vol. I.—EARLY ENGLAND, 449-1071—Foreign Kings, 1071-1214—The Charter, 1214-1291—The Parliament, 1307-1461. With eight Coloured Maps. 8vo. 16*s.*
Vol. II.—THE MONARCHY, 1461-1540—The Reformation, 1540-1603. 8vo. 16*s.*
Vol. III.—PURITAN ENGLAND, 1603-1660—The Revolution, 1660-1688. With four Maps. 8vo. 16*s.*
THE MAKING OF ENGLAND. With Maps. 8vo. 16*s.*
THE CONQUEST OF ENGLAND. With Maps and Portrait. 8vo. 18*s.*
ANALYSIS OF ENGLISH HISTORY, based on Green's "Short History of the English People." By C. W. A. TAIT, M.A., Assistant-Master, Clifton College. Crown 8vo. 3*s.* 6*d.*
READINGS FROM ENGLISH HISTORY. Selected and Edited by JOHN RICHARD GREEN. Three Parts. Globe 8vo. 1*s.* 6*d.* each. I. Hengist to Cressy. II. Cressy to Cromwell. III. Cromwell to Balaklava.

Green.—A SHORT GEOGRAPHY OF THE BRITISH ISLANDS. By JOHN RICHARD GREEN and ALICE STOPFORD GREEN. With Maps. Fcap. 8vo. 3*s.* 6*d.*

Grove.—A PRIMER OF GEOGRAPHY. By Sir GEORGE GROVE, D.C.L. With Illustrations. 18mo. 1*s.* (*Science Primers.*)

Guest.—LECTURES ON THE HISTORY OF ENGLAND. By M. J. GUEST. With Maps. Crown 8vo. 6*s.*

Historical Course for Schools—Edited by EDWARD A. FREEMAN, D.C.L., LL.D., late Fellow of Trinity College, Oxford, Regius Professor of Modern History in the University of Oxford.
I.—GENERAL SKETCH OF EUROPEAN HISTORY. By EDWARD A. FREEMAN, D.C.L. New Edition, revised and enlarged, with Chronological Table, Maps, and Index. 18mo. 3*s.* 6*d.*
II.—HISTORY OF ENGLAND. By EDITH THOMPSON. New Ed., revised and enlarged, with Coloured Maps. 18mo. 2*s.* 6*d.*
III.—HISTORY OF SCOTLAND. By MARGARET MACARTHUR. New Edition. 18mo. 2*s.*
IV.—HISTORY OF ITALY. By the Rev. W. HUNT, M.A. New Edition, with Coloured Maps. 18mo. 3*s.* 6*d.*

Historical Course for Schools—*continued.*

V.—HISTORY OF GERMANY. By J. SIME, M.A. New Edition Revised. 18mo. 3*s.*

VI.—HISTORY OF AMERICA. By JOHN A. DOYLE. With Maps. 18mo. 4*s.* 6*d.*

VII.—EUROPEAN COLONIES. By E. J. PAYNE, M.A. With Maps. 18mo. 4*s.* 6*d.*

VIII.—FRANCE. By CHARLOTTE M. YONGE. With Maps. 18mo. 3*s.* 6*d.*

GREECE. By EDWARD A. FREEMAN, D.C.L. [*In preparation.*

ROME. By EDWARD A. FREEMAN, D.C.L. [*In preparation.*

History Primers—Edited by JOHN RICHARD GREEN, M.A., LL.D., Author of "A Short History of the English People."

ROME. By the Rev. M. CREIGHTON, M.A., Dixie Professor of Ecclesiastical History in the University of Cambridge. With Eleven Maps. 18mo. 1*s.*

GREECE. By C. A. FYFFE, M.A., Fellow and late Tutor of University College, Oxford. With Five Maps. 18mo. 1*s.*

EUROPEAN HISTORY. By E. A. FREEMAN, D.C.L., LL.D. With Maps. 18mo. 1*s.*

GREEK ANTIQUITIES. By the Rev. J. P. MAHAFFY, M.A. Illustrated. 18mo. 1*s.*

CLASSICAL GEOGRAPHY. By H. F. TOZER, M.A. 18mo. 1*s.*

GEOGRAPHY. By Sir G. GROVE, D.C.L. Maps. 18mo. 1*s.*

ROMAN ANTIQUITIES. By Professor WILKINS. Illustrated. 18mo. 1*s.*

FRANCE. By CHARLOTTE M. YONGE. 18mo. 1*s.*

Hole.—A GENEALOGICAL STEMMA OF THE KINGS OF ENGLAND AND FRANCE. By the Rev. C. HOLE. On Sheet. 1*s.*

Jennings—CHRONOLOGICAL TABLES. A synchronistic arrangement of the events of Ancient History (with an Index). By the Rev. ARTHUR C. JENNINGS, Rector of King's Stanley, Gloucestershire, Author of "A Commentary on the Psalms," "Ecclesia Anglicana," "Manual of Church History," &c. 8vo. [*Immediately.*

Kiepert.—A MANUAL OF ANCIENT GEOGRAPHY. From the German of Dr. H. KIEPERT. Crown 8vo. 5*s.*

Labberton.—NEW HISTORICAL ATLAS AND GENERAL HISTORY. By R. H. LABBERTON, Litt.Hum.D. 4to. New Edition Revised and Enlarged. 15*s.*

Lethbridge.—A SHORT MANUAL OF THE HISTORY OF INDIA. With an Account of INDIA AS IT IS. The Soil, Climate, and Productions; the People, their Races, Religions, Public Works, and Industries; the Civil Services, and System of Administration. By Sir ROPER LETHBRIDGE, M.A., C.I.E., late Scholar of Exeter College, Oxford, formerly Principal of Kishnaghur College, Bengal, Fellow and sometime Examiner of the Calcutta University. With Maps. Crown 8vo. 5*s.*

Macmillan's Geographical Series. Edited by ARCHIBALD GEIKIE, F.R.S., Director-General of the Geological Survey of the United Kingdom.

THE TEACHING OF GEOGRAPHY. A Practical Handbook for the use of Teachers. By ARCHIBALD GEIKIE, F.R.S. Crown 8vo. 2*s*.

*** The aim of this volume is to advocate the claims of geography as an educational discipline of a high order, and to show how these claims may be practically recognized by teachers.

AN ELEMENTARY GEOGRAPHY OF THE BRITISH ISLES. By ARCHIBALD GEIKIE, F.R.S. 18mo. 1*s*.

MAPS AND MAP MAKING. By ALFRED HUGHES, M.A., late Scholar of Corpus Christi College, Oxford, Assistant Master at Manchester Grammar School. Crown 8vo. [*In the press.*

AN ELEMENTARY GENERAL GEOGRAPHY. By HUGH ROBERT MILL, D.Sc. Edin. Crown 8vo. [*In the press.*

Michelet.—A SUMMARY OF MODERN HISTORY. Translated from the French of M. MICHELET, and continued to the Present Time, by M. C. M. SIMPSON. Globe 8vo. 4*s*. 6*d*.

Norgate.—ENGLAND UNDER THE ANGEVIN KINGS. By KATE NORGATE. With Maps and Plans. 2 vols. 8vo. 32*s*.

Otté.—SCANDINAVIAN HISTORY. By E. C. OTTÉ. With Maps. Globe 8vo. 6*s*.

Ramsay.—A SCHOOL HISTORY OF ROME. By G. G. RAMSAY, M.A., Professor of Humanity in the University of Glasgow. With Maps. Crown 8vo. [*In preparation.*

Seeley.—Works by J. R. SEELEY, M.A., Regius Professor of Modern History in the University of Cambridge.

THE EXPANSION OF ENGLAND. Crown 8vo. 4*s*. 6*d*.

OUR COLONIAL EXPANSION. Extracts from the above. Crown 8vo. Sewed. 1*s*.

Tait.—ANALYSIS OF ENGLISH HISTORY, based on Green's "Short History of the English People." By C. W. A. TAIT, M.A., Assistant-Master, Clifton College. Crown 8vo. 3*s*. 6*d*.

Wheeler.—A SHORT HISTORY OF INDIA AND OF THE FRONTIER STATES OF AFGHANISTAN, NEPAUL, AND BURMA. By J. TALBOYS WHEELER. With Maps. Crown 8vo. 12*s*.

COLLEGE HISTORY OF INDIA, ASIATIC AND EUROPEAN. By the same. With Maps. Crown 8vo. 3*s*. 6*d*.

Yonge (Charlotte M.).—CAMEOS FROM ENGLISH HISTORY. By CHARLOTTE M. YONGE, Author of "The Heir of Redclyffe," Extra fcap. 8vo. New Edition. 5*s*. each. (1) FROM ROLLO TO EDWARD II. (2) THE WARS IN FRANCE. (3) THE WARS OF THE ROSES. (4) REFORMATION TIMES. (5) ENGLAND AND SPAIN. (6) FORTY YEARS OF STUART RULE (1603—1643).

Young.—Works by CHARLOTTE M., *continued.*

EUROPEAN HISTORY. Narrated in a Series of Historical Selections from the Best Authorities. Edited and arranged by E. M. SEWELL and C. M. YONGE. First Series, 1003—1154. New Edition. Crown 8vo. 6*s.* Second Series, 1088—1228. New Edition. Crown 8vo. 6*s.*

THE VICTORIAN HALF CENTURY—A JUBILEE BOOK. With a New Portrait of the Queen. Crown 8vo., paper covers, 1*s.* Cloth, 1*s.* 6*d.*

MODERN LANGUAGES AND LITERATURE.

(1) English, (2) French, (3) German, (4) Modern Greek, (5) Italian, (6) Spanish.

ENGLISH.

Abbott.—A SHAKESPEARIAN GRAMMAR. An attempt to illustrate some of the Differences between Elizabethan and Modern English. By the Rev. E. A. ABBOTT, D.D., Head Master of the City of London School. New Edition. Extra fcap. 8vo. 6*s.*

Bacon.—ESSAYS. Edited by F. G. SELBY, M.A., Professor of Logic and Moral Philosophy, Deccan College, Poona. Globe 8vo. [*In preparation.*

Burke.—REFLECTIONS ON THE FRENCH REVOLUTION. Edited by F. G. SELBY, M.A., Professor of Logic and Moral Philosophy, Deccan College, Poona. Globe 8vo. [*In preparation.*

Brooke.—PRIMER OF ENGLISH LITERATURE. By the Rev. STOPFORD A. BROOKE, M.A. 18mo. 1*s.* (*Literature Primers.*)

Butler.—HUDIBRAS. Edited, with Introduction and Notes, by ALFRED MILNES, M.A. Lon., late Student of Lincoln College, Oxford. Extra fcap 8vo. Part I. 3*s.* 6*d.* Parts II. and III. 4*s.* 6*d.*

Cowper's TASK: AN EPISTLE TO JOSEPH HILL, ESQ.; TIROCINIUM, or a Review of the Schools; and THE HISTORY OF JOHN GILPIN. Edited, with Notes, by WILLIAM BENHAM, B.D. Globe 8vo. 1*s.* (*Globe Readings from Standard Authors.*)

THE TASK. Edited by W. T. WEBB, M.A., Professor of English Literature, Presidency College, Calcutta. [*In preparation.*

Dowden.—SHAKESPEARE. By Professor DOWDEN. 18mo. 1*s.* (*Literature Primers.*)

Dryden.—SELECT PROSE WORKS. Edited, with Introduction and Notes, by Professor C. D. YONGE. Fcap. 8vo. 2*s.* 6*d.*

ENGLISH CLASSICS FOR INDIAN STUDENTS.

A SERIES OF SELECTIONS FROM THE WORKS OF THE GREAT ENGLISH CLASSICS, with Introductions and Notes, specially written for the use of Native Students preparing for the Examinations of the Universities of Bombay, Calcutta, Madras, and the Punjab. The books are also adapted for the use of English Students.

The following Volumes are ready or in preparation.

Bacon.—ESSAYS. Edited by F. G. SELBY, M.A., Professor of Logic and Moral Philosophy, Deccan College, Poona. [*In preparation.*

Burke.—REFLECTIONS ON THE FRENCH REVOLUTION. By the same Editor. [*In preparation.*

Cowper.—THE TASK. Edited by W. T. WEBB, M.A., Professor of English Literature, Presidency College, Calcutta. Globe 8vo. [*In preparation.*

Goldsmith.—THE TRAVELLER AND THE DESERTED VILLAGE. Edited by ARTHUR BARRETT, B.A., Professor of English Literature, Elphinstone College, Bombay. Globe 8vo. 1*s*. 6*d*. [*Ready.*

THE VICAR OF WAKEFIELD. Edited by HAROLD LITTLEDALE, B.A., Professor of History and English Literature, Baroda College. [*In preparation.*

Helps.—ESSAYS WRITTEN IN THE INTERVALS OF BUSINESS. Edited by F. J. ROWE, M.A., Professor of English Literature, Presidency College, Calcutta. [*In preparation.*

Milton.—PARADISE LOST, Books I. and II. Edited by MICHAEL MACMILLAN, B.A., Professor of Logic and Moral Philosophy, Elphinstone College, Bombay. Globe 8vo. 2*s*. 6*d*. [*Ready.*

Scott.—THE LADY OF THE LAKE. Edited by G. H. STUART, Professor of English Literature, Presidency College, Madras. [*In preparation.*

THE LAY OF THE LAST MINSTREL. By the same Editor. [*In preparation.*

MARMION. Edited by MICHAEL MACMILLAN, B.A. Globe 8vo. 3*s*. 6*d*. [*Ready.*

ROKEBY. By the same Editor. [*In preparation.*

Shakespeare.—MUCH ADO ABOUT NOTHING. Edited by K. DEIGHTON, M.A., late Principal of Agra College. Globe 8vo. 2*s*.

HENRY V. By the same Editor. [*In preparation.*

THE WINTER'S TALE. By the same Editor. [*In preparation.*

RICHARD III. Edited by C. H. TAWNEY, M.A., Principal and Professor of English Literature, Elphinstone College, Calcutta. [*In preparation.*

Tennyson.—SELECTIONS. Edited by F. J. ROWE, M.A., and W. T. WEBB, M.A., Professors of English Literature, Presidency College, Calcutta. [*In the press.*

This Volume contains:—Recollections of the Arabian Nights—The Lady of Shalott—Œnone—The Lotos-Eaters—A Dream of Fair Women—Morte D'Arthur—Dora—Ulysses—Tithonus—Sir Galahad—The Lord of Burleigh—Ode on the Death of the Duke of Wellington—The Revenge.

Wordsworth.—SELECTIONS. Edited by WILLIAM WORDSWORTH, B.A., Principal and Professor of History and Political Economy, Elphinstone College, Bombay. [*In preparation.*

Gladstone.—SPELLING REFORM FROM AN EDUCATIONAL POINT OF VIEW. By J. H. GLADSTONE, Ph.D., F.R.S., Member of the School Board for London. New Edition. Crown 8vo. 1*s.* 6*d.*

Globe Readers. For Standards I.—VI. Edited by A. F. MURISON. Sometime English Master at the Aberdeen Grammar School. With Illustrations. Globe 8vo.

Primer I.	(48 pp.)	3*d.*	Book III.	(232 pp.)	1*s.* 3*d.*
Primer II.	(48 pp.)	3*d.*	Book IV.	(328 pp.)	1*s.* 9*d.*
Book I.	(96 pp.)	6*d.*	Book V.	(416 pp.)	2*s.*
Book II.	(136 pp.)	9*d.*	Book VI.	(448 pp.)	2*s.* 6*d.*

"Among the numerous sets of readers before the public the present series is honourably distinguished by the marked superiority of its materials and the careful ability with which they have been adapted to the growing capacity of the pupils. The plan of the two primers is excellent for facilitating the child's first attempts to read. In the first three following books there is abundance of entertaining reading. . . . . Better food for young minds could hardly be found."—THE ATHENÆUM.

***The Shorter Globe Readers.**—With Illustrations. Globe 8vo.

Primer I.	(48 pp.)	3*d.*	Standard III.	(178 pp.)	1*s.*
Primer II.	(48 pp.)	3*d.*	Standard IV.	(182 pp.)	1*s.*
Standard I.	(92 pp.)	6*d.*	Standard V.	(216 pp.)	1*s.* 3*d.*
Standard II.	(124 pp.)	9*d.*	Standard VI.	(228 pp.)	1*s.* 6*d.*

* This Series has been abridged from "The Globe Readers" to meet the demand for smaller reading books.

GLOBE READINGS FROM STANDARD AUTHORS.

Cowper's TASK: AN EPISTLE TO JOSEPH HILL, ESQ.; TIROCINIUM, or a Review of the Schools; and THE HISTORY OF JOHN GILPIN. Edited, with Notes, by WILLIAM BENHAM, B.D. Globe 8vo. 1*s.*

Goldsmith's VICAR OF WAKEFIELD. With a Memoir of Goldsmith by Professor MASSON. Globe 8vo. 1*s.*

Lamb's (Charles) TALES FROM SHAKESPEARE. Edited, with Preface, by the Rev. CANON AINGER, M.A. Globe 8vo. 2*s.*

Scott's (Sir Walter) LAY OF THE LAST MINSTREL; and THE LADY OF THE LAKE. Edited, with Introductions and Notes, by FRANCIS TURNER PALGRAVE. Globe 8vo. 1s.

MARMION; and the LORD OF THE ISLES. By the same Editor. Globe 8vo. 1s.

The Children's Garland from the Best Poets.—Selected and arranged by COVENTRY PATMORE. Globe 8vo. 2s.

Yonge (Charlotte M.).—A BOOK OF GOLDEN DEEDS OF ALL TIMES AND ALL COUNTRIES. Gathered and narrated anew by CHARLOTTE M. YONGE, the Author of "The Heir of Redclyffe." Globe 8vo. 2s.

Goldsmith.—THE TRAVELLER, or a Prospect of Society; and THE DESERTED VILLAGE. By OLIVER GOLDSMITH. With Notes, Philological and Explanatory, by J. W. HALES, M.A. Crown 8vo. 6*d.*

THE VICAR OF WAKEFIELD. With a Memoir of Goldsmith by Professor MASSON. Globe 8vo. 1s. (*Globe Readings from Standard Authors.*)

SELECT ESSAYS. Edited, with Introduction and Notes, by Professor C. D. YONGE. Fcap. 8vo. 2s. 6*d.*

THE TRAVELLER AND THE DESERTED VILLAGE. Edited by ARTHUR BARRETT, B.A., Professor of English Literature, Elphinstone College, Bombay. Globe 8vo. 1s. 6*d.*

THE VICAR OF WAKEFIELD. Edited by HAROLD LITTLEDALE, B.A., Professor of History and English Literature, Baroda College. Globe 8vo. [*In preparation.*

Gosse.—A HISTORY OF ENGLISH LITERATURE IN THE REIGN OF QUEEN ANNE. By EDMUND GOSSE. Crown 8vo. [*In the press.*

Hales.—LONGER ENGLISH POEMS, with Notes, Philological and Explanatory, and an Introduction on the Teaching of English, Chiefly for Use in Schools. Edited by J. W. HALES, M.A., Professor of English Literature at King's College, London. New Edition. Extra fcap. 8vo. 4s. 6*d.*

Helps.—ESSAYS WRITTEN IN THE INTERVALS OF BUSINESS. Edited by F. J. ROWE, M.A., Professor of English Literature, Presidency College, Calcutta. Globe 8vo. [*In preparation.*

Johnson's LIVES OF THE POETS. The Six Chief Lives (Milton, Dryden, Swift, Addison, Pope, Gray), with Macaulay's "Life of Johnson." Edited with Preface and Notes by MATTHEW ARNOLD. New and cheaper edition. Crown 8vo. 4s. 6*d.*

Lamb (Charles).—TALES FROM SHAKESPEARE. Edited, with Preface, by the Rev. CANON AINGER, M.A. Globe 8vo. 2s. (*Globe Readings from Standard Authors.*)

Literature Primers—Edited by JOHN RICHARD GREEN, M.A., LL.D., Author of "A Short History of the English People."

ENGLISH COMPOSITION. By Professor NICHOL. 18mo. 1*s.*

ENGLISH GRAMMAR. By the Rev. R. MORRIS, LL.D., sometime President of the Philological Society. 18mo. 1*s.*

ENGLISH GRAMMAR EXERCISES. By R. MORRIS, LL.D., and H. C. BOWEN, M.A. 18mo. 1*s.*

EXERCISES ON MORRIS'S PRIMER OF ENGLISH GRAMMAR. By JOHN WETHERELL, of the Middle School, Liverpool College. 18mo. 1*s.*

ENGLISH LITERATURE. By STOPFORD BROOKE, M.A. New Edition. 18mo. 1*s.*

SHAKSPERE. By Professor DOWDEN. 18mo. 1*s.*

THE CHILDREN'S TREASURY OF LYRICAL POETRY. Selected and arranged with Notes by FRANCIS TURNER PALGRAVE. In Two Parts. 18mo. 1*s.* each.

PHILOLOGY. By J. PEILE, M.A. 18mo. 1*s.*

A History of English Literature in Four Volumes. Crown 8vo.

EARLY ENGLISH LITERATURE. By STOPFORD BROOKE, M.A. [*In preparation.*

ELIZABETHAN LITERATURE. By GEORGE SAINTSBURY. 7*s.* 6*d.*

THE AGE OF QUEEN ANNE. By EDMUND GOSSE. [*In the press.*

THE MODERN PERIOD. By PROFESSOR E. DOWDEN. [*In prep.*

Macmillan's Reading Books.—Adapted to the English and Scotch Codes. Bound in Cloth.

PRIMER. 18mo. (48 pp.) 2*d.*

BOOK I. for Standard I. 18mo. (96 pp.) 4*d.*

BOOK II. for Standard II. 18mo. (144 pp.) 5*d.*

BOOK III. for Standard III. 18mo. (160 pp.) 6*d.*

BOOK IV. for Standard IV. 18mo. (176 pp.) 8*d.*

BOOK V. for Standard V. 18mo. (380 pp.) 1*s.*

BOOK VI. for Standard VI. Cr. 8vo. (430 pp.) 2*s.*

Book VI. is fitted for higher Classes, and as an Introduction to English Literature.

Macmillan's Copy-Books—

Published in two sizes, viz.:—

1. Large Post 4to. Price 4*d.* each.
2. Post Oblong. Price 2*d.* each.

1. INITIATORY EXERCISES AND SHORT LETTERS.
2. WORDS CONSISTING OF SHORT LETTERS

*3. LONG LETTERS. With Words containing Long Letters—Figures.

*4. WORDS CONTAINING LONG LETTERS.

4a. PRACTISING AND REVISING COPY-BOOK. For Nos. 1 to 4.

*5. CAPITALS AND SHORT HALF-TEXT. Words beginning with a Capital.

*6. HALF-TEXT WORDS beginning with Capitals—Figures.

*7. SMALL-HAND AND HALF-TEXT. With Capitals and Figures.

*8. SMALL-HAND AND HALF-TEXT. With Capitals and Figures.

8a. PRACTISING AND REVISING COPY-BOOK. For Nos. 5 to 8.

Macmillan's Copy Books (*continued*)

*9. SMALL-HAND SINGLE HEADLINES—Figures.
10. SMALL-HAND SINGLE HEADLINES—Figures.
11. SMALL-HAND DOUBLE HEADLINES—Figures.
12. COMMERCIAL AND ARITHMETICAL EXAMPLES, &c.
12a. PRACTISING AND REVISING COPY-BOOK. For Nos. 8 to 12.

* *These numbers may be had with Goodman's Patent Sliding Copies.* Large Post 4to. Price 6*d.* each.

Martin.—THE POET'S HOUR: Poetry selected and arranged for Children. By FRANCES MARTIN. New Edition. 18mo. 2*s.* 6*d.*

SPRING-TIME WITH THE POETS: Poetry selected by FRANCES MARTIN. New Edition. 18mo. 3*s* 6*d.*

Milton.—By STOPFORD BROOKE, M.A. Fcap. 8vo. 1*s.* 6*d* (*Classical Writers Series.*)

Milton.—PARADISE LOST. Books I. and II. Edited, with Introduction and Notes, by MICHAEL MACMILLAN, B.A. Oxon, Professor of Logic and Moral Philosophy, Elphinstone College, Bombay. Globe 8vo. 2*s.* 6*d.*

Morley.—ON THE STUDY OF LITERATURE. The Annual Address to the Students of the London Society for the Extension of University Teaching. Delivered at the Mansion House, February 26, 1887. By JOHN MORLEY. Globe 8vo. Cloth. 1*s.* 6*d.*

* *Also a Popular Edition in Pamphlet form for Distribution, price* 2*d.*

Morris.—Works by the Rev. R. MORRIS, LL.D.

HISTORICAL OUTLINES OF ENGLISH ACCIDENCE, comprising Chapters on the History and Development of the Language, and on Word-formation. New Edition. Extra fcap. 8vo. 6*s.*

ELEMENTARY LESSONS IN HISTORICAL ENGLISH GRAMMAR, containing Accidence and Word-formation. New Edition. 18mo. 2*s.* 6*d.*

PRIMER OF ENGLISH GRAMMAR. 18mo. 1*s.* (See also *Literature Primers.*)

Oliphant.—THE OLD AND MIDDLE ENGLISH. A New Edition of "THE SOURCES OF STANDARD ENGLISH," revised and greatly enlarged. By T. L. KINGTON OLIPHANT. Extra fcap. 8vo. 9*s.*

THE NEW ENGLISH. By the same Author. 2 vols. Cr. 8vo. 21*s.*

Palgrave.—THE CHILDREN'S TREASURY OF LYRICAL POETRY. Selected and arranged, with Notes, by FRANCIS TURNER PALGRAVE. 18mo. 2*s.* 6*d.* Also in Two Parts. 1*s.* each.

Patmore.—THE CHILDREN'S GARLAND FROM THE BEST POETS. Selected and arranged by COVENTRY PATMORE. Globe 8vo. 2*s.* (*Globe Readings from Standard Authors.*)

Plutarch.—Being a Selection from the Lives which Illustrate Shakespeare. North's Translation. Edited, with Introductions, Notes, Index of Names, and Glossarial Index, by the Rev. W. W. SKEAT, M.A. Crown 8vo. 6*s.*

Saintsbury.—A HISTORY OF ELIZABETHAN LITERATURE. By GEORGE SAINTSBURY. Cr. 8vo. 7*s.* 6*d.*

Scott's (Sir Walter) LAY OF THE LAST MINSTREL, and THE LADY OF THE LAKE. Edited, with Introduction and Notes, by FRANCIS TURNER PALGRAVE. Globe 8vo. 1*s.* (*Globe Readings from Standard Authors.*)

MARMION; and THE LORD OF THE ISLES. By the same Editor. Globe 8vo. 1*s.* (*Globe Readings from Standard Authors.*)

MARMION. Edited, with Introduction and Notes, by M. MACMILLAN, B.A. Oxon, Professor of Logic and Moral Philosophy, Elphinstone College, Bombay. Globe 8vo. 3*s.* 6*d.*

THE LADY OF THE LAKE. Edited by G. H. STUART, M.A., Professor of English Literature, Presidency College, Madras. Globe 8vo. [*In preparation.*

THE LAY OF THE LAST MINSTREL. By the same Editor. Globe 8vo. [*In preparation.*

ROKEBY. By MICHAEL MACMILLAN, B.A. Globe 8vo. [*In preparation.*

Shakespeare.—A SHAKESPEARIAN GRAMMAR. By Rev. E. A. ABBOTT, D.D., Head Master of the City of London School. Globe 8vo. 6*s.*

A SHAKESPEARE MANUAL. By F. G. FLEAY, M.A., late Head Master of Skipton Grammar School. Second Edition. Extra fcap. 8vo. 4*s.* 6*d.*

PRIMER OF SHAKESPEARE. By Professor DOWDEN. 18mo. 1*s.* (*Literature Primers.*)

MUCH ADO ABOUT NOTHING. Edited by K. DEIGHTON, M.A., late Principal of Agra College. Globe 8vo. 2*s.*

HENRY V. By the same Editor. Globe 8vo. [*In preparation.*

THE WINTER'S TALE. By the same Editor. Globe 8vo. [*In preparation.*

RICHARD III. Edited by C. H. TAWNEY, M.A., Principal and Professor of English Literature, Elphinstone College, Calcutta. Globe 8vo. [*In preparation.*

Sonnenschein and Meiklejohn. — THE ENGLISH METHOD OF TEACHING TO READ. By A. SONNENSCHEIN and J. M. D. MEIKLEJOHN, M.A. Fcap. 8vo.

COMPRISING:

THE NURSERY BOOK, containing all the Two-Letter Words in the Language. 1*d.* (Also in Large Type on Sheets for School Walls. 5*s.*)

THE FIRST COURSE, consisting of Short Vowels with Single Consonants. 6*d.*

THE SECOND COURSE, with Combinations and Bridges, consisting of Short Vowels with Double Consonants. 6*d.*

THE THIRD AND FOURTH COURSES, consisting of Long Vowels, and all the Double Vowels in the Language. 6*d.*

"These are admirable books, because they are constructed on a principle, and that the simplest principle on which it is possible to learn to read English."—SPECTATOR.

Taylor.—WORDS AND PLACES; or, Etymological Illustrations of History, Ethnology, and Geography. By the Rev. ISAAC TAYLOR, M.A., Litt. D., Hon. LL.D., Canon of York. Third and Cheaper Edition, revised and compressed. With Maps. Globe 8vo. 6*s.*

Tennyson.—The COLLECTED WORKS of LORD TENNYSON, Poet Laureate. An Edition for Schools. In Four Parts. Crown 8vo. 2*s.* 6*d.* each.

SELECTIONS FROM LORD TENNYSON'S POEMS. Edited with Notes for the Use of Schools. By the Rev. ALFRED AINGER, M.A., LL.D., Canon of Bristol. [*In preparation*.

SELECT POEMS OF LORD TENNYSON. With short Introduction and Notes. By W. T. WEBB, M.A., Professor of History and Political Economy, and F. J. ROWE, Professor of English Literature, Presidency College, Calcutta. Globe 8vo. [*In the press*.

This selection contains:—"Recollections of the Arabian Nights," "The Lady of Shalott," "Oenone," "The Lotos Eaters," "Ulysses," "Tithonus," "Morte d'Arthur," "Sir Galahad," "Dora," "The Ode on the Death of the Duke of Wellington," and the Ballad of the "Revenge."

Thring.—THE ELEMENTS OF GRAMMAR TAUGHT IN ENGLISH. By EDWARD THRING, M.A., late Head Master of Uppingham. With Questions. Fourth Edition. 18mo. 2*s.*

Vaughan (C.M.).—WORDS FROM THE POETS. By C. M. VAUGHAN. New Edition. 18mo, cloth. 1*s.*

Ward.—THE ENGLISH POETS. Selections, with Critical Introductions by various Writers and a General Introduction by MATTHEW ARNOLD. Edited by T. H. WARD, M.A. 4 Vols. Vol. I. CHAUCER TO DONNE.—Vol. II. BEN JONSON TO DRYDEN.—Vol. III. ADDISON TO BLAKE.—Vol. IV. WORDSWORTH TO ROSSETTI. Crown 8vo. Each 7*s.* 6*d.*

Wetherell.—EXERCISES ON MORRIS'S PRIMER OF ENGLISH GRAMMAR. By JOHN WETHERELL, M.A. 18mo. 1*s.* (*Literature Primers.*)

Woods.—A FIRST SCHOOL POETRY BOOK. Compiled by M. A. WOODS, Head Mistress of the Clifton High School for Girls. Fcap. 8vo. 2*s.* 6*d.*

A SECOND SCHOOL POETRY BOOK. By the same Author. Fcap. 8vo. 4*s.* 6*d.*

A THIRD SCHOOL POETRY BOOK. By the same Author. Fcap. 8vo. [*In preparation*.

Wordsworth.—SELECTIONS. Edited by WILLIAM WORDSWORTH, B.A., Principal and Professor of History and Political Economy, Elphinstone College, Bombay. [*In preparation*.

Yonge (Charlotte M.).—THE ABRIDGED BOOK OF GOLDEN DEEDS. A Reading Book for Schools and general readers. By the Author of "The Heir of Redclyffe." 18mo, cloth. 1*s.*

GLOBE READINGS EDITION. Globe 8vo. 2*s.* (See p. 60.)

FRENCH.

Beaumarchais.—LE BARBIER DE SEVILLE. Edited, with Introduction and Notes, by L. P. BLOUET, Assistant Master in St. Paul's School. Fcap. 8vo. 3*s*. 6*d*.

Bowen.—FIRST LESSONS IN FRENCH. By H. COURTHOPE BOWEN, M.A., Principal of the Finsbury Training College for Higher and Middle Schools. Extra fcap. 8vo. 1*s*.

Breymann.—Works by HERMANN BREYMANN, Ph.D., Professor of Philology in the University of Munich.

A FRENCH GRAMMAR BASED ON PHILOLOGICAL PRINCIPLES. Second Edition. Extra fcap. 8vo. 4*s*. 6*d*.

FIRST FRENCH EXERCISE BOOK. Extra fcap. 8vo. 4*s*. 6*d*.

SECOND FRENCH EXERCISE BOOK. Extra fcap. 8vo. 2*s*. 6*d*.

Fasnacht.—Works by G. EUGÈNE FASNACHT, Author of "Macmillan's Progressive French Course," Editor of "Macmillan's Foreign School Classics," &c.

THE ORGANIC METHOD OF STUDYING LANGUAGES. Extra fcap. 8vo. I. French. 3*s*. 6*d*.

A SYNTHETIC FRENCH GRAMMAR FOR SCHOOLS. Crown 8vo. 3*s*. 6*d*.

GRAMMAR AND GLOSSARY OF THE FRENCH LANGUAGE OF THE SEVENTEENTH CENTURY. Crown 8vo. [*In preparation.*

Macmillan's Primary Series of French and German Reading Books.—Edited by G. EUGÈNE FASNACHT, Assistant-Master in Westminster School. With Illustrations. Globe 8vo.

DE MAISTRE—LA JEUNE SIBÉRIENNE ET LE LÉPREUX DE LA CITÉ D'AOSTE. Edited, with Introduction, Notes, and Vocabulary. By STÉPHANE BARLET, B.Sc. Univ. Gall. and London; Assistant-Master at the Mercers' School, Examiner to the College of Preceptors, the Royal Naval College, &c. 1*s*. 6*d*.

FLORIAN—FABLES. Selected and Edited, with Notes, Vocabulary, Dialogues, and Exercises, by the Rev. CHARLES YELD, M.A., Head Master of University School, Nottingham. Illustrated. 1*s*. 6*d*.

GRIMM—KINDER UND HAUSMÄRCHEN. Selected and Edited, with Notes, and Vocabulary, by G. E. FASNACHT. 2*s*.

HAUFF.—DIE KARAVANE. Edited, with Notes and Vocabulary, by HERMAN HAGER, Ph.D. Lecturer in the Owens College, Manchester. 2*s*. 6*d*.

LA FONTAINE—A SELECTION OF FABLES. Edited, with Introduction, Notes, and Vocabulary, by L. M. MORIARTY, B.A., Professor of French in King's College, London. 2*s*.

PERRAULT—CONTES DE FÉES. Edited, with Introduction, Notes, and Vocabulary, by G. E. FASNACHT. 1*s*.

G. SCHWAB—ODYSSEUS. With Introduction, Notes, and Vocabulary, by the same Editor. [*In preparation.*

Macmillan's Progressive French Course.—By G. EUGÈNE FASNACHT, Assistant-Master in Westminster School.

I.—FIRST YEAR, containing Easy Lessons on the Regular Accidence. New and thoroughly revised Edition. Extra fcap. 8vo. 1*s.*

II.—SECOND YEAR, containing an Elementary Grammar with copious Exercises, Notes, and Vocabularies. A new Edition, enlarged and thoroughly revised. Extra fcap. 8vo. 2*s.*

III.—THIRD YEAR, containing a Systematic Syntax, and Lessons in Composition. Extra fcap. 8vo. 2*s.* 6*d.*

EXERCISES IN FRENCH COMPOSITION. Part I. Elementary. Part II. Advanced. By G. E. FASNACHT. [*Part I. in the press.*

THE TEACHER'S COMPANION TO MACMILLAN'S PROGRESSIVE FRENCH COURSE. With Copious Notes, Hints for Different Renderings, Synonyms, Philological Remarks, &c. By G. E. FASNACHT. Globe 8vo. *First Year* 4*s.* 6*d.*, *Second Year* 4*s.* 6*d.*, *Third Year* 4*s.* 6*d.*

Macmillan's Progressive French Readers. By G. EUGÈNE FASNACHT.

I.—FIRST YEAR, containing Fables, Historical Extracts, Letters, Dialogues, Ballads, Nursery Songs, &c., with Two Vocabularies: (1) in the order of subjects; (2) in alphabetical order. Extra fcap. 8vo. 2*s.* 6*d.*

II.—SECOND YEAR, containing Fiction in Prose and Verse, Historical and Descriptive Extracts, Essays, Letters, Dialogues, &c. Extra fcap. 8vo. 2*s.* 6*d.*

Macmillan's Foreign School Classics. Edited by G. EUGÈNE FASNACHT. 18mo.

FRENCH.

CORNEILLE—LE CID. Edited by G. E. FASNACHT. 1*s.*

DUMAS—LES DEMOISELLES DE ST. CYR. Edited by VICTOR OGER, Lecturer in University College, Liverpool. 1*s.* 6*d.*

LA FONTAINE'S FABLES. Books I.—VI. Edited by L. M. MORIARTY, B.A., Professor of French in King's College, London. [*In preparation.*

MOLIÈRE—L'AVARE. By the same Editor. 1*s.*

MOLIÈRE—LE BOURGEOIS GENTILHOMME. By the same Editor. 1*s.* 6*d.*

MOLIÈRE—LES FEMMES SAVANTES. By G. E. FASNACHT. 1*s.*

MOLIÈRE—LE MISANTHROPE. By the same Editor. 1*s.*

MOLIERE—LE MÉDECIN MALGRE LUI. By the same Editor. 1*s.*

RACINE—BRITANNICUS. Edited by EUGÈNE PELLISSIER, Assistant-Master in Clifton College, and Lecturer in University College, Bristol. 2*s.*

Macmillan's Foreign School Classics (*continued*)—

FRENCH READINGS FROM ROMAN HISTORY. Selected from Various Authors and Edited by C. COLBECK, M.A., late Fellow of Trinity College, Cambridge; Assistant-Master at Harrow. 4*s.* 6*d.*

SAND, GEORGE—LA MARE AU DIABLE. Edited by W. E. RUSSELL, M.A., Assistant-Master in Haileybury College. 1*s.*

SANDEAU, JULES—MADEMOISELLE DE LA SEIGLIERE. Edited by H. C. STEEL, Assistant-Master in Winchester College. 1*s.* 6*d.*

THIERS'S HISTORY OF THE EGYPTIAN EXPEDITION. Edited by Rev. H. A. BULL, M.A. Assistant-Master in Wellington College. [*In preparation.*

VOLTAIRE—CHARLES XII. Edited by G. E. FASNACHT. 3*s.* 6*d.*

*** *Other volumes to follow.*

(See also *German Authors*, page 69.

Masson (Gustave).—A COMPENDIOUS DICTIONARY OF THE FRENCH LANGUAGE (French-English and English-French). Adapted from the Dictionaries of Professor ALFRED ELWALL. Followed by a List of the Principal Diverging Derivations, and preceded by Chronological and Historical Tables. By GUSTAVE MASSON, Assistant-Master and Librarian, Harrow School. New Edition. Crown 8vo. 6*s.*

Molière.—LE MALADE IMAGINAIRE. Edited, with Introduction and Notes, by FRANCIS TARVER, M.A., Assistant-Master at Eton. Fcap. 8vo. 2*s.* 6*d.*

(See also *Macmillan's Foreign School Classics.*)

Pellissier.—FRENCH ROOTS AND THEIR FAMILIES. A Synthetic Vocabulary, based upon Derivations, for Schools and Candidates for Public Examinations. By EUGÈNE PELLISSIER, M.A., B.Sc., LL.B., Assistant-Master at Clifton College, Lecturer at University College, Bristol. Globe 8vo. 6*s.*

GERMAN.

Huss.—A SYSTEM OF ORAL INSTRUCTION IN GERMAN, by means of Progressive Illustrations and Applications of the leading Rules of Grammar. By HERMANN C. O. HUSS, Ph.D. Crown 8vo. 5*s.*

Macmillan's Progressive German Course. By G. EUGÈNE FASNACHT.

PART I.—FIRST YEAR. Easy Lessons and Rules on the Regular Accidence. Extra fcap. 8vo. 1*s.* 6*d.*

Part II.—SECOND YEAR. Conversational Lessons in Systematic Accidence and Elementary Syntax. With Philological Illustrations and Etymological Vocabulary. New Edition, enlarged and thoroughly recast. Extra fcap. 8vo. 3*s.* 6*d.*

Part III.—THIRD YEAR. [*In preparation,*

Macmillan's Progressive German Course (*continued*).

TEACHER'S COMPANION TO MACMILLAN'S PROGRESSIVE GERMAN COURSE. With copious Notes, Hints for Different Renderings, Synonyms, Philological Remarks, &c. By G. E. FASNACHT. Extra Fcap. 8vo. FIRST YEAR. 4*s*. 6*d*. SECOND YEAR. 4*s*. 6*d*.

Macmillan's Progressive German Readers. By G. E. FASNACHT.

I.—FIRST YEAR, containing an Introduction to the German order of Words, with Copious Examples, extracts from German Authors in Prose and Poetry; Notes, and Vocabularies. Extra Fcap. 8vo., 2*s*. 6*d*.

Macmillan's Primary German Reading Books.

(See page 66.)

Macmillan's Foreign School Classics. Edited by G. EUGÈNE FASNACHT, 18mo.

GERMAN.

FREYTAG (G.).—DOKTOR LUTHER. Edited by FRANCIS STORR, M.A., Head Master of the Modern Side, Merchant Taylors' School. [*In preparation.*

GOETHE—GÖTZ VON BERLICHINGEN. Edited by H. A. BULL, M.A., Assistant Master at Wellington College. 2*s*.

GOETHE—FAUST. PART I., followed by an Appendix on PART II. Edited by JANE LEE, Lecturer in German Literature at Newnham College, Cambridge. 4*s*. 6*d*.

HEINE—SELECTIONS FROM THE REISEBILDER AND OTHER PROSE WORKS. Edited by C. COLBECK, M.A., Assistant-Master at Harrow, late Fellow of Trinity College, Cambridge. 2*s*. 6*d*.

LESSING.—MINNA VON BARNHELM. Edited by JAMES SIME. [*In preparation.*

SCHILLER—SELECTIONS FROM SCHILLER'S LYRICAL POEMS. Edited, with Notes and a Memoir of Schiller, by E. J. TURNER, B.A., and E. D. A. MORSHEAD, M.A. Assistant-Masters in Winchester College. 2*s*. 6*d*.

SCHILLER—DIE JUNGFRAU VON ORLEANS. Edited by JOSEPH GOSTWICK. 2*s*. 6*d*.

SCHILLER—MARIA STUART. Edited by C. SHELDON, M.A., D.Lit., of the Royal Academical Institution, Belfast. 2*s*. 6*d*.

SCHILLER—WILHELM TELL. Edited by G. E. FASNACHT. 2*s*. 6*d*.

SCHILLER.—WALLENSTEIN. Part I. DAS LAGER. Edited by H. B. COTTERILL, M.A. 2*s*.

UHLAND—SELECT BALLADS. Adapted as a First Easy Reading Book for Beginners. With Vocabulary. Edited by G. E. FASNACHT. 1*s*.

*** *Other Volumes to follow.*

(See also *French Authors*, page 67.)

Pylodet.—NEW GUIDE TO GERMAN CONVERSATION; containing an Alphabetical List of nearly 800 Familiar Words; followed by Exercises; Vocabulary of Words in frequent use; Familiar Phrases and Dialogues; a Sketch of German Literature, Idiomatic Expressions, &c. By L. PYLODET. 18mo, cloth limp. 2*s*. 6*d*.

Whitney.—Works by W. D. WHITNEY, Professor of Sanskrit and Instructor in Modern Languages in Yale College.

A COMPENDIOUS GERMAN GRAMMAR. Crown 8vo. 4*s*. 6*d*.

A GERMAN READER IN PROSE AND VERSE. With Notes and Vocabulary. Crown 8vo. 5*s*.

Whitney and Edgren.—A COMPENDIOUS GERMAN AND ENGLISH DICTIONARY, with Notation of Correspondences and Brief Etymologies. By Professor W. D. WHITNEY, assisted by A. H. EDGREN. Crown 8vo. 7*s*. 6*d*.

THE GERMAN-ENGLISH PART, separately, 5*s*.

MODERN GREEK.

Vincent and Dickson. — HANDBOOK TO MODERN GREEK. By Sir EDGAR VINCENT, K.C.M.G. and T. G. DICKSON, M.A. Second Edition, revised and enlarged, with Appendix on the relation of Modern and Classical Greek by Professor JEBB. Crown 8vo. 6*s*.

ITALIAN.

Dante. — THE PURGATORY OF DANTE. Edited, with Translation and Notes, by A. J. BUTLER, M.A., late Fellow of Trinity College, Cambridge. Crown 8vo. 12*s*. 6*d*.

THE PARADISO OF DANTE. Edited, with Translation and Notes, by the same Author. Crown 8vo. 12*s*. 6*d*.

SPANISH.

Calderon.—FOUR PLAYS OF CALDERON. Edited, with Introduction and Notes, by NORMAN MACCOLL, M.A., Fellow of Downing College, Cambridge. Crown 8vo. [*In the press.*

DOMESTIC ECONOMY.

Barker.—FIRST LESSONS IN THE PRINCIPLES OF COOKING. By LADY BARKER. New Edition. 18mo. 1*s*.

Berners.—FIRST LESSONS ON HEALTH. By J. BERNERS. New Editlon. 18mo. 1*s*.

Fawcett.—TALES IN POLITICAL ECONOMY. By MILLICENT GARRETT FAWCETT. Globe 8vo. 3*s*.

Frederick.—HINTS TO HOUSEWIVES ON SEVERAL POINTS, PARTICULARLY ON THE PREPARATION OF ECONOMICAL AND TASTEFUL DISHES. By Mrs. FREDERICK. Crown 8vo. 1*s*.

"This unpretending and useful little volume distinctly supplies a desideratum . . . . The author steadily keeps in view the simple aim of 'making every-day meals at home, particularly the dinner, attractive,' without adding to the ordinary household expenses."—SATURDAY REVIEW.

Grand'homme.— CUTTING-OUT AND DRESSMAKING. From the French of Mdlle. E. GRAND'HOMME. With Diagrams. 18mo. 1*s*.

Jex-Blake.—THE CARE OF INFANTS. A Manual for Mothers and Nurses. By SOPHIA JEX-BLAKE, M.D., Member of the Irish College of Physicians; Lecturer on Hygiene at the London School of Medicine for Women. 18mo. 1*s*.

Tegetmeier.—HOUSEHOLD MANAGEMENT AND COOKERY. With an Appendix of Recipes used by the Teachers of the National School of Cookery. By W. B. TEGETMEIER. Compiled at the request of the School Board for London. 18mo. 1*s*.

Thornton.—FIRST LESSONS IN BOOK-KEEPING. By J. THORNTON. New Edition. Crown 8vo. 2*s*. 6*d*.

The object of this volume is to make the theory of Book-keeping sufficiently plain for even children to understand it.

A KEY TO THE ABOVE FOR THE USE OF TEACHERS AND PRIVATE STUDENTS. Containing all the Exercises worked out, with brief Notes. By J. THORNTON. Oblong 4to. 10*s*. 6*d*.

Wright.—THE SCHOOL COOKERY-BOOK. Compiled and Edited by C. E. GUTHRIE WRIGHT, Hon Sec. to the Edinburgh School of Cookery. 18mo. 1*s*.

ART AND KINDRED SUBJECTS.

Anderson.—LINEAR PERSPECTIVE, AND MODEL DRAWING. A School and Art Class Manual, with Questions and Exercises for Examination, and Examples of Examination Papers. By LAURENCE ANDERSON. With Illustrations. Royal 8vo. 2*s*.

Collier.—A PRIMER OF ART. With Illustrations. By JOHN COLLIER. 18mo. 1*s*.

Delamotte.—A BEGINNER'S DRAWING BOOK. By P. H. DELAMOTTE, F.S.A. Progressively arranged. New Edition improved. Crown 8vo. 3*s*. 6*d*.

Ellis.—SKETCHING FROM NATURE. A Handbook for Students and Amateurs. By TRISTRAM J. ELLIS. With a Frontispiece and Ten Illustrations, by H. STACY MARKS, R.A., and Thirty Sketches by the Author. New Edition, revised and enlarged. Crown 8vo. 3*s*. 6*d*.

Hunt.—TALKS ABOUT ART. By WILLIAM HUNT. With a Letter from Sir J. E. MILLAIS, Bart., R.A. Crown 8vo. 3*s.* 6*d.*

Taylor.—A PRIMER OF PIANOFORTE PLAYING. By FRANKLIN TAYLOR. Edited by Sir GEORGE GROVE. 18mo. 1*s.*

WORKS ON TEACHING.

Ball.—THE STUDENT'S GUIDE TO THE BAR. By WALTER W. R. BALL, M.A., of the Inner Temple, Barrister-at-Law; Fellow and Assistant Tutor of Trinity College, Cambridge, and Fellow of University College, London. Fourth Edition Revised. Crown 8vo. 2*s.* 6*d.*

Blakiston—THE TEACHER. Hints on School Management. A Handbook for Managers, Teachers' Assistants, and Pupil Teacheres. By J. R. BLAKISTON, M.A. Crown 8vo. 2*s.* 6*d.* (Recommended by the London, Birmingham, and Leicester School Boards.)

"Into a comparatively small book he has crowded a great deal of exceedingly useful and sound advice. It is a plain, common-sense book, full of hints to the teacher on the management of his school and his children."—SCHOOL BOARD CHRONICLE.

Calderwood.—ON TEACHING. By Professor HENRY CALDERWOOD. New Edition. Extra fcap. 8vo. 2*s.* 6*d.*

Carter.—EYESIGHT IN SCHOOLS. A Paper read before the Association of Medical Officers of Schools on April 15th, 1885. By R. BRUDENELL CARTER, F.R.C.S., Ophthalmic Surgeon to St. George's Hospital. Crown 8vo. Sewed. 1*s.*

Fearon.—SCHOOL INSPECTION. By D. R. FEARON, M.A., Assistant Commissioner of Endowed Schools. New Edition. Crown 8vo. 2*s.* 6*d.*

Gladstone.—OBJECT TEACHING. A Lecture delivered at the Pupil-Teacher Centre, William Street Board School, Hammersmith. By J. H. GLADSTONE, Ph.D., F.R.S., Member of the London School Board. With an Appendix. Crown 8vo. 3*d.*

"It is a short but interesting and instructive publication, and our younger teachers will do well to read it carefully and thoroughly. There is much in these few pages which they can learn and profit by."—THE SCHOOL GUARDIAN.

Hertel.—OVERPRESSURE IN HIGH SCHOOLS IN DENMARK. By Dr. HERTEL, Municipal Medical Officer, Copenhagen. Translated from the Danish by C. GODFREY SÖRENSEN. With Introduction by Sir J. CRICHTON-BROWNE, M.D., LL.D., F.R.S. Crown 8vo. 3*s.* 6*d.*

DIVINITY.

. For other Works by these Authors, see THEOLOGICAL CATALOGUE.

Abbott (Rev. E. A.)—BIBLE LESSONS. By the Rev. E. A. ABBOTT, D.D., Head Master of the City of London School. New Edition. Crown 8vo. 4*s.* 6*d.*

"Wise, suggestive, and really profound initiation into religious thought." —GUARDIAN.

Abbott—Rushbrooke.—THE COMMON TRADITION OF THE SYNOPTIC GOSPELS, in the Text of the Revised Version. By EDWIN A. ABBOTT, D.D., formerly Fellow of St. John's College, Cambridge, and W. G. RUSHBROOKE, M.L., formerly Fellow of St. John's College, Cambridge. Cr. 8vo. 3*s*. 6*d*.

The Acts of the Apostles.— Being the Greek Text as revised by Professors WESTCOTT and HORT. With Explanatory Notes for the Use of Schools, by T. E. PAGE, M.A., late Fellow of St. John's College, Cambridge; Assistant Master at the Charterhouse. Fcap. 8vo. 4*s*. 6*d*.

Arnold. — A BIBLE-READING FOR SCHOOLS. — THE GREAT PROPHECY OF ISRAEL'S RESTORATION (Isaiah, Chapters xl.—lxvi.). Arranged and Edited for Young Learners. By MATTHEW ARNOLD, D.C.L., formerly Professor of Poetry in the University of Oxford, and Fellow of Oriel. New Edition. 18mo, cloth. 1*s*.

Arnold.—ISAIAH XL.—LXVI. With the Shorter Prophecies allied to it. Arranged and Edited, with Notes, by MATTHEW ARNOLD. Crown 8vo. 5*s*.

ISAIAH OF JERUSALEM, IN THE AUTHORISED ENGLISH VERSION. With Introduction, Corrections, and Notes. By MATTHEW ARNOLD. Crown 8vo. 4*s*. 6*d*.

Benham.—A COMPANION TO THE LECTIONARY. Being a Commentary on the Proper Lessons for Sundays and Holy Days. By Rev. W. BENHAM, B.D., Rector of S. Edmund with S. Nicholas Acons, &c. New Edition. Crown 8vo. 4*s*. 6*d*.

Calvert.—GREEK TESTAMENT, School Readings in the. A Course of thirty-six Lessons mainly following upon the Narrative of St. Mark. Edited and Arranged with Introduction, Notes and Vocabulary, by the Rev. A. CALVERT, M.A., late Fellow of St. John's College, Cambridge. Fcap. 8vo. 4*s*. 6*d*.

Cassel.—MANUAL OF JEWISH HISTORY AND LITERATURE; preceded by a BRIEF SUMMARY OF BIBLE HISTORY. By DR. D. CASSEL. Translated by Mrs. HENRY LUCAS. Fcap. 8vo. 2*s*. 6*d*.

Cheetham.—A CHURCH HISTORY OF THE FIRST SIX CENTURIES. By the Ven. ARCHDEACON CHEETHAM, Crown 8vo. [*In the press.*

Cross.—BIBLE READINGS SELECTED FROM THE PENTATEUCH AND THE BOOK OF JOSHUA. By the Rev. JOHN A. CROSS. Second Edition enlarged, with Notes. Globe 8vo. 2*s*. 6*d*.

Curteis.—MANUAL OF THE THIRTY-NINE ARTICLES. By G. H. CURTEIS, M.A., Principal of the Lichfield Theological College. [*In preparation.*

Davies.—THE EPISTLES OF ST. PAUL TO THE EPHESIANS, THE COLOSSIANS, AND PHILEMON; with Introductions and Notes, and an Essay on the Traces of Foreign Elements in the Theology of these Epistles. By the Rev. J. LLEWELYN DAVIES, M.A., Rector of Christ Church, St. Marylebone; late Fellow of Trinity College, Cambridge. Second Edition. Demy 8vo. 7*s.* 6*d.*

Drummond.—THE STUDY OF THEOLOGY, INTRODUCTION TO. By JAMES DRUMMOND, LL.D., Professor of Theology in Manchester New College, London. Crown 8vo. 5*s.*

Gaskoin.—THE CHILDREN'S TREASURY OF BIBLE STORIES. By Mrs. HERMAN GASKOIN. Edited with Preface by Rev. G. F. MACLEAR, D.D. PART I.—OLD TESTAMENT HISTORY. 18mo. 1*s.* PART II.—NEW TESTAMENT. 18mo. 1*s.* PART III.—THE APOSTLES: ST. JAMES THE GREAT, ST. PAUL, AND ST JOHN THE DIVINE. 18mo. 1*s.*

Golden Treasury Psalter.—Students' Edition. Being an Edition of "The Psalms Chronologically arranged, by Four Friends," with briefer Notes. 18mo. 3*s.* 6*d.*

Greek Testament.—Edited, with Introduction and Appendices, by CANON WESTCOTT and Dr. F. J. A. HORT. Two Vols. Crown 8vo. 10*s.* 6*d.* each.

Vol. I. The Text.

Vol. II. Introduction and Appendix.

Greek Testament.—Edited by Canon WESTCOTT and Dr. HORT. School Edition of Text. 12mo. cloth. 4*s.* 6*d.* 18mo. roan, red edges. 5*s.* 6*d.*

GREEK TESTAMENT, SCHOOL READINGS IN THE. Being the outline of the life of our Lord, as given by St. Mark, with additions from the Text of the other Evangelists. Arranged and Edited, with Notes and Vocabulary, by the Rev. A. CALVERT, M.A., late Fellow of St. John's College, Cambridge. Fcap. 8vo. 4*s.* 6*d.*

THE ACTS OF THE APOSTLES. Being the Greek Text as revised by Drs. WESTCOTT and HORT. With Explanatory Notes by T. E. PAGE, M.A., Assistant Master at the Charterhouse. Fcap. 8vo. 4*s.* 6*d.*

THE GOSPEL ACCORDING TO ST. MARK. Being the Greek Text as revised by Drs. WESTCOTT and HORT. With Explanatory Notes by Rev. J. O. F. MURRAY, M.A., Lecturer in Emmanuel College, Cambridge. Fcap. 8vo. [*In preparation.*

Hardwick.—Works by Archdeacon HARDWICK:—

A HISTORY OF THE CHRISTIAN CHURCH. Middle Age. From Gregory the Great to the Excommunication of Luther. Edited by WILLIAM STUBBS, M.A., Regius Professor of Modern History in the University of Oxford. With Four Maps. New Edition. Crown 8vo. 10*s.* 6*d.*

Hardwick.—Works by Archdeacon HARDWICK, *continued.*

A HISTORY OF THE CHRISTIAN CHURCH DURING THE REFORMATION. Eighth Edition. Edited by Professor STUBBS. Crown 8vo. 10*s.* 6*d.*

Jennings and Lowe.—THE PSALMS, WITH INTRODUCTIONS AND CRITICAL NOTES. By A. C. JENNINGS, M.A.; assisted in parts by W. H. LOWE, M.A. In 2 vols. Second Edition Revised. Crown 8vo. 10*s.* 6*d.* each.

Kay.—T. PAUL'S TWO EPISTLES TO THE CORINTHIANS, A COMMENTARY ON. By the late Rev. W. KAY, D.D., Rector of Great Leghs, Essex, and Hon. Canon of St. Albans; formerly Principal of Bishop's College, Calcutta; and Fellow and Tutor of Lincoln College. Demy 8vo. 9*s.*

Kuenen.—PENTATEUCH AND BOOK OF JOSHUA: an Historico-Critical Inquiry into the Origin and Composition of the Hexateuch. By A. KUENEN, Professor of Theology at Leiden. Translated from the Dutch, with the assistance of the Author, by PHILLIP H. WICKSTEED, M.A. 8vo. 14*s.*

The OXFORD MAGAZINE says:—"The work is absolutely indispensable to all special students of the Old Testament."

Lightfoot.—Works by the Right Rev. J. B. LIGHTFOOT, D.D., D.C.L., LL.D., Lord Bishop of Durham.

ST. PAUL'S EPISTLE TO THE GALATIANS. A Revised Text, with Introduction, Notes, and Dissertations. Ninth Edition, revised. 8vo. 12*s.*

ST. PAUL'S EPISTLE TO THE PHILIPPIANS. A Revised Text, with Introduction, Notes, and Dissertations. Ninth Edition, revised. 8vo. 12*s.*

ST. CLEMENT OF ROME—THE TWO EPISTLES TO THE CORINTHIANS. A Revised Text, with Introduction and Notes. 8vo. 8*s.* 6*d.*

ST. PAUL'S EPISTLES TO THE COLOSSIANS AND TO PHILEMON. A Revised Text, with Introductions, Notes, and Dissertations. Eighth Edition, revised. 8vo. 12*s.*

THE APOSTOLIC FATHERS. Part II. S. IGNATIUS—S. POLYCARP. Revised Texts, with Introductions, Notes, Dissertations, and Translations. 2 volumes in 3. Demy 8vo. 48*s.*

Maclear.—Works by the Rev. G. F. MACLEAR, D.D., Canon of Canterbury, Warden of St. Augustine's College, Canterbury, and late Head-Master of King's College School, London:—

A CLASS-BOOK OF OLD TESTAMENT HISTORY. New Edition, with Four Maps. 18mo. 4*s.* 6*d.*

A CLASS-BOOK OF NEW TESTAMENT HISTORY, including the Connection of the Old and New Testaments. With Four Maps. New Edition. 18mo. 5*s.* 6*d.*

A SHILLING BOOK OF OLD TESTAMENT HISTORY, for National and Elementary Schools. With Map. 18mo, cloth. New Edition.

Maclear.—Works by the Rev. G. F., *continued.*

A SHILLING BOOK OF NEW TESTAMENT HISTORY, for National and Elementary Schools. With Map. 18mo, cloth. New Edition.

These works have been carefully abridged from the Author's large manuals.

CLASS-BOOK OF THE CATECHISM OF THE CHURCH OF ENGLAND. New Edition. 18mo. 1*s.* 6*d.*

A FIRST CLASS-BOOK OF THE CATECHISM OF THE CHURCH OF ENGLAND. With Scripture Proofs, for Junior Classes and Schools. New Edition. 18mo. 6*d.*

A MANUAL OF INSTRUCTION FOR CONFIRMATION AND FIRST COMMUNION. WITH PRAYERS AND DEVOTIONS. 32mo. cloth extra, red edges. 2*s.*

Maurice.—THE LORD'S PRAYER, THE CREED, AND THE COMMANDMENTS. A Manual for Parents and Schoolmasters. To which is added the Order of the Scriptures. By the Rev. F. DENISON MAURICE, M.A. 18mo, cloth, limp. 1*s.*

Pentateuch and Book of Joshua: an Historico-Critical Inquiry into the Origin and Composition of the Hexateuch. By A. KUENEN, Professor of Theology at Leiden. Translated from the Dutch, with the assistance of the Author, by PHILIP H. WICKSTEED, M.A. 8vo. 14*s.*

Procter.—A HISTORY OF THE BOOK OF COMMON PRAYER, with a Rationale of its Offices. By Rev. F. PROCTER. M.A. 17th Edition, revised and enlarged. Crown 8vo. 10*s.* 6*d.*

Procter and Maclear.—AN ELEMENTARY INTRODUCTION TO THE BOOK OF COMMON PRAYER. Rearranged and supplemented by an Explanation of the Morning and Evening Prayer and the Litany. By the Rev. F. PROCTER and the Rev. Dr. MACLEAR. New and Enlarged Edition, containing the Communion Service and the Confirmation and Baptismal Offices. 18mo. 2*s.* 6*d.*

The Psalms, with Introductions and Critical Notes.—By A. C. JENNINGS, M.A., Jesus College, Cambridge, Tyrwhitt Scholar, Crosse Scholar, Hebrew University Prizeman, and Fry Scholar of St. John's College, Carus and Scholefield Prizeman, Vicar of Whittlesford, Cambs.; assisted in Parts by W. H. LOWE, M.A., Hebrew Lecturer and late Scholar of Christ's College, Cambridge, and Tyrwhitt Scholar. In 2 vols. Second Edition Revised. Crown 8vo. 10*s.* 6*d.* each.

Ramsay.—THE CATECHISER'S MANUAL; or, the Church Catechism Illustrated and Explained, for the Use of Clergymen, Schoolmasters, and Teachers. By the Rev. ARTHUR RAMSAY, M.A. New Edition. 18mo. 1*s.* 6*d.*

Ryle.—AN INTRODUCTION TO THE CANON OF THE OLD TESTAMENT. By Rev. H. E. RYLE, M.A., Fellow of King's College, and Hulsean Professor of Divinity in the University of Cambridge. Crown 8vo. [*In preparation.*

St. James' Epistle.—The Greek Text with Introduction and Notes. By Rev. JOSEPH MAYOR, M.A., Professor of Moral Philosophy in King's College, London. 8vo. [*In preparation.*

St. John's Epistles.—The Greek Text with Notes and Essays, by BROOKE FOSS WESTCOTT, D.D., Regius Professor of Divinity and Fellow of King's College, Cambridge, Canon of Westminster, &c. Second Edition Revised. 8vo. 12*s.* 6*d.*

St. Paul's Epistles.—Greek Text, with Introduction and Notes.

THE EPISTLE TO THE GALATIANS. Edited by the Right Rev. J. B. LIGHTFOOT, D.D., Bishop of Durham. Ninth Edition. 8vo. 12*s.*

THE EPISTLE TO THE PHILIPPIANS. By the same Editor. Ninth Edition. 8vo. 12*s.*

THE EPISTLE TO THE COLOSSIANS AND TO PHILEMON. By the same Editor. Eighth Edition. 8vo. 12*s.*

THE EPISTLE TO THE ROMANS. Edited by the Very Rev. C. J. VAUGHAN, D.D., Dean of Llandaff, and Master of the Temple. Fifth Edition. Crown 8vo. 7*s.* 6*d.*

THE EPISTLE TO THE PHILIPPIANS, with Translation, Paraphrase, and Notes for English Readers. By the same Editor. Crown 8vo. 5*s.*

THE EPISTLE TO THE THESSALONIANS, COMMENTARY ON THE GREEK TEXT. By JOHN EADIE, D.D., LL.D. Edited by the Rev. W. YOUNG, M.A., with Preface by Professor CAIRNS. 8vo. 12*s.*

THE EPISTLES TO THE EPHESIANS, THE COLOSSIANS, AND PHILEMON; with Introductions and Notes, and an Essay on the Traces of Foreign Elements in the Theology of these Epistles. By the Rev. J. LLEWELYN DAVIES, M.A., Rector of Christ Church, St. Marylebone; late Fellow of Trinity College, Cambridge. Second Edition, revised. Demy 8vo. 7*s.* 6*d.*

THE TWO EPISTLES TO THE CORINTHIANS, A COMMENTARY ON. By the late Rev. W. KAY, D.D., Rector of Great Leghs, Essex, and Hon. Canon of St. Albans; formerly Principal of Bishop's College, Calcutta; and Fellow and Tutor of Lincoln College. Demy 8vo. 9*s.*

The Epistle to the Hebrews. In Greek and English. With Critical and Explanatory Notes. Edited by Rev. FREDERIC RENDALL, M.A., formerly Fellow of Trinity College, Cambridge, and Assistant-Master at Harrow School. Crown 8vo. 6*s.*

THE ENGLISH TEXT, WITH NOTES. By the same Editor. Crown 8vo. [*In the press.*

The Epistle to the Hebrews. The Greek Text with Notes and Essays by B. F. WESTCOTT, D.D. 8vo. [*In the press.*

Westcott.—Works by BROOKE FOSS WESTCOTT, D.D., Canon of Westminster, Regius Professor of Divinity, and Fellow of King's College, Cambridge.

Westcott.—Works by BROOKE FOSS, *continued.*

A GENERAL SURVEY OF THE HISTORY OF THE CANON OF THE NEW TESTAMENT DURING THE FIRST FOUR CENTURIES. Sixth Edition. With Preface on "Supernatural Religion." Crown 8vo. 10*s*. 6*d*.

INTRODUCTION TO THE STUDY OF THE FOUR GOSPELS. Sixth Edition. Crown 8vo. 10*s*. 6*d*.

THE BIBLE IN THE CHURCH. A Popular Account of the Collection and Reception of the Holy Scriptures in the Christian Churches. New Edition. 18mo, cloth. 4*s*. 6*d*.

THE EPISTLES OF ST. JOHN. The Greek Text, with Notes and Essays. Second Edition Revised. 8vo. 12*s*. 6*d*.

THE EPISTLE TO THE HEBREWS. The Greek Text Revised, with Notes and Essays. 8vo. [*In the press.*

SOME THOUGHTS FROM THE ORDINAL. Cr. 8vo. 1*s*. 6*d*.

Westcott and Hort.—THE NEW TESTAMENT IN THE ORIGINAL GREEK. The Text Revised by B. F. WESTCOTT, D.D., Regius Professor of Divinity, Canon of Westminster, and F. J. A. HORT, D.D., Lady Margaret Professor of Divinity; Fellow of Emmanuel College, Cambridge: late Fellows of Trinity College, Cambridge. 2 vols. Crown 8vo. 10*s*. 6*d*. each.

Vol. I. Text.

Vol. II. Introduction and Appendix.

THE NEW TESTAMENT IN THE ORIGINAL GREEK, FOR SCHOOLS. The Text Revised by BROOKE FOSS WESTCOTT, D.D., and FENTON JOHN ANTHONY HORT, D.D. 12mo. cloth. 4*s*. 6*d*. 18mo. roan, red edges. 5*s*. 6*d*.

Wilson.—THE BIBLE STUDENT'S GUIDE to the more Correct Understanding of the English Translation of the Old Testament, by reference to the original Hebrew. By WILLIAM WILSON, D.D., Canon of Winchester, late Fellow of Queen's College, Oxford. Second Edition, carefully revised. 4to. cloth. 25*s*.

Wright.—THE BIBLE WORD-BOOK: A Glossary of Archaic Words and Phrases in the Authorised Version of the Bible and the Book of Common Prayer. By W. ALDIS WRIGHT, M.A., Vice-Master of Trinity College, Cambridge. Second Edition, Revised and Enlarged. Crown 8vo. 7*s*. 6*d*.

Yonge (Charlotte M.).—SCRIPTURE READINGS FOR SCHOOLS AND FAMILIES. By CHARLOTTE M. YONGE. Author of "The Heir of Redclyffe." In Five Vols.

FIRST SERIES. GENESIS TO DEUTERONOMY. Extra fcap. 8vo. 1*s*. 6*d*. With Comments, 3*s*. 6*d*.

SECOND SERIES. From JOSHUA to SOLOMON. Extra fcap. 8vo. 1*s*. 6*d*. With Comments, 3*s*. 6*d*.

THIRD SERIES. The KINGS and the PROPHETS. Extra fcap. 8vo. 1s. 6d. With Comments, 3s. 6d.

FOURTH SERIES. The GOSPEL TIMES. 1s. 6d. With Comments. Extra fcap. 8vo, 3s. 6d.

FIFTH SERIES. APOSTOLIC TIMES. Extra fcap. 8vo. 1s. 6d. With Comments, 3s. 6d.

Zechariah—Lowe.—THE HEBREW STUDENT'S COMMENTARY ON ZECHARIAH, HEBREW AND LXX. With Excursus on Syllable-dividing, Metheg, Initial Dagesh, and Siman Rapheh. By W. H. LOWE, M.A., Hebrew Lecturer at Christ's College, Cambridge. Demy 8vo. 10s. 6d.

MACMILLAN'S GEOGRAPHICAL SERIES.

Edited by ARCHIBALD GEIKIE, F.R.S., Director-General of the Geological Survey of the United Kingdom.

The following List of Volumes is contemplated:—

THE TEACHING OF GEOGRAPHY. A Practical Handbook for the use of Teachers. By ARCHIBALD GEIKIE, F.R.S., Director-General of the Geological Survey of the United Kingdom, and Director of the Museum of Practical Geology, Jermyn Street, London; formerly Murchison Professor of Geology and Mineralogy in the University of Edinburgh. Crown 8vo. 2s. [*Ready.*

*** The aim of this volume is to advocate the claims of geography as an educational discipline of a high order, and to show how these claims may be practically recognised by teachers.

AN ELEMENTARY GEOGRAPHY OF THE BRITISH ISLES. By ARCHIBALD GEIKIE, F.R.S., &c. 18mo. 1s. [*Ready.*

THE ELEMENTARY SCHOOL ATLAS. By JOHN BARTHOLOMEW, F.R.G.S. Designed to illustrate the principal Text-Books on Elementary Geography. Price 1s.

MAPS AND MAP MAKING. By ALFRED HUGHES, M.A., late Scholar of Corpus Christi College, Oxford, Assistant Master at Manchester Grammar School. Crown 8vo.

AN ELEMENTARY GENERAL GEOGRAPHY. By HUGH ROBERT MILL, D.Sc. Edin. Crown 8vo.

A GEOGRAPHY OF THE BRITISH COLONIES.

A GEOGRAPHY OF EUROPE.

A GEOGRAPHY OF AMERICA.

A GEOGRAPHY OF ASIA.

A GEOGRAPHY OF AFRICA.

A GEOGRAPHY OF THE OCEANS AND OCEANIC ISLANDS.

ADVANCED CLASS-BOOK OF THE GEOGRAPHY OF BRITAIN.

GEOGRAPHY OF AUSTRALIA AND NEW ZEALAND.

GEOGRAPHY OF BRITISH NORTH AMERICA.

GEOGRAPHY OF INDIA.

GEOGRAPHY OF THE UNITED STATES.

ADVANCED CLASS-BOOK OF THE GEOGRAPHY OF EUROPE.

RICHARD CLAY AND SONS, LONDON AND BUNGAY.

www.ingramcontent.com/pod-product-compliance
Lightning Source LLC
Chambersburg PA
CBHW021403210326
41599CB00011B/991
* 9 7 8 3 3 3 7 2 5 6 4 1 8 *